Theory and Applications of
Higher-Dimensional Hadamard Matrices

Combinatorics and Computer Science

Editor-in-Chief

Liu Yanpei (Beijing)

Co-Editor-in-Chief

David M. Jackson (Waterloo)

Pierre Hansen (Montreal)

Fred S. Roberts (New Brunswick)

Theory and Applications of Higher-Dimensional Hadamard Matrices

By

Yang Yi Xian

Department of Information Engineering,
Beijing University of Posts and Telecommunications,
Beijing, People's Republic of China

Science Press
Beijing/New York,

Kluwer Academic Publishers
Dordrecht/Boston/London

A C.I.P Catalogue record for this book is available from the Library of Congress.

ISBN 978-90-481-5730-3

Published by Kluwer Academic Publishers,
P. O. Box 17, 3300 AA Dordrecht, The Netherlands.

Sold and distributed in North, Central and South America
by Kluwer Academic Publishers,
101 Philip Drive, Norwell, MA 02061, U.S.A.

Sold and distributed in the People's Republic of China
by Science Press, Beijing.

In all other countries, sold and distributed
by Kluwer Academic Publishers,
P. O. Box 322, 3300 AH Dordrecht, The Netherlands.

Printed on acid-free paper

Contents

Preface

Just over one hundred years ago, in 1893, Jacques Hadamard found 'binary' (± 1) matrices of orders 12 and 20 whose rows (resp. columns) were pairwise orthogonal. These matrices satisfy the determinantal upper bound for 'binary' matrices. Hadamard actually proposed the question of seeking the maximal determinant of matrices with entries on the unit circle, but his name has become associated with the question concerning real (binary) matrices. Hadamard was not the first person to study these matrices. For example, J. J. Sylvester had found, in 1857, such row (column) pairwise orthogonal binary matrices of all orders of powers of two. Nevertheless, Hadamard proved that binary matrices with a maximal determinant could exist only for orders 1, 2, and $4t$, t a positive integer.

With regard to the practical applications of Hadamard matrices, it was M. Hall, Jr., L. Baumert, and S. Golomb who sparked the interest in Hadamard matrices over the past 30 years. They made use of the Hadamard matrix of order 32 to design an eight bit error-correcting code for two reasons. First, error-correcting codes based on Hadamard matrices have good error correction capability and good decoding algorithms. Second, because Hadamard matrices are (± 1)-valued, all the computer processing can be accomplished using additions and subtractions rather than multiplication.

Walsh matrices are the simplest and most popular special kinds of Hadamard matrices. Walsh matrices are generated by sampling the Walsh functions, which are families of orthogonal complete functions. Based on the Walsh matrices, a very efficient orthogonal transform, called Walsh–Hadamard transform, was developed. The Walsh–Hadamard transform is now playing a more and more important role in signal processing and image coding.

P. J. Shlichta discovered in 1971 that there exist higher-dimensional binary arrays which possess a range of orthogonality properties. In particular, P. J. Shlichta constructed 3-dimensional arrays with the property that any sub-array obtained by fixing one index is a 2-dimensional Hadamard matrix. The study of higher-dimensional Hadamard matrices was mainly motivated by another important paper of P. J. Shlichta 'Higher-Dimensional Hadamard Matrices', which was published in *IEEE Trans. on Inform.*, in 1979. Since then a lot of papers on the existence, construction, and enumeration of higher-dimensional Hadamard matrices have been reported. For example, J.Hammer and J. Seberry, found, in 1982, that higher-dimensional orthogonal designs can be used to construct higher-dimensional Hadamard matrices. To the author's knowledge much of the research achievements on higher-dimensional Hadamard matrices have been accomplished by S. S. Agaian, W. De Launey, J. Hammer, J. Seberry, Yi Xian Yang, K.J. Horadam, P. J. Shlichta, J. Jedwab, C. Lin, Y. Q. Chen, and others. Many new papers have been published, thus none can collect together all of the newest results in this area.

The book divides naturally into three parts according to the dimensions of Hadamard matrices processed.

The first part, Chapter 1 and Chapter 2, lay stress upon the classical 2-dimensional cases. Because quite a few books (or chapters in them) have been published which introduce the progress of (2-dimensional) Hadamard matrices, we prefer to present an introductory survey rather than to restate many known long proofs. Chapter 1 introduces Walsh matrices and Walsh transforms, which have been widely used in engineering fields. Fast algorithms for Walsh transforms and various useful properties of Walsh matrices are also stated. Chapter 2 is about (2-dimensional) Hadamard matrices, especially their construction, existence, and their generalized forms. The updated strongest Hadamard construction theorems presented in this chapter are helpful for readers to understand how difficult it is to prove or disprove the famous Hadamard conjecture.

The second part, Chapters 3 and 4, deals with the lower-dimensional cases, e.g., 3-, 4-, and 6-dimensional Walsh and Hadmard matrices and transforms. One of the aims of this part is to make it easier to smoothly move from 2-dimensional cases to the general higher-dimensional cases. Chapter 3 concentrates on the 3-dimensional Hadamard and Walsh matrices. Constructions based upon direct multiplication, and upon recursive methods, perfect binary arrays are introduced. Another important topic of this chapter is the existence and construction of 3-dimensional Hadamard matrices of orders $4k$ and $4k + 2$, respectively. Chapter 4 introduces a group of transforms based on 2-, 3-, 4-, and 6-dimensional Walsh–Hadamard matrices and their corresponding fast algorithms. The algebraic theory of higher-dimensional Walsh–Hadamard matrices is presented also.

Finally, the third part, which is the key part of the book, consists of the last two chapters (Chapter 5 and 6). To the author's knowledge, the contents in this part (and the previous second part) have never been included in any published books. This part is divided into chapters according to the orders of the matrices (arrays) processed. Chapter 5 investigates the N-dimensional Hadamard matrices of order 2, which have been proved equivalent to the well known H–Boolean functions and the perfect binary arrays of order 2. This equivalence motivates a group of perfect results about the enumeration of higher-dimensional Hadamard matrices of order 2. Applications of these matrices to feed forward networking, stream cipher, Bent functions and error correcting codes are presented in turn. Chapter 6, which is the longest chapter of the book, aims at introducing Hadamard matrices of general dimension and order. After introducing the definitions of the regular, proper, improper, and generalized higher-dimensional Hadamard matrices, many theorems about the existence and constructions are presented. Perfect binary arrays, generalized perfect arrays, and the orthogonal designs are also used to construct new higher-dimensional Hadamard matrices. The last chapter of the book is a concluding chapter of questions, which includes a list of open problems in the study of the theory of higher-dimensional Hadamard matrices. We hope that these research problems will motivate further developments.

In order to satisfy readers with this special interest, we list, at the end of each chapter, as many up to date references as possible.

I would like to thank my supervisors, Professors. Zhen Ming Hu and Jiong Pang Zhou for their guidance during my academic years at the Information Security Center of Beijing University of Posts and Telecommunications (BUPT). During my research years in higher-dimensional Hadamard matrices I benefited from Professors W. De Launey, J. Hammer, J. Seberry, K. J. Horadam, P. J. Shlichta, J. Jedwab. My thanks go to many of their papers, theses and communications. I was attracted into the area of higher-dimensional Hadamard matrices by P.J. Shlichta's paper 'Higher-Dimensional Hadamard Matrices' published in *IEEE Trans. on Inform. Theory*. My first journal paper was motivated by J. Hammer and J. Seberry's paper 'Higher-Dimensional Orthogonal Designs and Applications' published in *IEEE Trans. on Inform. Theory*. It is Dr. J. Jedwab's wonderful Ph.D thesis 'Perfect Arrays, Barker Arrays and Difference Sets' that motivated me to finish the first book on higher-dimensional Hadamard matrices. One of my main aims in this book is to motivate other authors to begin to publish more books on higher-dimensional Hadamard matrices and their applications, so that the readers in other areas can know what has been done in the area of higher-dimensional Hadamard matrices.

I specially thank my wife, Xin Xin Niu, and my son, Mu Long Yang, for their support. It is not hard to imagine how much they have sacrificed in family life during the past years. I would like to dedicate this book to my wife and son. Finally, I also dedicate this book to my parents, Mr. Zhong Quan Yang and Mrs. De Lian Wei for their love.

Part I
Two-Dimensional Cases

Chapter 1
Walsh Matrices

Walsh matrices are the simplest and most popular special kind of Hadamard matrices, which is defined as the (± 1)-valued orthogonal matrix. Walsh matrices are generated by sampling the Walsh functions, which are families of orthogonal complete functions. The orders of Walsh matrices are always equal to 2^n, where n is a non-negative integer. If the $+1$s in a Walsh matrix are replaced by -1s and -1s by 1s, then a good error correcting code with Hamming distance $m/2$, where m is the order of the matrix, is constructed. Walsh matrices are widely used in communications, signal processing, and physics, and have an extensive and widely scattered literature. This chapter concentrates on the definitions, generations and ordering of Walsh matrices, and on Walsh transforms with fast algorithms.

1.1   Walsh Functions and Matrices

Walsh functions belong to the class of piecewise constant basis functions which were developed in the nineteen twenties and have played an important role in scientific and engineering applications. The foundations of the field of Walsh functions were made by Rademacher (in 1922), Walsh (in 1923), Fine (in 1945), Paley (in 1952), and Kaczmarz and Steinhaus (in 1951). The engineering approach to the study and utilization of these functions was originated by Harmuth (in 1969), who introduced the concept of sequence to represent the associated, generalized frequency defined as one half the mean rate of zero crossings. Possible applications of Walsh functions to signal multiplexing, bandwidth compression, digital filtering, pattern recognition, statistical analysis, function approximation, and others are suggested and extensively examined.

1.1.1 Definitions

In order to define the Walsh functions we introduce, at first, a family of important orthogonal (but incomplete) functions which are called Rademacher functions ([1]):

$$\text{RAD}(n, t) = \text{sign}[\sin(2^n \pi t)], n = 0, 1, \ldots, \tag{1.1}$$

where sign$[x]$ is the sign function of x, i.e., sign$[x] = 1$ if $x > 0$ and sign$[x] = -1$ if $x < 0$.

Clearly, Rademacher functions are derived from sinusoidal functions which have identical zero crossing positions. Rademacher functions have two arguments n and t such that $R(n, t)$ has 2^{n-1} periods of square wave over a normalized time base $0 \le t \le 1$. The amplitudes of the functions are $+1$ and -1. The first function $R(0, t)$ is equal to one for the entire interval $0 \le t \le 1$. The next and subsequent functions are square waves having odd symmetry.

Rademacher functions are periodic with period 1, i.e.,

$$\text{RAD}(n, t) = \text{RAD}(n, t + 1).$$

They are also periodic over shorter intervals such that

$$\text{RAD}(n, t + m2^{1-n}) = \text{RAD}(n, t), \quad n = 1, 2, \ldots; \quad m = \pm 1, \pm 2, \ldots$$

Rademacher functions can also be generated using the recurrence relation

$$\text{RAD}(n, t) = \text{RAD}(1, 2^{n-1}t)$$

with

$$\text{RAD}(1, t) = \begin{cases} 1, t \in [0, 1/2) \\ -1, t \in [1/2, 1). \end{cases}$$

In order to define the Walsh functions we note that each integer n, $0 \le n \le 2^m - 1$, has a unique binary extension of the form

$$n = \sum_{k=0}^{m-1} n_k 2^k, \quad \text{where } n_k = 0 \text{ or } 1. \tag{1.2}$$

Then the n-th Paley-ordered Walsh function is defined by

$$\mathrm{Wal_P}(n,t) = \prod_{k=0}^{m-1} [\mathrm{RAD}(k+1,t)]^{n_k}. \tag{1.3}$$

For example, because $7 = 1 \times 2^2 + 1 \times 2^1 + 1 \times 2^0$,

$$\mathrm{Wal_P}(7,t) = \mathrm{RAD}(3,t)\mathrm{RAD}(2,t)\mathrm{RAD}(1,t)$$
$$= \mathrm{sign}[\sin(2^3\pi t)] \times \mathrm{sign}[\sin(2^2\pi t)] \times \mathrm{sign}[\sin(2^1\pi t)].$$

Thus Walsh functions form an ordered set of rectangular waveforms taking only two amplitude values $+1$ and -1. Unlike the Rademacher functions the Walsh rectangular waveforms do not have unit mark–space ratio. Like the sine and cosine functions, two arguments are required for complete definition, a time period, t, and an ordering number, n, related to frequency in a way which is described later.

The Walsh functions can also be defined by their time argument. In fact, each non-negative real number t, $0 \le t < 1$, can be uniquely decomposed as

$$t = \sum_{k=1}^{\infty} t_k 2^{-k}, \quad \text{with } t_k = 0 \text{ or } 1. \tag{1.4}$$

Then the Rademacher functions are derived by

$$\begin{cases} \mathrm{RAD}(0,t) = 1 \\ \mathrm{RAD}(k,t) = (-1)^{t_k}, \ k = 1, 2, \dots. \end{cases} \tag{1.5}$$

Hence by setting Equation (1.5) into Equation (1.3), we have the following equivalent definition of the Paley-ordered Walsh functions ([1]):

$$\mathrm{Wal_P}(n,t) = (-1)^{\sum_{k=0}^{m-1} n_k t_{k+1}}. \tag{1.6}$$

A straightforward consequence of Equation (1.6) is the following identity:

Lemma 1.1.1 ([2] , [3], [4]) *Let q and n be two non-negative integers. Thus*

$$\mathrm{Wal_P}(n,t)\mathrm{Wal_P}(q,t) = \mathrm{Wal_P}(q \oplus n, t) \tag{1.7}$$

where $q \oplus n$ is the dyadic summation of q and n, i.e., $q \oplus n = k$ if and only if their binary extensions $q = (q_0, q_1, \dots, q_{m-1})$, $n = (n_0, n_1, \dots, n_{m-1})$, and $k = (k_0, k_1, \dots, k_{m-1})$ satisfy $(n_i + q_i) \bmod 2 = k_i$ for all $0 \le i \le m-1$.

Thus the set of Paley-ordered Walsh functions forms an Abelian group under the multiplication operation.

Let k, $0 \leq k \leq 2^m - 1$, be an integer with its binary extension being $k = \sum_{i=0}^{m-1} k_i 2^i$. Then, by Equation (1.6), the discrete sampling of a Walsh function $\text{Wal}_P(n, t)$ at the point $t = k/2^m$ is

$$W_P(n, k) = (-1)^{\sum_{i=0}^{m-1} n_i k_{m-1-i}}. \tag{1.8}$$

Therefore by sampling the continuous Walsh functions with the unit space $1/2^m$ we have the following (± 1)-valued matrix of size $2^m \times 2^m$:

$$W_P = [W_P(n, k)] = [(-1)^{\sum_{i=0}^{m-1} n_i k_{m-1-i}}]. \tag{1.9}$$

This matrix is called the Paley ordered Walsh matrix.

For example, if $m = 3$, then the corresponding Paley ordered Walsh matrix is

$$W_P = \begin{bmatrix} + & + & + & + & + & + & + & + \\ + & + & + & + & - & - & - & - \\ + & + & - & - & + & + & - & - \\ + & + & - & - & - & - & + & + \\ + & - & + & - & + & - & + & - \\ + & - & + & - & - & + & - & + \\ + & - & - & + & + & - & - & + \\ + & - & - & + & - & + & + & - \end{bmatrix}.$$

Theorem 1.1.1 ([2] , [3], [4]) *Let $W_P = [W_P(n, k)]$, $0 \leq n, k \leq 2^m - 1$, be a Paley-ordered Walsh matrix. Then*

1. $W_P(n, k) = W_P(k, n)$ *for all $0 \leq n, k \leq 2^m - 1$, i.e. the Paley ordered Walsh matrix is symmetrical;*

2. $W_P(n, k) W_P(n, q) = W_P(n, k \oplus q)$ *for all $0 \leq n, k, q \leq 2^m - 1$, i.e. the set of columns of the Paley-ordered Walsh matrix is also closed under the bit-wise multiplication;*

3. *The matrix W_P is (± 1)-valued and orthogonal. In other words, W_P is an Hadamard matrix.*

Proof. The first two statements are direct consequences of Equation (1.9). The third statement is owed to the second statement and the following identity:

$$\sum_{k=0}^{2^m-1} W_p(n,k) = \sum_{k=0}^{2^m-1} (-1)^{\sum_{i=0}^{m-1} n_i k_{m-1-i}}$$

$$= \sum_{k_0=0}^{1} \cdots \sum_{k_{m-1}=0}^{1} (-1)^{\sum_{i=0}^{m-1} n_i k_{m-1-i}}$$

$$= \prod_{i=0}^{m-1} \sum_{k_i=0}^{1} (-1)^{n_i k_{m-1-i}}$$

$$= 0 \ (\text{ provided that } n_i \neq 0 \text{ for some } i).$$

In other words, except for the all ones row (the 0-th row), the rows of the matrix W_p are balanced by 1 and -1.   **Q.E.D.**

The set of Walsh function series produced by Equation (1.5) or equivalently by Equation (1.3) can also be obtained in several other different ways, each of which has its own particular advantages. The methods considered in the following context are:

1. By means of a difference equation;

2. Through the Hadamard matrices;

Both of these derivations are, of course, mathematical processes for which computational algorithms can be developed and the series produced using the digital computer or obtained directly by using digital logic.

From Difference Equations: ([2], [3], [4], [5]) This method gives the function directly in sequence order. Sequence is a term used for describing a periodic repetition rate which is independent of waveform. It is defined as, 'One half of the average number of zero crossings per unit time interval'. From this we see that frequency can be regarded as a special measure of sequency applicable to sinusoidal waveforms only. Applying the definition of sequency to periodic and a periodic function, we obtain:

1. The sequency of a periodic function equals one half the number of sign changes per period;

2. The sequency of an a periodic function equals one half the number of sign changes per unit time, if this limit exists.

Assume that the normalized time base is $0 \le t \le 1$. A given sequency-ordered Walsh function is defined from its preceding harmonic function so that, we commence with a definition of $\mathrm{Wal}(0,t) = 1$ within the time-base and 0 outside the time-base. Then the entire set of sequency-ordered Walsh functions $\{\mathrm{Wal}(n,t) : n = 0,1,\ldots\}$ can be obtained by an iterative process. The difference equation is given as

$$\mathrm{Wal}(2j+i,t) = (-1)^{\lfloor j/2 \rfloor + i}[\mathrm{Wal}(j,2t) + (-1)^{j+i}\mathrm{Wal}(j,2(t-1/2))], \quad (1.10)$$

where $i = 0$ or 1, $j = 0,1,\ldots$ and $\lfloor x \rfloor$ is the floor function.

For $N = 2^m$ equally spaced discrete points, Equation (1.10) can be written as

$$\mathrm{Wal}(2j+i,n) = (-1)^{\lfloor j/2 \rfloor + i}[\mathrm{Wal}(j,2n) + (-1)^{j+i}\mathrm{Wal}(j,2(n-N/2))],$$
$$(1.11)$$

where $0 \le n \le 2^m - 1$.

Commencing with the known $\mathrm{Wal}(0,t) = 1$ within the time base (i.e., $j = 0$, and $i = 1$) for $n \le N/2$ its value for $\mathrm{Wal}(j,2n)$ will be 1 and for $n > N/2$ the function falls outside the time base and will be 0. Similarly to $\mathrm{Wal}(j,2(n-N/2))$ for $n < N/2$ the function again falls outside the time base and will become 0, whilst for $n > N/2$ the value is 1. The sign of these functions will be modified by the factors $(-1)^{j+i}$ and $(-1)^{\lfloor j/2 \rfloor + i}$ in accordance with Equation (1.11).

The operation of difference Equation (1.10) may be considered as equivalent to compressing the previous Walsh function $\mathrm{Wal}(j,2n)$ into the left hand part of the time base by selection of alternate points and, after left adjustment, adding to these on the right hand side a similar valued set of points but all having an opposite sign.

From Walsh–Hadamard Matrices: ([2] , [3], [4], [6]) A Walsh–Hadamard matrix is a (± 1)-valued square array with its rows (and columns) are orthogonal to one another. The smallest non-trivial Walsh–Hadamard matrix is

$$H_2 = \begin{bmatrix} 1 & 1 \\ 1 & -1 \end{bmatrix}$$

The other higher matrices of size $2^m \times 2^m$ can be obtained from the recursive relationship

$$H_{2^m} = \begin{bmatrix} H_{2^{m-1}} & H_{2^{m-1}} \\ H_{2^{m-1}} & -H_{2^{m-1}} \end{bmatrix} \quad (1.12)$$

i.e., the direct product of H_{2m-1} and H_2. (In general, the direct product of two matrices A and B, represented by $A \otimes B$, is another matrix obtained by replacing the (i,j)-th entry a_{ij} by the matrix $a_{ij}B$). For example

$$H_4 = H_2 \otimes H_2 = \begin{bmatrix} H_2 & H_2 \\ H_2 & -H_2 \end{bmatrix} = \begin{bmatrix} 1 & 1 & 1 & 1 \\ 1 & -1 & 1 & -1 \\ 1 & 1 & -1 & -1 \\ 1 & -1 & -1 & 1 \end{bmatrix}$$

and

$$H_8 = H_4 \otimes H_2 = \begin{bmatrix} H_4 & H_4 \\ H_4 & -H_4 \end{bmatrix}$$

$$= \begin{bmatrix} 1 & 1 & 1 & 1 & 1 & 1 & 1 & 1 \\ 1 & -1 & 1 & -1 & 1 & -1 & 1 & -1 \\ 1 & 1 & -1 & -1 & 1 & 1 & -1 & -1 \\ 1 & -1 & -1 & 1 & 1 & -1 & -1 & 1 \\ 1 & 1 & 1 & 1 & -1 & -1 & -1 & -1 \\ 1 & -1 & 1 & -1 & -1 & 1 & -1 & 1 \\ 1 & 1 & -1 & -1 & -1 & -1 & 1 & 1 \\ 1 & -1 & -1 & 1 & -1 & 1 & 1 & -1 \end{bmatrix}.$$

When the n-th row of the matrix H_{2m} is denoted by a function $\text{Wal}_h(n, t)$, $0 \leq n \leq 2^m - 1$, we obtain the Hadamard-ordered Walsh function series.

Replacing each row of the matrix in Equation (1.12) by its equivalent Paley-ordered Walsh functions we can form a series of functions which will indicate the ordering obtained through this derivation. For example, from the above matrix H_8 it is easy to see that

$$\text{Wal}_h(0,t) = \text{Wal}_P(0,t), \quad \text{Wal}_h(1,t) = \text{Wal}_P(4,t),$$

$$\text{Wal}_h(2,t) = \text{Wal}_P(2,t), \quad \text{Wal}_h(3,t) = \text{Wal}_P(6,t),$$

$$\text{Wal}_h(4,t) = \text{Wal}_P(1,t), \quad \text{Wal}_h(5,t) = \text{Wal}_P(5,t),$$

$$\text{Wal}_h(6,t) = \text{Wal}_P(3,t), \quad \text{Wal}_h(7,t) = \text{Wal}_P(7,t).$$

The relationship between these Walsh–Hadamard matrices and a sampled set Paley-ordered Walsh functions is now clear. They simply express a Walsh function series having positive phasing and arranged in bit-reversed Paley order. This ordering is sometimes referred to as the 'lexicographic ordering'.

1.1.2   Ordering

From the last subsection we know that the generation of a Walsh function series can be carried out in a number of ways, and that the order of the functions produced can be different to each other. So it is necessary now to consider the ordering of Walsh functions in some details. There are three main ordering conventions in common use:

Sequency Order([1]): This is Walsh's original order for his function $\mathrm{Wal}(n,t)$, see Equations (1.10) and (1.11). In this order the components are arranged in ascending order of zero crossings. It is directly related to frequency, in which we find that Fourier components are also arranged in increasing harmonic number (zero crossings divided by two). Sequency order is most close to our practical experience with other orthogonal functions (e.g., sinusoidal functions).

Paley Order([1]): This is the order obtained by generation from successive Rademacher functions, see Equation (1.3). The Paley ordering has certain analytical and computational advantages and is used for most mathematical discussions. The Paley-ordered functions may be defined as the eigenfunctions of a logical differential operator and that this definition is of great value in the mathematical development of the theory.

Hadamard Order([1]): This ordering follows the Walsh–Hadamard matrices, see Equation (1.12). It is the ordering that is obtained if one computes fast Walsh transforms without sorting in the manner of the Cooley–Tukey fast Fourier transform algorithm. Hence it is computationally advantageous.

Paley order and Hadamard order are used in theoretical mathematical work in image transmission and for computational efficiency. Sequency order is favored for communications and signal processing work such as spectral analysis and filtering.

In order to show the relationships amongst the Walsh functions of different orders, we now introduce the concept of 'binary to Gray code' conversion and 'Gray code to binary' conversion as follows:

Binary to Gray code conversion: Let $0 \leq n \leq 2^m - 1$ be an integer with its binary extension being

$$(n_{m-1}, n_{m-2}, \ldots, n_0).$$

Its Binary to Gray code is

$$(g_{m-1}, g_{m-2}, \ldots, g_0),$$

where $g_{m-1} = n_{m-1}$ and $g_i = (n_i + n_{i+1})\mathrm{mod}2$, $0 \leq i \leq m - 2$.

Gray code to Binary conversion: converting Gray code to binary, we start with the left most digit and move to the right, making $n_i = g_i$ if the number of 1s preceding g_i is even, and making $n_i = 1 - g_i$ if the number of 1s preceding g_i is odd. During this process, zero 1s is treated as an even number of 1s.

Theorem 1.1.2 ([1]) *Let* t, $0 \leq t < 1$, *be a number and* $t = \sum_{k=1}^{\infty} t_k 2^{-k}$, *where* $t_k = 0$ *or* 1. *And let* $(g_{m-1}, g_{m-2}, \ldots, g_0)$ *be the Binary to Gray code of* $(n_{m-1}, n_{m-2}, \ldots, n_0)$. *Then the sequency-ordered Walsh functions can be equivalently defined by*

$$\mathrm{Wal}(n, t) = (-1)^{\sum_{k=0}^{m-1} g_k t_{k+1}} = (-1)^{\sum_{k=0}^{m-1} (n_k + n_{k+1}) t_{k+1}}. \qquad (1.13)$$

Proof. The functions defined by Equation (1.13) satisfy the recursive relationship in Equations (1.10) and (1.11).   **Q.E.D.**

From Theorem 1.13 and Equation (1.6) we find that the Walsh functions in Paley order and sequency order are related by

$$\mathrm{Wal}_P(n, t) = \mathrm{Wal}(b(n), t), \qquad (1.14)$$

where $b(n)$ represents the Gray code to Binary conversion of the integer n.

Theorem 1.1.3 ([1]) *The function* $\mathrm{Wal}(n, t)$ *has* n *zero-crossings. Or, equivalently, the sequency* s_n *of* $\mathrm{Wal}(n, t)$ *is given by*

$$S_n = \begin{cases} 0 & n = 0, \\ n/2 & n = \mathrm{even}, \\ (n+1)/2 & n = \mathrm{odd}. \end{cases}$$

Proof. By induction on m: The case of $m = 0$ is trivial. Assume that the theorem is true for $n < 2^m$. It is now necessary to prove that the theorem is true for $n < 2^{m+1}$.

If $2^m \leq n < 2^{m+1}$ then $n_{m+1} = 1$. Hence by Equation (1.13) we have:

$$
\begin{aligned}
\mathrm{Wal}(n,t) &= (-1)^{\sum_{k=0}^{m}(n_k+n_{k+1})t_{k+1}} \\
&= (-1)^{[\sum_{k=0}^{m-2}(n_k+n_{k+1})t_{k+1}+n_{m-1}t_m]+n_m t_m+n_m t_{m-1}} \\
&= (-1)^{n_m(t_m+t_{m+1})}\mathrm{Wal}(n-2^m,t) \\
&= (-1)^{(t_m+t_{m+1})}\mathrm{Wal}(n-2^m,t).
\end{aligned}
\tag{1.15}
$$

While the $(-1)^{t_m+t_{m+1}}$ produces a zero-crossing only if (t_m,t_{m+1}) is changed from $(0,0)$ to $(0,1)$ or from $(1,0)$ to $(1,1)$. In other words, the zero-crossing points are of the form $t = k/2^{m+1}$, where k is odd. Thus $(-1)^{t_m+t_{m+1}}$ has 2^m zero-crossing points.

From Equation (1.13) the function $\mathrm{Wal}(n-2^m,t)$ is unchanged at the zero-crossing points of $(-1)^{t_m+t_{m+1}}$. Thus, by Equation (1.15), the zero-crossing points of $(-1)^{t_m+t_{m+1}}$ are also those of $\mathrm{Wal}(n,t)$.

Outside the zero-crossing points of $(-1)^{t_m+t_{m+1}}$, by the induction assumption and Equation (1.15), the function $\mathrm{Wal}(n-2^m,t)$ has $n-2^m$ zero-crossing points.

Thus, the total zero-crossing points of $\mathrm{Wal}(n,t)$ is equal to $2^m + (n - 2^m) = n$.   **Q.E.D.**

An alternative notation of $\mathrm{Wal}(n,t)$ is to classify the Walsh functions in terms of even and odd waveform symmetry, viz.,

$$
\begin{cases}
\mathrm{Wal}(2k,t) = \mathrm{Cal}(k,t), & \text{called the Walsh cosine} \\
\mathrm{Wal}(2k-1,t) = \mathrm{Sal}(k,t), & \text{called the Walsh sine },
\end{cases}
\tag{1.16}
$$

where $0 \leq k \leq 2^{m-1}$. This classification defines two further Walsh series having close similarities with the cosine and sine series. It can be indicated that the normalized Walsh functions are symmetrical about their mid or zero time point. Defining the range of the function as $-1/2 \leq t \leq 1/2$, the functions are either directly symmetrical (Cal functions) or inversely symmetrical (Sal functions). In the latter case the ones found in the left hand side are mirrored by zeros in the right hand side and *viceversa*. Thus a symmetry relationship can be stated as $\mathrm{Wal}(n,t) = \mathrm{Wal}(t,n)$.

From Equations (1.5) and (1.13) we obtain the following equivalent definition of sequency-ordered Walsh functions:

$$
\mathrm{Wal}(n,t) = \prod_{k=0}^{m-1} \mathrm{RAD}(k+1,t)^{g_k},
\tag{1.17}
$$

where $g = (g_{m-1}, g_{m-2}, \ldots, g_0)$ is the binary to Gray code of the integer $n = (n_{m-1}, n_{m-2}, \ldots, n_0)$.

By sampling the functions in Equation (1.17) with the unit space $1/2^m$, we get the sequency-ordered Walsh matrix:

$$W = [W(n, i)] = [(-1)^{\sum_{k=0}^{m-1} (n_k + n_{k+1}) i_{m-1-k}}], \qquad (1.18)$$

where $0 \leq n, i \leq 2^m - 1$. This matrix W is also a (± 1)-valued orthogonal matrix of size $2^m \times 2^m$.

Similarly, the Hadamard-ordered Walsh functions are equal to

$$\text{Wal}_h(n, t) = \prod_{k=0}^{m-1} \text{RAD}(k+1, t)^{n_{m-1-k}}. \qquad (1.19)$$

And its matrix form is

$$W_h = [W_h(n, i)] = [(-1)^{\sum_{k=0}^{m-1} n_k i_k}], \qquad (1.20)$$

where $0 \leq n, i \leq 2^m - 1$. This matrix W_h is also a (± 1)-valued orthogonal matrix of size $2^m \times 2^m$.

From Equations (1.19) and (1.17), it is clear that the Hadamard-ordered Walsh functions are related to those in sequency order by the relation

$$\text{Wal}_h(n, t) = \text{Wal}(b(\langle n \rangle), t), \qquad (1.21)$$

where $\langle n \rangle$ is obtained by the bit-reversal of n, and $b(\langle n \rangle)$ is the Gray code to binary conversion of $\langle n \rangle$.

Similarly, from Equations (1.19) and (1.3), we have

$$\text{Wal}_h(n, t) = \text{Wal}_P(\langle n \rangle, t). \qquad (1.22)$$

Up to now, the relationships amongst Walsh functions and matrices of Paley, sequency, and Hadamard orders have been completed by Equations (1.22), (1.21), and (1.14).

Besides the Paley order, Hadamard order, and sequency order, there are many other orders for Walsh functions. It has been proved that the universal order of Walsh functions, denoted by $\text{UW} = [\text{UW}(i, j)]$, $0 \leq i, j \leq 2^n - 1$, can be defined by

$$\text{UW}(i, j) = (-1)^{iAj'} \qquad (1.23)$$

for some non-singular matrix $A = [A(i,j)]$ of size $n \times n$, where $\mathbf{i} := (i_0, i_1, \ldots, i_{n-1})$ and $\mathbf{j} := (j_0, j_1, \ldots, j_{n-1})$ are the binary expressions of the integers i and j, respectively.

For example, if

$$A = \begin{bmatrix} 0 & 0 & 0 & 0 & \ldots & 0 & 0 & 1 & 1 \\ 0 & 0 & 0 & 0 & \ldots & 0 & 1 & 1 & 0 \\ 0 & 0 & 0 & 0 & \ldots & 1 & 1 & 0 & 0 \\ & & & & \ldots & & & & \\ 0 & 1 & 1 & 0 & \ldots & 0 & 0 & 0 & 0 \\ 1 & 1 & 0 & 0 & \ldots & 0 & 0 & 0 & 0 \end{bmatrix},$$

$$\begin{bmatrix} 0 & 0 & 0 & \ldots & 0 & 0 & 1 \\ 0 & 0 & 0 & \ldots & 0 & 1 & 0 \\ 0 & 0 & 0 & \ldots & 1 & 0 & 0 \\ & & & \ldots & & & \\ 0 & 0 & 1 & \ldots & 0 & 0 & 0 \\ 0 & 1 & 0 & \ldots & 0 & 0 & 0 \\ 1 & 0 & 0 & \ldots & 0 & 0 & 0 \end{bmatrix}$$

or

$$\begin{bmatrix} 1 & 0 & 0 & \ldots & 0 & 0 & 0 \\ 0 & 1 & 0 & \ldots & 0 & 0 & 0 \\ 0 & 0 & 1 & \ldots & 0 & 0 & 0 \\ & & & \ldots & & & \\ 0 & 0 & 0 & \ldots & 1 & 0 & 0 \\ 0 & 0 & 0 & \ldots & 0 & 1 & 0 \\ 0 & 0 & 0 & \ldots & 0 & 0 & 1 \end{bmatrix},$$

then the Walsh functions in Equation (1.23) are sequency ordered, Paley ordered, and Hadamard ordered, respectively.

Because the number of non-singular matrices, in $GF(2)$ and of size $n \times n$, is enumerated by

$$S(n) := \begin{cases} \prod_{i=1}^{n/2}(2^{n+1} - 2^{2i}) & n \text{ even}, \\ \\ \prod_{i=0}^{(n-1)/2}(2^n - 2^{2i}) & n \text{ odd}, \end{cases}$$

there are at all $S(n)$ different orders for the Walsh functions defined by binary matrices of size $n \times n$.

1.2   Orthogonality and Completeness

The mathematical techniques of studying functions, signals, and systems through series expansions in orthogonal complete sets of basis functions are now a standard tool in all branches of science and engineering. This section introduces the orthogonality and completeness of Walsh functions and matrices.

1.2.1   Orthogonality

A family of real-valued functions $S_n(t)$, $n = 0, 1, \ldots$, is said to be orthogonal with weight K, a non-negative constant, over the interval $0 \le t \le T$ if each pair of its member functions, say $S_n(t)$ and $S_m(t)$, are orthogonal with each other, i.e.,

$$\int_0^T K S_n(t) S_m(t) dt = \begin{cases} K \text{ if } n = m \\ 0 \text{ if } n \ne m. \end{cases} \tag{1.24}$$

If the constant K is one, then the family is normalized and is called an orthonormal family of functions. An orthogonal non-normalized family can always be converted into an orthonormal family.

Theorem 1.2.1 ([1] – [4], [7]) *The family of Walsh functions* $\{\text{Wal}_P(n, t) : 0 \le n \le 2^m - 1, 0 \le t < 1\}$ *is orthonormal over the interval* $[0, 1)$. *In other words,*

$$\int_0^1 \text{Wal}_P(u, t) \text{Wal}_P(v, t) dt = 0, \quad 0 \le u \ne v \le 2^m - 1 \tag{1.25}$$

and

$$\int_0^1 [\text{Wal}_P(n, t)]^2 = 1, \quad n = 0, 1, 2, \ldots. \tag{1.26}$$

Proof. The equation (1.26) is trivial, because of the identity $[\text{Wal}_P(n, t)]^2 = 1$. The equation (1.25) is based on the facts that (1): the nontrivial rows of the discrete sampling of $\text{Wal}_P(n, t)$, or equivalently the Walsh matrix W_p, are balanced by 1s and -1s; and (2): the functions $\{\text{Wal}_P(n, t) : 0 \le n \le 2^m - 1, 0 \le t < 1\}$ are of waveforms with their space being larger than or equal to $1/2^m$.   **Q.E.D.**

From Theorem 1.2.1 and Equations (1.22), (1.21), and (1.14), it is easy to see that the Walsh functions in both Hadamard and sequency orders are also orthonormal over the interval $[0, 1)$.

1.2.2   Completeness

Let $f(t)$ be a signal time function defined over a time interval $(0, T)$, and $S_n(t)$, $n = 0, 1, \ldots$, be a family of orthogonal functions. When we use the following series

$$\sum_{n=0}^{N-1} C_n S_n(t), \quad N = 1, 2, \ldots,$$

to represent the function $f(t)$, the resultant mean square approximation error(MSE) is

$$\text{MSE} = \int_0^T \left[f(t) - \sum_{n=0}^{N-1} C_n S_n(t) \right]^2 dt. \tag{1.27}$$

The family $S_n(t)$, $n = 0, 1, \ldots$, is called a complete orthogonal function series, if the MSE in Equation (1.27) monotonically decreases to zero as the integer N increases to infinite, where the coefficient C_n, $n = 0, 1, \ldots$, is defined by

$$C_n = \frac{1}{T} \int_0^T f(t) S_n(t) dt. \tag{1.28}$$

A function series is called incomplete if it is not complete. For example, the Rademacher functions defined by Equation (1.1) are incomplete.

A complete orthogonal function series $S_n(t)$ is always a closed series, or equivalently, there exists no $f(t)$ that satisfies both $0 < \int_0^T f^2(t) dt < \infty$ and $\int_0^T f(t) S_n(t) dt = 0$ for each integer n. More generally, a function series is said to be complete (or closed) if no function exists which is orthogonal to every function of the series unless the integral of the square of the function is itself zero.

One of the most important complete orthogonal function series is the trigonometric functions sine and cosine defined by

$$S_n(t) = \sqrt{2} \cos(2\pi n t) \quad \text{or} \quad \sqrt{2} \sin(2\pi n t). \tag{1.29}$$

Thus every function can be expressed as a sum of sinusoidal components (Fourier series), viz.,

$$f(t) = \frac{a_0}{2} + \sum_{k=1}^{\infty} [a_k \cos(k\omega_0 t) + b_k \sin(k\omega_0 t)].$$

Similarly to the sine–cosine functions, we have the following:

Theorem 1.2.2 ([1]–[4],[7]) *The family of Walsh functions* $\{\text{Walp}(n,t) :$ $0 \leq n \leq 2^m - 1, 0 \leq t < 1\}$ *is complete and orthogonal. In other words, if* $f(t)$ *is a function satisfying both* $f(t) \in L^2[0,1)$, *i.e.,* $f^2(t)$ *is integrable over* $[0,1)$, *and*

$$\int_0^1 f(t)\text{Walp}(n,t)dt = 0, \quad n = 0,1,2,\ldots,$$

then the $f(t)$ *is a zero function.*

This theorem together with Equations (1.22), (1.21), and (1.14) implies that the Walsh functions in both Hadamard and sequency orders are also complete orthogonal.

Moreover, it has been proved that:

Theorem 1.2.3 ([1], [2], [3], [4],[7]) *Every function* $f(t) \in L^2[0,1)$ *can be decomposed as*

$$f(t) = a_0\text{Cal}(0,t) + \sum_{k=1}^{\infty}[a_k\text{Cal}(k,t) + b_k\text{Sal}(k,t)]$$

$$= \sum_{k=0}^{\infty}[a_k\text{Cal}(k,t) + b_{k+1}\text{Sal}(k+1,t)]$$

where the coefficients a_k *and* b_k *are defined by*

$$a_k = \int_0^1 f(t)\text{Cal}(k,t)dt, \quad k = 0,1,2,\ldots,$$

and

$$b_k = \int_0^1 f(t)\text{Sal}(k,t)dt, \quad k = 1,2,3,\ldots.$$

The Parseval Identity is now stated as:

Theorem 1.2.4 *If* $f(t) \in L^2[0,1)$, *then*

$$\int_0^1 [f(t)]^2 dt = \sum_{k=0}^{\infty}[a_k^2 + b_{k+1}^2],$$

where a_k *and* b_k *are the same as those in Theorem 1.2.3.*

1.3   Walsh Transforms and Fast Algorithms ([8])

The famous Fourier transforms are based on the orthogonal and complete families of functions, sine and cosine functions. Walsh matrices are sampled from another orthogonal and complete families of functions called Walsh functions, thus a great deal of research in the field of Walsh functions has been devoted to the study and development of easy and fast algorithms for computing Walsh transforms. The sampling principle in the framework of Walsh signals and transforms were studied by Kak (in 1970) and Cheng and Johnson (1973). Their sequency-based sampling theorem states that: 'A causal time function $x(t)$ sequency-band limited to $B = 2^k$ zeros per second, can be uniquely reconstructed from its samples at every $T = 1/B$ seconds for all positive time'. The simplest version of a Walsh transform which provides the coefficients in Hadamard ordering, was given by Whechel and Guinn (in 1968). They actually derived the so called fast Hadamard transform (FHT) by using the recursive structure of the Hadamard matrix and applying a modified form of the Cooley–Tukey FFT algorithm. In 1970 Henderson showed how to circumvent the difficulty of obtaining the Walsh coefficients in bit-reversed order when the data are arranged in the Hadamard forward order. Andrews and Caspari (1970) presented a generalized approach for fast digital Walsh–Hadamard spectral analysis using the direct multiplication of matrices. Another method for computing the Walsh–Hadamard transform and the R–transform was presented in 1970 by Ulman, where the transform coefficients were produced in increasing sequency order.

A real time technique for computing the digital Walsh–Hadamard transform of two-dimensional pictures was developed by Alexandridis and Klinger (in 1972) by using the matrix formulation. A hardware implementation for real-time, parallel Walsh–Hadamard transformation was also provided. In 1972, another sequency-ordered fast Walsh transform (FWT) computation algorithm was presented. This algorithm was based on the Cooley–Tukey type of FHT and required an equal computational effort as the FHT.

This section is devoted to the study of the Walsh–Hadamard transform, which is perhaps the most well known of the non-sinusoidal orthogonal transforms. Fast algorithms to compute these transforms are developed and the notion of Walsh spectra is introduced.

1.3.1   Walsh Ordered Walsh–Hadamard Transforms ([8])

For details about the contents of this subsection the readers are recommended to [1].

Before considering the development of Walsh–Hadamard transforms, it is instructive to study some aspects of representing a given continuous signal $f(t)$ in the form of a Walsh series. For the purposes of discussion we will assume that $f(t)$ is defined on the half-open unit interval $t \in [0,1)$. From Theorem 1.2.3 it is known that every integrable function $f(t)$ may be represented by the following Walsh series

$$f(t) = a_0 + a_1 \mathrm{Wal}(1,t) + a_2 \mathrm{Wal}(2,t) + \ldots , \qquad (1.30)$$

where $t \in (0,1)$ and for each integer n the coefficient a_n is

$$a_n = \int_0^1 f(t)\mathrm{Wal}(n,t)dt. \qquad (1.31)$$

In other words, we have the following pair of transforms:

1. The forward transform is

$$f(t) = \sum_{n=0}^{\infty} F_n \mathrm{Wal}(n,t); \qquad (1.32)$$

2. The reverse transform is

$$F_n = \int_0^1 f(t)\mathrm{Wal}(n,t)dt. \qquad (1.33)$$

If the integration in Equation (1.33) is replaced by summation through the use of the trapezium rule of $N = 2^m$ sampling points f_i, then we obtain the following finite discrete Walsh transform pair

$$F_n = \frac{1}{N} \sum_{i=0}^{N-1} f_i \mathrm{Wal}(n,i), \quad 0 \le n \le N-1; \quad \text{(Forward Transform)} \quad (1.34)$$

and

$$f_i = \sum_{n=0}^{N-1} F_n \mathrm{Wal}(n,i), \quad 0 \le i \le N-1; \quad \text{(Inverse Transform)}. \quad (1.35)$$

Let $W = [\mathrm{Wal}(n,i)]$, $0 \le n, i \le N - 1$, be the sequency-ordered Walsh matrix of size $N \times N$. Then the transforms in Equations (1.34) and (1.35) can be equivalently stated as

$$F = \frac{1}{N} fW \qquad (1.36)$$

and

$$f = FW, \qquad (1.37)$$

where $f := (f_0, \ldots, f_{N-1})$ and $F := (F_0, \ldots, F_{N-1})$. The transforms defined by Equations (1.36) and (1.37) are called the Walsh-ordered Walsh–Hadamard transforms (WHT$_\mathrm{W}$). It is evident from this definition that N^2 additions and subtractions are required to compute the WHT$_\mathrm{W}$ coefficients F. In the coming text, we will develop a fast algorithm which yields the F in only $N \log_2 N$ additions and subtractions.

The fast Walsh-ordered Walsh–Hadamard transform (FWHT$_\mathrm{W}$) is illustrated in the following four steps:

Step 1: Bit-reverse the input and order it in ascending index order. If $f = : (f_0, \ldots, f_{N-1})$ is the given data sequence, then we denote the bit-reversed sequence in ascending index order by $f' = : (f'_0, \ldots, f'_{N-1})$;

Step 2: Define a reversal, which is illustrated by: without a reversal, we would have

$$\left. \begin{array}{l} f'_{k+1}(s) = f'_k(s) + f'_k(s+2) \\ f'_{k+1}(s+1) = f'_k(s+1) + f'_k(s+3) \end{array} \right\} \text{ additions}$$

and

$$\left. \begin{array}{l} f'_{k+1}(s+2) = f'_k(s) - f'_k(s+2) \\ f'_{k+1}(s+3) = f'_k(s+1) - f'_k(s+3) \end{array} \right\} \text{ subtractions.}$$

However, with a reversal we have

$$\left. \begin{array}{l} f'_{k+1}(s) = f'_k(s) - f'_k(s+2) \\ f'_{k+1}(s+1) = f'_k(s+1) - f'_k(s+3) \end{array} \right\} \text{ subtractions}$$

and

$$\left. \begin{array}{l} f'_{k+1}(s+2) = f'_k(s) + f'_k(s+2) \\ f'_{k+1}(s+3) = f'_k(s+1) + f'_k(s+3) \end{array} \right\} \text{ additions.}$$

Step 3: Define a block, which is a group of additions/subtractions which are disconnected from their neighbors, above or below;

Step 4: List the rules for locating the blocks where reversals occur.

1.3.2   Hadamard Ordered Walsh–Hadamard Transforms

For details of the contents of this subsection the readers are recommended to the paper [8] .

Let H_N be the Walsh–Hadamard matrix of size $N \times N = 2^n \times 2^n$ that is recursively constructed by Equation (1.12). The forward Hadamard-ordered Walsh–Hadamard transform (WHT$_H$) of the data sequence $g = (g_0, g_1, \ldots, g_{N-1})$ is defined by

$$G := (G_0, \ldots, G_{N-1}) = \frac{1}{N} g H_N. \tag{1.38}$$

Because the Walsh–Hadamard matrix H_N is symmetric (i.e. $H_N' = H_N$) and orthogonal (i.e., $H_N' H_N = N I_N$) and the inverse matrix of H_N is proportional to itself (i.e., $(H_N)^{-1} = H_N / N$), the inverse transform of Equation (1.38) is

$$g = G H_N, \tag{1.39}$$

which is called the inverse Hadamard-ordered Walsh–Hadamard transform (IWHT$_H$).

The other equivalent definition of the Walsh–Hadamard matrix H_N, $N = 2^n$, is

$$H_N = \overbrace{H_2 \otimes H_2 \otimes \ldots \otimes H_2}^{n},$$

where $\otimes$ stands for the direct multiplication of matrices.

Thus the matrix H_N can be decomposed as the multiplication of the following n sparse matrices

$$H_N = \prod_{k=0}^{n-1} (I_{2^k} \otimes H_2 \otimes I_{2^{n-1-k}}). \tag{1.40}$$

Setting Equation (1.40) into (1.38) and (1.39), a fast Hadamard-ordered Walsh–Hadamard transform is obtained which requires only $N \log_2 N$ additions and subtractions.

Bibliography

[1] N.Ahmed and K.R.Rao, *Orthogonal Transforms for Digital Signal Processing*, Springer-Verlag, Berlin, 1975.

[2] K.G. Beauchamp, *Applications of Walsh and Related Functions*, Academic Press, London, 1984.

[3] S.G. Tzafestas, *Walsh Functions in Signal and Systems Analysis and Design*, Van Nostrand Reinhold Co., New York, 1985.

[4] Y.X. Yang, and X.D.Lin, *Coding and Cryptography*, PPT Press, Beijing, 1992.

[5] H.Y.Song and S.W.Golomb, *On the Existence of Cyclic Hadamard Difference Sets*, IEEE Trans. IT, Vol.40, No.4, pp1266-1270, 1994.

[6] C.Gotsman, *A Note on Functions Governed by Walsh Expressions*, IEEE Trans. on Inform. Theory, Vol.37, No.3, pp694-695, 1991.

[7] K.G. Beauchamp, *Walsh Functions and Their Applications*, Academic Press, London, 1975.

[8] E.D. Banta, *Coding and Decoding Nonlinear Block Codes Using Logical Hadamard Transform*, IEEE Trans. IT, Vol.24, No.4, pp761-763, 1978.

Chapter 2
Hadamard Matrices

Hadamard matrices were originally investigated as a family of orthogonal matrices by Sylvester in 1857. In 1893. Hadamard proved that if $A = [A_{i,j}]$, $0 \leq i,j \leq n-1$, is a matrix of size $n \times n$ (or called 'of order n'), then

$$| \det A |^2 \leq \prod_{i=0}^{n-1} \sum_{j=0}^{n-1} | A_{i,j} |^2 . \qquad (2.1)$$

Furthermore, Hadamard discovered that if $| A_{i,j} | \leq 1$ and the equality (2.1) is satisfied then the matrix A is (± 1)-valued and its size m is equal to 1, 2, or $4t$. Subsequently matrices satisfying the equality (2.1) came to be called Hadamard matrices, which were further studied by Scarpis in 1898. The next major milestone was owed to Paley, in 1933. In 1944 and 1947 Williamson obtained more results of considerable interest. With regard to the practical applications of Hadamard matrices, M. Hall, Jr., L. Baumert, and S. Golomb working at the U.S. Jet Propulsion Laboratories (JPL) sparked the interest over Hadamard matrices over the past 30 years([1]). In the sixties the JPL was working toward building the Mariner and Voyager space probes to visit Mars and the other planets of the solar system. In order to obtain high quality color pictures of the backs of the planets, the space probes have to take three black and white pictures through red, green, and blue filters. Each picture is then divided into a 1000×1000 array of black and white pixels. Each pixel is graded on a scale of 1 to 16. These grades are then used to choose a code word in an eight error correcting code based on the Hadamard matrix of order 32. The code word is transmitted to the earth. After error correction, the three black and white pictures are reconstructed and then a computer is used to obtain the colored pictures. Hadamard matrices are used for these code words for two reasons. First, error-correcting codes based on Hadamard matrices have good error correction capability and good decoding algorithms. Second, because Hadamard matrices are (± 1)-valued,

all the computer processing can be accomplished using additions and subtractions rather than multiplication.

Up to now, besides more and more applications, much of the research
work is devoted to the proof of the following famous Hadamard conjecture:

There exists at least one Hadamard matrix of size 1×1, 2×2 *and*
$4t \times 4t$, *for each positive integer* t.

This chapter concentrates on the basic definitions, constructions and
existence results about the Hadamard matrices. Many references are listed
at the end of this chapter so that interested readers can trace the up to
date results related to Hadamard matrices.

2.1   Definition

2.1.1   Hadamard Matrices

Although the Hadamard matrices can be defined by the equality (2.1),
in the following text we will use another equivalent and more intuitive
definition:

Definition 2.1.1 ([2]) *Let* $H = [H_{i,j}]$, $0 \leq i, j \leq n - 1$, *be a* (± 1)-*valued*
matrix of size $n \times n$. *Then* H *is called an Hadamard matrix if and only if*

$$HH' = nI_n, \tag{2.2}$$

where H' *stands for the transpose of* H *and* I_n *is the unit matrix of order*
n.

In other words, an Hadamard matrix is an orthogonal (± 1)-valued matrix.
Thus $H'H = HH'$. Up to now many Hadamard matrices have been
constructed. For example, the Hadamard matrices of order less than or
equal to 200 have been proved to exist for the following orders: 4, 8, 12,
16, 20, 24, 28, 32, 36, 40, 44, 48, 52, 56, 60, 64, 68, 72, 76, 80, 84, 88, 92,
96, 100, 104, 108, 112, 116, 120, 124, 128, 132, 136, 140, 144, 148, 152,
156, 160, 164, 168, 172, 176, 180, 184, 192, 196, 200. ([3])

It is obvious that the set of Hadamard matrices is closed under the
following five transforms:

1. Permuting the rows;

2. Permuting the columns;

3. Multiplying some rows by -1;

4. Multiplying some columns by -1;

5. Transposing.

By using these five transforms in suitable order, every Hadamard matrix can be changed to an Hadamard matrix such that its first row (resp. column) is all 1 which is called row-normal (resp. column-normal).

Definition 2.1.2 ([4]) *An Hadamard matrix is called normal iff it is both row-normal and column-normal.*

For example, the Hadamard-ordered Walsh matrices introduced in the last chapter are normal Hadamard matrices of order 2^m, $m = 1, 2, \ldots$. If H is a normal Hadamard matrix of order $4n$, then every row (column) except the first has $2n$ '+1's in each row (column); further, n '-1's in any row (column) overlap n '-1's in each other row (column).

An important necessary condition for Hadamard matrices is that their sizes should be the multiple of 4, precisely,

Theorem 2.1.1 ([1]) *A matrix of size $m \times m$, $m > 2$, is an Hadamard matrix only if $4 \mid m$.*

Proof. Let the matrix be $A = [A_{ij}]$. Because of the orthogonality we have

$$\sum_{i=0}^{m-1} (A_{1i} + A_{2i})(A_{1i} + A_{3i}) = \sum_{i=0}^{m-1} A_{1i}^2 = m. \qquad (2.3)$$

Whilst, on the other hand,

$$A_{1i} + A_{2i} = \pm 2, 0, \quad \text{and} \quad A_{1i} + A_{3i} = \pm 2, 0.$$

Thus, the summation in Equation (2.3) is of the form $4t$, because each of its terms is a multiple of 4. **Q.E.D.**

Clearly, the famous Hadamard conjecture is the inverse of this theorem. Another subclass of Hadamard matrices are the regular subclasses which are defined by:

Definition 2.1.3 ([5]) *An Hadamard matrix H is called regular if there exists some integer, say t, such that*

$$HJ = tJ.$$

Equivalently, the sum of each row of the regular Hadamard matrix H is equal to the same number t.

For example ([1]), the following matrix H is a regular Hadamard matrix of order 16 with $t = -4$:

$$
H = \begin{bmatrix}
-1 & 1 & 1 & 1 & 1 & 1 & -1 & -1 & -1 & -1 & 1 & -1 & -1 & -1 & -1 & -1 \\
1 & -1 & 1 & -1 & -1 & 1 & 1 & 1 & -1 & -1 & -1 & -1 & -1 & 1 & -1 & -1 \\
1 & 1 & -1 & -1 & -1 & 1 & -1 & -1 & 1 & 1 & -1 & -1 & -1 & -1 & 1 & -1 \\
1 & -1 & -1 & -1 & 1 & -1 & 1 & -1 & 1 & 1 & 1 & -1 & -1 & -1 & -1 & -1 \\
1 & -1 & -1 & 1 & -1 & -1 & -1 & 1 & -1 & 1 & 1 & -1 & 1 & -1 & -1 & -1 \\
1 & 1 & 1 & -1 & -1 & -1 & -1 & -1 & -1 & -1 & -1 & 1 & 1 & -1 & -1 & 1 \\
-1 & 1 & -1 & 1 & -1 & -1 & -1 & 1 & 1 & -1 & -1 & 1 & -1 & 1 & -1 & -1 \\
-1 & 1 & -1 & -1 & 1 & -1 & 1 & -1 & -1 & 1 & -1 & -1 & 1 & 1 & -1 & -1 \\
-1 & -1 & 1 & 1 & -1 & -1 & 1 & 1 & -1 & -1 & 1 & -1 & 1 & -1 & -1 & 1 \\
-1 & -1 & 1 & -1 & 1 & -1 & -1 & 1 & 1 & -1 & -1 & -1 & -1 & 1 & -1 & 1 \\
1 & -1 & -1 & 1 & 1 & -1 & -1 & -1 & -1 & -1 & -1 & -1 & -1 & 1 & 1 & 1 \\
-1 & -1 & -1 & 1 & -1 & 1 & 1 & 1 & -1 & 1 & -1 & -1 & -1 & 1 & -1 & 1 \\
-1 & -1 & -1 & -1 & 1 & 1 & -1 & 1 & -1 & 1 & -1 & 1 & -1 & -1 & -1 & 1 \\
-1 & 1 & -1 & -1 & -1 & -1 & 1 & 1 & -1 & -1 & 1 & -1 & -1 & -1 & -1 & 1 & 1 \\
-1 & -1 & 1 & -1 & -1 & -1 & -1 & -1 & 1 & 1 & 1 & -1 & -1 & 1 & -1 & 1 & 1 \\
-1 & -1 & -1 & -1 & -1 & 1 & -1 & -1 & -1 & -1 & -1 & 1 & 1 & 1 & 1 & 1 & -1
\end{bmatrix}.
$$

An Hadamard matrix H is called a symmetric Hadamard matrix iff $H' = H$, exactly one of the top left corners of H and $-H$ is 1. Rename by B the matrix with 1 as its top left corner. Multiplying by -1 those columns of B where the top most row is -1ed. Similarly, multiplying by -1 those rows of B where the left most column is -1ed. Then the matrix B is changed into the following normal form

$$
A = \begin{pmatrix} 1 & e \\ e' & C \end{pmatrix}, \tag{2.4}
$$

where $e = (1, 1, \ldots, 1)$ is the all-ones vector of length $m-1$. The matrix C in Equation (2.4) is called the kernel of the symmetric Hadamard matrix H.

Lemma 2.1.1 *Let C be the kernel of a symmetric Hadamard matrix H of order m. Then*

$$C' = C; \quad CJ_{m-1} = J_{m-1}C = -J_{m-1}; \quad C^2 = mI_{m-1} - J_{m-1}. \quad (2.5)$$

Proof. Without loss of generality we assume that

$$H = \begin{pmatrix} 1 & e \\ e' & C \end{pmatrix}. \qquad (2.6)$$

Because $H = H'$ we have $C = C'$. From the equation

$$mI_m = HH'$$
$$= \begin{pmatrix} 1 & e \\ e' & C \end{pmatrix}\begin{pmatrix} 1 & e \\ e' & C \end{pmatrix}$$
$$= \begin{pmatrix} m & e + eC \\ e' + Ce' & e'e + C^2 \end{pmatrix},$$

we obtain the equations

$$eC = -e, \quad C^2 = mI_{m-1} - e'e = mI_{m-1} - J_{m-1}.$$

The lemma follows.   **Q.E.D.**

Definition 2.1.4 *An Hadamard matrix H of order m is called a skew Hadamard matrix iff*

$$H = S + I, \quad \text{and} \quad S' = -S. \qquad (2.7)$$

Lemma 2.1.2 *Let H be a skew Hadamard matrix of order m, see Equation (2.7). Then*
$$SS' = (m - 1)I. \qquad (2.8)$$

Proof. The lemma follows the identity:

$$mI = HH'$$
$$= (S + I)(S' + I)$$
$$= SS' + I.$$

Q.E.D.

The condition $S' = -S$ implies that all of the diagonal elements of S are zero. Multiplying '-1's to some rows and columns, the matrix S can be changed to

$$\begin{pmatrix} 0 & e \\ -e' & W \end{pmatrix}. \tag{2.9}$$

The matrix W in Equation (2.9) is called the kernel of the skew Hadamard matrix H.

Lemma 2.1.3 *Let W be the kernel of a skew Hadamard matrix H of order m. Then*

$$WW' = (m-1)I_{m-1} - J_{m-1}, \tag{2.10}$$

and

$$WJ_{m-1} = 0, \quad W' = -W. \tag{2.11}$$

Proof. Without loss of generality we assume that

$$H = I + \begin{pmatrix} 0 & e \\ -e' & W \end{pmatrix} := I + S.$$

The equation $S' = -S$ implies $W' = -W$. From Lemma 2.1.2 we have

$$(m-1)I = \begin{pmatrix} 0 & e \\ -e' & W \end{pmatrix} \begin{pmatrix} 0 & -e \\ e' & W' \end{pmatrix}$$

$$= \begin{pmatrix} ee' & eW' \\ We' & e'e + WW' \end{pmatrix}.$$

Hence

$$We' = 0, \quad J_{m-1} + WW' = (m-1)I_{m-1}.$$

The lemma follows.   **Q.E.D.**

Definition 2.1.5 *Let M be a skew Hadamard matrix, and N a symmetric Hadamard matrix of the same order. The pair M and N are called amicable Hadamard matrices if*

$$MN = NM'. \tag{2.12}$$

The equation (2.12) implies that MN is also a symmetric matrix.

Lemma 2.1.4 *If $M = S + I$ and N are a pair of amicable Hadamard matrices, then*

$$SN = NS'. \tag{2.13}$$

Proof. Because of $MN = (S+I)N = SN + N$, and $NM' = N(S'+I) = NS' + N$, the proof is finished by Equation (2.12).   **Q.E.D.**

2.1.2   Hadamard Design

Let $S = \{s_1, s_2, \ldots, s_v\}$ be a v-element set and B$=\{B_1, B_2, \ldots, B_b\}$ a family consisted of b subsets of S. The family B is called a (b, v, r, k, λ)-balanced incomplete block design, or a (b, v, r, k, λ)-BIBD in short, if and only if all of the following three conditions are satisfied.

1. Each subset B_i, $1 \leq i \leq b$, contains the same number, k, of elements of S;

2. Each element $s \in S$ appears in exactly r subsets of B;

3. Each pair $\{s_i, s_j\}$ is contained in just λ subsets of B.

The family B can also be uniquely described by its associated matrix $A = [A(i, j)]$, which is a $(0, 1)$-valued matrix of size $b \times v$ and for each $1 \leq i \leq b$, $1 \leq j \leq v$,

$$A(i, j) = \begin{cases} 1 \text{ if } s_j \in B_i, \\ 0 \text{ if } s_j \notin B_i. \end{cases} \tag{2.14}$$

In particular, if $b = v$ and $r = k$, then a (b, v, r, k, λ)-BIBD is called a symmetrical balanced incomplete block design, or shortly, a (v, k, λ)-design. It is known that the associated matrix of every (v, k, λ)-design satisfies the necessary conditions presented in the following lemma.

Lemma 2.1.5 *If $A = [A(i, j)]$ is the associated matrix of some (v, k, λ)-design, then,*

$$\lambda(v - 1) = k(k - 1); \tag{2.15}$$

$$A'A = (k - \lambda)I_v + \lambda J_v; \tag{2.16}$$

$$Aw_v = kw_v; \tag{2.17}$$

$$\frac{(v + \lambda)}{2} \geq k \geq \sqrt{\lambda v}; \tag{2.18}$$

If v is even, then

$$k - \lambda = c^2 \tag{2.19}$$

for some integer c, where I_v is the unit matrix of order v; J_v the all-one matrix of order v; and w_v the all-one column vector of length v.

One of the useful corollaries of Lemma 2.1.5 is

Corollary 2.1.1 Let $A = [A(i, j)]$ be the associated matrix of some (v, k, λ)-design. Then,

$$A'A = (k - \lambda)I_v + \lambda J_v; \tag{2.20}$$

$$AA' = (k - \lambda)I_v + \lambda J_v; \tag{2.21}$$

$$J_v A = kJ_v; \tag{2.22}$$

$$AJ_v = kJ_v. \tag{2.23}$$

Lemma 2.1.6 Let $A = [A(i, j)]$ be a $(0, 1)$-valued non-singular matrix of order v. Then Equation (2.20) is true iff Equation (2.21) is true. More-over, the A is an associate matrix of some (v, k, λ)-design iff Equation (2.20) (or (2.21)) is satisfied.

One of the most important such designs is the Hadamard design, which is defined by:

Definition 2.1.6 A $(4t - 1, 2t - 1, t - 1)$-design is called an Hadamard design.

It will be clear soon that Hadamard designs and Hadamard matrices are uniquely generated by each other. In fact:

Theorem 2.1.2 There exists a $(4t - 1, 2t - 1, t - 1)$-Hadamard design if and only if there exists an Hadamard matrix of order 4t.

Proof. At first, we prove that each Hadamard matrix of order $4t$ produces a $(4t - 1, 2t - 1, t - 1)$-Hadamard design.

Without the loss of generality we assume that $H = [H_{ij}]$ is a normal Hadamard matrix of order $4t$. Let A_1 be the sub-matrix of H which is produced by canceling the first row and column of H. Thus A_1 is of order $4t - 1$. Let A be the matrix defined by

$$A = \frac{1}{2}(A_1 + J), \tag{2.24}$$

where J is the all-one matrix of order $4t - 1$. In other words, A is transformed from A_1 by changing -1 to 0.

Let $e = (1, 1, \ldots, 1)$ be an all-one vector of length $4t - 1$. Then the equation (2.2) is equivalent to

$$\begin{pmatrix} 1 & e \\ e' & A_1 \end{pmatrix} \begin{pmatrix} 1 & e \\ e' & A_1' \end{pmatrix} = \begin{pmatrix} 4t & e + eA_1' \\ e' + A_1e' & e'e + A_1A_1' \end{pmatrix} = 4tI. \tag{2.25}$$

Because of $e'e = J_{4t-1}$, we have

$$\begin{cases} A_1A_1' = 4tI - J, \\ A_1J = -J, \\ JA_1' = -J. \end{cases} \tag{2.26}$$

Thus the following two equations are obtained:

$$\begin{aligned} AA' &= \frac{1}{4}(A_1 + J)(A_1' + J) \\ &= tI + (t - 1)J \end{aligned} \tag{2.27}$$

and

$$\begin{aligned} AJ &= \frac{1}{2}(A_1 + J)J \\ &= \frac{1}{2}(A_1J + J^2) \\ &= (2t - 1)J. \end{aligned} \tag{2.28}$$

Equation (2.27) implies that $\det A \neq 0$, for nontrivial t. Thus from Equations (2.27) and (2.28) we know that A is the associated matrix of some $(4t - 1, 2t - 1, t - 1)$-design, which results in the existence of a $(4t - 1, 2t - 1, t - 1)$-Hadamard design.

We then try to prove that an Hadamard matrix of order $4t$ can be constructed by a $(4t - 1, 2t - 1, t - 1)$-Hadamard design.

If there exists a $(4t-1, 2t-1, t-1)$-Hadamard design with its associated matrix be A, then from Corollary 2.1.1 we know that the equations (2.27) and (2.28) are satisfied. Let

$$A_1 = 2A - J. \tag{2.29}$$

Then the (± 1)-valued matrix A_1 satisfies

$$\begin{aligned} A_1 J &= (2A - J)J \\ &= 2AJ - (4t - 1)J \\ &= -J, \end{aligned}$$

$$JA_1' = (A_1 J)' = (-J)' = -J,$$

and

$$\begin{aligned} A_1 A_1' &= (2A - J)(2A' - J) \\ &= 4AA' - 2AJ - 2JA' + J^2 \\ &= 4tI - J. \end{aligned}$$

Hence the matrix $\begin{pmatrix} 1 & e \\ e' & A_1 \end{pmatrix}$ is the desired Hadamard matrix of order $4t$, where e is the all-one row of length $4t - 1$.   **Q.E.D.**

Besides Theorem 2.1.2 there are some other relationships between (v, k, λ)-designs and Hadamard matrices. For example, by making use of the design theory we can prove the following theorems, which provide with us some necessary and/or sufficient conditions of regular Hadamard matrices.

Theorem 2.1.3 *If H is a regular Hadamard matrix of order $4m$ then there exists a positive integer s such that $m = s^2$ and*

$$HJ = JH = \pm 2sJ.$$

Proof. Because H is regular,

$$HH' = H'H = 4mI, \qquad \text{and} \qquad HJ = tJ, \tag{2.30}$$

for some integer t. Let A be the $(0, 1)$-valued matrix defined by

$$A := \frac{1}{2}(H + J).$$

Then, by using Equation (2.30), we have

$$\begin{aligned}
AA' &= \frac{1}{4}(H + J)(H' + J) \\
&= mI + (m + t/2)J
\end{aligned} \tag{2.31}$$

and

$$AJ = (2m + t/2)J. \tag{2.32}$$

Hence t is even. Note that $\det A \neq 0$. Then from Lemma 2.1.6, it is known that the matrix A is the associated matrix of some $(4m, 2m+t/2, m+t/2)$-design. Hence Equation (2.15) becomes

$$(m + t/2)(4m - 1) = (2m + t/2)(2m + t/2 - 1)$$

or equivalently

$$m = (t/2)^2 := s^2 \qquad \text{and} \qquad t = \pm 2s. \tag{2.33}$$

On the other hand, Equation (2.30) and the identity $JH'H = 4mJ$ imply

$$JH = \frac{4m}{tJ} = tJ.$$

Q.E.D.

Theorem 2.1.4 *There exists a $(4u^2, 2u^2 \pm u, u^2 \pm u)$-design if and only if there exists a regular Hadamard matrix of order $4u^2$.*

Proof. On the one hand, if H is a regular Hadamard matrix of order $4u^2$, then by Equation (2.33) we have $t = \pm 2u$. Equations (2.31), (2.32), and Lemma 2.1.6 imply that the matrix $A := (H + J)/2$ is the associated matrix of some $(4u^2, 2u^2 \pm u, u^2 \pm u)$-design.

On the other hand, if there is a $(4u^2, 2u^2 \pm u, u^2 \pm u)$-design with its associated matrix A, then

$$AA' = u^2I + (u^2 \pm u)J$$

and

$$AJ = JA = (2u^2 \pm u)J.$$

Let H be the (± 1)-valued matrix defined by $H := 2A - J$. Then

$$HH' = (2A - J)(2A' - J) = 4u^2 I,$$

and

$$HJ = JH = \pm 2uJ.$$

Therefore H is the regular Hadamard matrix of order $4u^2$. **Q.E.D.**

Definition 2.1.7 *Let v and k be two positive integers, and $D = \{a_1, a_2, \dots, a_k\}$ (mod v) a set of k different (mod v) integers. The set D is called a (v, k, λ)-cyclic difference set, abbreviated as (v, k, λ)-CDS, iff for each $d \not\equiv 0$ (mod v), there are exactly λ pairs (a_i, a_j) such that*

$$d \equiv a_i - a_j (\mathrm{mod} v).$$

For example, let $v = 11$ and $k = 5$. Then the set $D = \{1, 3, 4, 5, 9\}$ is a $(11, 5, 2)$-CDS.

 From the following lemma it is clear that the cyclic difference sets are, in fact, subclasses of (v, k, λ)-designs.

Lemma 2.1.7 *A set $D = \{a_1, a_2, \dots, a_k\}$ (mod v) of k different (mod v) integers is a (v, k, λ)-CDS if and only if the family $\mathcal{B} = \{B_1, B_2, \dots, B_v\}$ forms a (v, k, λ)-design. Where $B_1 := D$, $B_2 := \alpha(B_1)$, $\dots$, $B_v := \alpha(B_{v-1})$, and α the one-to-one mapping from the ring Z_v to itself defined by*

$$\alpha : i \longrightarrow i + 1 (\mathrm{mod} v).$$

A direct corollary of Lemma 2.1.7 and Theorem 2.1.2 is the following

Corollary 2.1.2 *If there is a $(4t - 1, 2t - 1, t - 1)$-CDS, then there exists an Hadamard matrix of order $4t$.*

 Two of the important cyclic difference sets are stated in the following lemma.

Lemma 2.1.8 *1. Let p be a prime, e a positive integer, and $p^e = 4t - 1$. Let G be the additive group of the field $GF(p^e)$. Then the set $Q := \{g^2 : g \in GF(p^e)\}$ forms a $(4t - 1, 2t - 1, t - 1)$-CDS in G;*

2. Let p and q be two different primes such that $p^e = q^f - 2$. Let G be the group $G := \{(a,b) : a \in GF(p^e), b \in GF(q^f)\}$ with its addition $+$ and production $.$ defined by

$$(a,b) + (c,d) := (a+c, b+d); \text{ and } (a,b).(c,d) = (ac, bd).$$

Let $D := D_1 \cup D_2 \cup D_3$, where

$$D_1 := \{(a,b) : a = x^2 \neq 0, b = y^2 \neq 0, x \in GF(p^e), y \in GF(q^f)\},$$

$$D_2 := \{(c,d) : c \text{ and } d \text{ are non-quadratic in } GF(p^e) \text{ and } GF(q^f),$$
$$\text{respectively } \}$$

and

$$D_3 := \{(g,0) : g \in GF(p^e)\}.$$

Then the set D is a $(p^e q^f, (p^e q^f - 1)/2, (p^e q^f - 3)/4)$-CDS in the group G.

Applying Lemma 2.1.8 to Theorem 2.1.2, the following theorem is obtained.

Theorem 2.1.5 *There exist Hadamard matrices of orders m if $m = p^e + 1 \equiv 0 \pmod 4$ or $m = p^e q^f + 1$, $q^f = p^e + 2$.*

Definition 2.1.8 ([6]) *Let V be an additive Abelian group of order v. Let $S_1, S_2, \ldots, S_n$, $S_i = \{s_{i1}, s_{i2}, \ldots, s_{ik_i}\}$, $1 \leq i \leq n$, be subsets of V. If, for each nonzero element $g \in V$, the equation*

$$g = s_{ij} - s_{im}$$

has exactly λ solutions, then $S_1, S_2, \ldots, S_n$ is called an n-$\{v; k_1, \ldots, k_n; \lambda\}$ supplementary difference sets (SDS). If $k_1 = k_2 = \ldots = k_n = k$, we abbreviate it as n-$(v; k; \lambda)$ SDS.

2.1.3 Williamson Matrices

Theorem 2.1.6 *If there are four $m \times m$ (± 1)-valued matrices A_1, A_2, A_3, and A_4 such that the following three conditions are satisfied:*

C1: *They are symmetric;*

C2: $A_i A_j = A_j A_i$, *for all $1 \leq i, j \leq 4$;*

C3: *The following equation is true:*

$$A_1^2 + A_2^2 + A_3^2 + A_4^2 = 4mI_m. \tag{2.34}$$

Then the matrix H defined by

$$H = \begin{pmatrix} A_1 & A_2 & A_2 & A_4 \\ -A_2 & A_1 & -A_4 & A_3 \\ -A_3 & A_4 & A_1 & -A_2 \\ -A_4 & -A_3 & A_2 & A_1 \end{pmatrix} \tag{2.35}$$

is an Hadamard matrix of order 4m.

This theorem can be proved by direct verification. The four matrices satisfying the conditions C1, C2 and C3 are called Williamson matrices. Thus constructions of Williamson matrices imply the constructions of Hadamard matrices. In the coming text we concentrate on Williamson matrices.

A matrix $B = [B(i,j)]$, $0 \le i,j \le m-1$, is called cyclic if it is recursively produced by the first row by

$$B(i,j) = B(0, (j-i) \bmod m).$$

This kind of matrix is denoted by $B = circ(B(0,0), B(0,1), \ldots, B(0,m-1))$.

Let $U = circ(0,1,0,\ldots,0)$ be a cyclic matrix. And let A, B, C, D be four (± 1)-valued matrices defined respectively by

$$A = a_0 I + a_1 U + \ldots + a_{m-1} U^{m-1}, \tag{2.36}$$

$$B = b_0 I + b_1 U + \ldots + b_{m-1} U^{m-1}, \tag{2.37}$$

$$C = c_0 I + c_1 U + \ldots + c_{m-1} U^{m-1}, \tag{2.38}$$

and

$$D = d_0 I + d_1 U + \ldots + d_{m-1} U^{m-1}, \tag{2.39}$$

where $a_i, b_i, c_i, d_i = \pm 1$, $0 \le i \le m-1$.

If the coefficients a_i, b_i, c_i, and d_i, $1 \le i \le m-1$, satisfy

$$a_{m-i} = a_i; \quad b_{m-i} = b_i; \quad c_{m-i} = c_i; \quad d_{m-i} = d_I, \tag{2.40}$$

then the matrices A, B, C, and D are all symmetric, i.e., the condition C1 in Theorem 2.1.6 is satisfied. It is not difficult to find that the condition C2 is also satisfied. Thus we obtain the following

Theorem 2.1.7 *The matrices A, B, C, and D respectively defined by Equations (2.36)–(2.39) are Williamson matrices if Equations (2.40) and (2.34) are simultaneously satisfied.*

Let m be odd integer. Without loss of generality we assume that

$$a_0 = b_0 = c_0 = d_0 = 1. \tag{2.41}$$

Then the matrices A, B, C, and D, respectively defined by Equations (2.36)–(2.39) can be rewritten as

$$A = P_1 - N_1; \quad B = P_2 - N_2; \quad C = P_3 - N_3; \quad \text{and} \quad D = P_4 - N_4, \tag{2.42}$$

where

$$\begin{cases} P_1 = \sum_{a_i=1} U^i, & N_1 = \sum_{a_i=-1} U^i, \\ P_2 = \sum_{b_i=1} U^i, & N_2 = \sum_{b_i=-1} U^i, \\ P_3 = \sum_{c_i=1} U^i, & N_3 = \sum_{c_i=-1} U^i, \\ P_4 = \sum_{d_i=1} U^i, & N_4 = \sum_{d_i=-1} U^i. \end{cases} \tag{2.43}$$

Because of the identity

$$P_i + N_i = J, 1 \le i \le 4, \tag{2.44}$$

the equation (2.34) is transformed to

$$\sum_{i=1}^{4} (2P_i - J)^2 = 4mI. \tag{2.45}$$

Note that each U^i is a permutation matrix. Then $U^i J = J U^i = J$. Hence

$$P_i J = J P_i = p_i J, \quad 1 \le i \le 4, \tag{2.46}$$

where p_i represents the number of terms in P_i, $1 \le i \le 4$.

By making use of Equation (2.46), the equation (2.45) can be rewritten as

$$4 \sum_{i=1}^{4} P_i^2 - 4 \sum_{i=1}^{4} P_i J + \sum_{i=1}^{4} J^2 = 4mI,$$

which is equivalent to

$$\sum_{i=1}^{4} P_i^2 = (\sum_{i=1}^{4} p_i - m)J + mI. \tag{2.47}$$

If Equation (2.40) is satisfied, then because of $a_0 = 1$, $m = $ odd, we know that p_1, p_2, p_3, and p_4 are all odd integers. Thus Equation (2.47) implies

$$\sum_{a_i=1} U^{2i} + \sum_{b_i=1} U^{2i} + \sum_{c_i=1} U^{2i} + \sum_{d_i=1} U^{2i} \equiv J + I \pmod{2}, \tag{2.48}$$

which implies further that $a_i + b_i + c_i + d_i = \pm 2$, for each $1 \le i \le m - 1$. Therefore, we have proved the following lemma.

Lemma 2.1.9 *If m is odd and the matrices A, B, C, and D, respectively defined by Equations (2.36)–(2.39), satisfy Equations (2.40), (2.41), and (2.34), then the coefficients a_i, b_i, c_i, and d_i satisfy $a_i + b_i + c_i + d_i = \pm 2$, for each $1 \le i \le m - 1$.*

Let

$$\begin{cases} W_1 = (A + B + C - D)/2, \\ W_2 = (A + B - C + D)/2, \\ W_3 = (A - B + C + D)/2, \\ W_4 = (-A + B + C + D)/2 \end{cases} \tag{2.49}$$

or equivalently,

$$\begin{cases} A = (W_1 + W_2 + W_3 - W_4)/2, \\ B = (W_1 + W_2 - W_3 + W_4)/2, \\ C = (W_1 - W_2 + W_3 + W_4)/2, \\ D = (-W_1 + W_2 + W_3 + W_4)/2. \end{cases} \tag{2.50}$$

Then Equation (2.34) is equivalent to

$$W_1^2 + W_2^2 + W_3^2 + W_4^2 = 4mI_m. \tag{2.51}$$

In conclusion, we have:

Lemma 2.1.10 *Let m be odd. There exist matrices A, B, C, and D, described in Lemma 2.1.9 if and only if there exist symmetric matrices*

1. *For $1 \leq i \leq 4$,*

   $$W_i = I \pm 2U^j \pm 2U^s \pm \ldots \pm 2U^t, \ 1 \leq j < s < \ldots < t \leq m-1; \quad (2.52)$$

2. *For each j, $1 \leq j \leq m-1$, the U^j appears once in just one of W_1, W_2, W_3, and W_4.*

If the matrices W_1, W_2, W_3, and W_4 described in Lemma 2.1.10 are constructed, then Williamson matrices A, B, C, and D can be generated by Equation (2.50). Hence an Hadamard matrix H is obtained by Equation (2.35).

Note that the characteristic equation of the cyclic matrix U is $\lambda^m - 1 = 0$, which has m different solutions. Thus there exists a non-singular matrix, say S, that

$$U = SVS^{-1}, \quad (2.53)$$

where V is a diagonal matrix $V = \mathrm{dia}(r_1, \ldots, r_m)$, and $r_1, \ldots, r_m$ is the m different solutions. Setting Equation (2.53) into Equation (2.52), one obtain that

$$W_i = S(I \pm 2V^j \pm 2V^s \pm \ldots \pm 2V^t)^{-1}.$$

Let $\overline{W}_i = I \pm 2V^j \pm 2V^s \pm \ldots \pm 2V^t$. Then Equation (2.51) is changed to

$$\overline{W_1}^2 + \overline{W_2}^2 + \overline{W_3}^2 + \overline{W_4}^2 = 4mI_m. \quad (2.54)$$

Because 1 is a solution of the characteristic equation, without loss of generality we assume that $r_j = 1$. Then by comparing the (j, j)-th elements of both left and right hand sides of Equation (2.54), we have

$$(1 \pm 2 \pm \ldots \pm 2)^2 + (1 \pm 2 \pm \ldots \pm 2)^2 + (1 \pm 2 \pm \ldots \pm 2)^2 + (1 \pm 2 \pm \ldots \pm 2)^2 = 4m. \quad (2.55)$$

Lemma 2.1.11 (Lagrange) *Every positive integer n can be decomposed into the sum of four squares, i.e., $n = a^2 + b^2 + c^2 + d^2$. Moreover, if m is a positive odd, then the number $4m$ can be decomposed into the summation of four odd squares.*

In conclusion, Williamson matrices can be constructed by the following steps:

Step 1: Decompose $4m$ into the sum of four odd squares:

where a, b, c, and d are odd.

Step 2: Compare Equations (2.55) with (2.56) to find the matrices W_i, $1 \leq i \leq 4$, that satisfy both Lemma 2.1.10 and Equation (2.51);

Step 3: Define the matrices A, B, C, and D from Equation (2.50);

Step 4: Construct the Hadamard matrix H by Equation (2.35).

In order to illustrate the above steps, we introduce a few examples here:

Example 1: Let $u^i + U^{23-i} := E_i$, $1 \leq i \leq 11$. Then because of

$$92 = 4 \times 23 = 9^2 + 3^2 + 1^2 + 1^2 \tag{2.57}$$

and

$$92I = (I + 2E_2 + 2E_6)^2 + (I - 2E_3 + 2E_1 - 2E_{10})^2$$
$$+ (I + 2E_5 - 2E_7)^2 + (I + 2E_{11} - 2E_8 + 2E_9 - 2E_4)^2,$$

the matrices W_i can be defined by $W_1 = I + 2E_2 + 2E_6$, $W_2 = I - 2E_3 + 2E_1 - 2E_{10}$, $W_3 = I + 2E_5 - 2E_7$, and $W_4 = I + 2E_{11} - 2E_8 + 2E_9 - 2E_4$. Hence Williamson matrices of order 23 can be constructed. Therefore there exists an Hadamard matrix of order 92.

Example 2: Because of $4 \times 43 = 172 = 13^2 + 1^2 + 1^2 + 1^2$ and

$$172I = (I + 2Y_0 - 2Y_2)^2 + (I + 2Y_3 - 2Y_1)^2$$
$$+ (I + 2Y_4 - 2Y_6)^2 + (I + 2Y_5)^2,$$

where

$$Y_i := E_{3i} + E_{37+i} + E_{314+i}, \text{ and } E_j := U^j + U^{43-j}.$$

Hence Williamson matrices A, B, C, and D of order 43 can be constructed by $W_1 := I + 2Y_0 - 2Y_2$, $W_2 := I + 2Y_3 - 2Y_1$, $W_3 := I + 2Y_4 - 2Y_6$, and $W_4 := I + 2Y_5$. Therefore there exists an Hadamard matrix of order 172.

Example 3: Because of $116 = 4 \times 29 = 9^2 + 5^2 + 3^2 + 1^2$ and

$$116I = (I + 2E_2 - 2E_4 + 2E_6 - 2E_9 - 2E_{11} + 2E_{12})^2$$
$$+ (I - 2E_3 - 2E_5 + 2E_7 - 2E_8 + 2E_{10})^2$$
$$(I + 2E_1)^2 + (I + 2E_{13} + 2E_{14})^2,$$

where $E_i := U^i + U^{29-i}$. Hence Williamson matrices A, B, C, and D of order 29 can be constructed by $W_1 := I + 2E_2 - 2E_4 + 2E_6 - 2E_9 - $

$2E_{11} + 2E_{12}$, $W_2 := I - 2E_3 - 2E_5 + 2E_7 - 2E_8 + 2E_{10}$, $W_3 := I + 2E_1$, and $W_4 := I + 2E_{13} + 2E_{14}$. Therefore there exists an Hadamard matrix of order 116.

The following theorem presents us more Hadamard matrices constructed by cyclic Williamson matrices.

Theorem 2.1.8 *If A, B, C, and D are cyclic Williamson matrices of order m, then the matrix H defined by*

$$
\begin{pmatrix}
A & A & A & B & -B & C & -C & -D & B & C & -D & -D \\
A & -A & B & -A & -B & -D & D & -C & -B & -D & -C & -C \\
A & -B & -A & A & -D & D & -B & B & -C & -D & C & -C \\
B & A & -A & -A & D & D & D & C & C & -B & -B & -C \\
B & -D & D & D & A & A & A & C & -C & B & -C & B \\
B & C & -D & D & A & -A & C & -A & -D & C & B & -B \\
D & -C & B & -B & A & -C & -A & A & B & C & D & -D \\
-C & -D & -C & -D & C & A & -A & -A & -D & B & -B & -B \\
D & -C & -B & -B & -B & C & C & -D & A & A & A & D \\
-D & -B & C & C & C & B & B & -D & A & -A & D & -A \\
C & -B & -C & C & D & -B & -D & -B & A & -D & -A & A \\
-C & -D & -D & C & -C & -B & B & B & D & A & -A & -A
\end{pmatrix}
$$

is an Hadamard matrix of order $12m$.

In order to construct more Hadamard matrices, we introduce here a new concept in the following definition.

Definition 2.1.9 ([6]) *A $(0, \pm 1)$-valued matrix $W := W(p, k)$, $p \geq k$, of order p is called a weighting matrix of order p and weight k if*

$$WW' = kI_p.$$

Thus, if $p = k$, a $W(p, p)$ is an Hadamard matrix of order p.

Lemma 2.1.12 ([6]) *Let p be a prime power. Then there exists a weighting matrix $W = W(p+1, p)$ satisfying $W' = (-1)^{(p-1)/2}W$. Furthermore, if $p \equiv 3 \pmod 4$, then $W + I_p$ is an Hadamard matrix of order $p + 1$.*

Proof. Arrange the elements of $GF(p) = \{a_0, a_1, \ldots, a_{p-1}\}$ so that $a_0 = 0$, and $a_{p-i} = -a_i$, $1 \leq i \leq p - 1$. Define a matrix $Q = [Q(i,j)]$ by

$$Q(i,j) = \chi(a_j - a_i) := \begin{cases} 0 & \text{if } i = j, \\ 1 & \text{if } a_j - a_i = y^2 \text{ for } y \in GF(p), \\ -1 & \text{otherwise.} \end{cases}$$

Then we have the next identities:

$$QQ' = pI - J, \qquad QJ = JQ = 0, \qquad Q' = (-1)^{(p-1)/2}Q,$$

since exactly half of $a_1, \ldots, a_{p-1}$ are squares, -1 is a square for $p \equiv 1 \pmod 4$ but not for $p \equiv 3 \pmod 4$, and $\sum_y \chi(y)\chi(y+c) = \sum_y \chi(y^2)\chi(1+cy^{-1}) = \sum_{x \neq 1} \chi(x) = -1$.

Let e be the all-one vector of length p. Then the matrix

$$W = \begin{bmatrix} 0 & e \\ (-1)^{(p-1)/2}e' & Q \end{bmatrix}$$

is the required weighting matrix $W(p+1,p)$.

If $p \equiv 3 \pmod 4$, then $W + I_p$ is an Hadamard matrix of order $p + 1$.
Q.E.D.

Theorem 2.1.9 ([6]) *Let $q \equiv 1 \pmod 4$ be a prime power. Then there exists a weighting matrix $W(q+1, q)$ of the form*

$$S = \begin{bmatrix} A & B \\ B & -A \end{bmatrix},$$

where A and B are cyclic and A is of zero diagonal.

Proof. Let α be a primitive element of $GF(q^2)$. Let V be a basis of the vector space $GF(q^2)$ over $GF(q)$. Based on this basis, we can define a matrix v by

$$(v) := \frac{1}{2} \begin{bmatrix} \alpha^{q-1} + \alpha^{1-q} & (\alpha^{q-1} - \alpha^{1-q})\alpha^{(q+1)/2} \\ (\alpha^{q-1} - \alpha^{1-q})\alpha^{-(q+1)/2} & \alpha^{q-1} + \alpha^{1-q} \end{bmatrix}.$$

This matrix v satisfies $\det(v) = 1$, and has two eigenvalues $\alpha_1 := \alpha^{q-1}$ and $\alpha_2 := \alpha^{1-q}$. Note that $\alpha_1^{(q+1)/2} = \alpha_2^{(q+1)/2} \in GF(q)$ and no $k < (q+1)/2$

satisfies $\alpha_1^k \in GF(q)$ or $\alpha_2^k \in GF(q)$. Hence v acts on the projective line $PG(1,q)$ as a permutation with period $(q+1)/2$ and without any fixed point. Thus the points of $PG(1,q)$ can be divided into two sets of transitivity, say v_1 and v_2, each containing $(q+1)/2$ points.

Let w be another matrix defined by

$$(w) := \begin{bmatrix} 0 & \alpha^{q+1} \\ 1 & 0 \end{bmatrix}.$$

Then $\chi\det(w) = -\chi(-1)$, $wv = vw$, and w has two eigenvalues $\beta_1 := \alpha^{(q+1)/2}$ and $\beta_2 := -\alpha^{(1+q)/2}$ satisfying $\beta_1^2 = \beta_2^2 \in GF(q)$. Hence w acts on $PG(1,q)$ as a permutation of period 2 which maps each point in v_1 into v_2. For each i, $1 \le i \le (q+1)/2$, the mapping $v^i w$ has no eigenvalue in $GF(q)$.

Represent the $q+1$ points of $PG(1,q) = \{x_0, x_1, \ldots, x_q\}$ by the following $q+1$ vectors in V: x, $v(x)$, $v^2(x)$, $\ldots$, $v^{(q-1)/2}$, $w(x)$, $vw(x)$, $\ldots$, $v^{(q-1)/2}w(x)$.

Let $S = [\chi\det(x_i, x_j)]$. Observe that each linear mapping $u : V \to V$ satisfies $\det(u(x), u(y)) = \det(w\det(x,y)$, $x, y \in V$. Thus

$$\det(v^i w(x), v^j w(x)) = \det(w)\det(v^i(x), v^j(x))$$
$$= \det(w)\det(x, v^{j-i}(x));$$

$$\det(v^i(x), v^j w(x)) = -\det(v^i w(x), v^j(x))$$
$$= \det(v^j(x), v^i w(x));$$

and

$$\det(v^i(x), v^j(x)) = -\det(v^{(q+1)/2+i}, v^j(x)).$$

In other words, this matrix S is the required matrix. The theorem follows. **Q.E.D.**

Theorem 2.1.10 ([6]) *Let $p \equiv 1 \pmod{4}$ be a prime power. Then there are Williamson matrices of order $p(p+1)/2$.*

Proof. Let A and B be the matrices in Theorem 2.1.9. Let Q be the matrix appeared in Lemma 2.1.12. Then the required Williamson matrices

X_1, X_2, X_3, and X_4 are:

$$X_1 = (I \otimes J) + (A \otimes (I + Q)),$$
$$X_2 = B \otimes (I + Q),$$
$$X_3 = (I \otimes J) + (A \otimes (I - Q)),$$
$$X_4 = B \otimes (I - Q).$$

Q.E.D.

2.2   Construction

2.2.1   General Constructions

We have seen in the last chapter that direct multiplication (or Kronecker Production) is a very useful approach for the construction of Walsh matrices. The following theorem shows that this approach works for Hadamard matrices too.

Theorem 2.2.1 *Let A and B be Hadamard matrices of orders m and n, respectively. Then their direct product $A \otimes B$ is an Hadamard matrix of order mn.*

Proof. At first the matrix $A \otimes B$ is clearly (± 1)-valued. Secondly,

$$
\begin{aligned}
(A \otimes B)(A \otimes B)' &= (A \otimes B)(A' \otimes B') \\
&= (AA') \otimes (BB') \\
&= (mI_m) \otimes (nI_n) \\
&= (mn)I_{mn}.
\end{aligned}
$$

The theorem follows.   **Q.E.D.**

A simple, but useful, consequence of this theorem is:

Corollary 2.2.1 *Let H_i, $1 \le i \le t$, be an Hadamard matrix of order m_i. Then $H_1 \otimes H_2 \otimes \ldots \otimes H_t$ is an Hadamard matrix of order $\prod_{i=1}^{t} m_i$.*

For example, Hadamard matrices of order 2^n can be formed as the direct product of some of order 2.

Theorem 2.2.2 *Let $H = U + I$ be a skew Hadamard matrix of order h, $M = W + I$ and $N = N'$ be a pair of amicable Hadamard matrices of order m. Let X, Y, Z be three (± 1)-valued matrices of order l and*

$$(XY')' = XY',$$
$$(YZ')' = YZ',$$
$$(ZX')' = ZX',$$
$$XX' = aI + (l - a)J,$$
$$YY' = cI + (l - c)J,$$
$$ZZ' = (l + 1)I - J,$$

where $(m - 1)c = m(l - h + 1) - a$. Then the matrix

$$D := U \otimes N \otimes Z + I_h \otimes W \otimes Y + I_h \otimes I_m \otimes X \qquad (2.58)$$

is an Hadamard matrix of order hlm.

Proof. On one hand, the matrix D defined by Equation (2.58) is clearly (± 1)-valued.

On the other hand, by the assumptions on H, M, and N, the following identities are obtained:

$$U' = -U,$$
$$UU' = (h - 1)I_h,$$
$$W' = -W,$$
$$WW' = (m - 1)I_m,$$
$$MN = NM',$$
$$MM' = N^2 = mI_m,$$
$$WN = NW'.$$

Thus we have

$$DD' = UU' \otimes N^2 \otimes ZZ' + I_h \otimes WW' \otimes YY' + I_h \otimes I_m \otimes XX'$$
$$+ U' \otimes WN \otimes YZ' + U \otimes NW' \otimes XY' + U' \otimes N \otimes XZ'$$
$$+ U \otimes N \otimes ZX' + I_h \otimes W' \otimes XY' + I_h \otimes W \otimes YX'$$
$$= UU' \otimes N^2 \otimes ZZ' + I_h \otimes WW' \otimes YY' + I_h \otimes I_m \otimes XX'$$
$$+ (U' + U) \otimes WN \otimes YZ' + (U' + U) \otimes N \otimes ZX'$$

$$+I_h \otimes (W + W') \otimes XY'$$
$$= (h-1)I_h \otimes mI_m \otimes ((l+1)I_l - J_l) + I_h \otimes I_m \otimes (c(m-1)I_l$$
$$+(l-c)(m-1)J_l) + I_h \otimes I_m \otimes (aI_l + (l-a)J_l)$$
$$= I_{mh} \otimes (m(h-1)(l+1) + c(m-1) + a)I_l$$
$$+I_{mh} \otimes (-m(h-1) + (l-c)(m-1) + l - a)J_l$$
$$= mlhI_{mlh}.$$

In other words, we have proved that D is an Hadamard matrix of order mhl.   **Q.E.D.**

Corollary 2.2.2 *If there exist:*

1. *A skew Hadamard matrix of order h;*

2. *Two (±1)-valued matrices Y and Z of order l such that*

$$(YZ')' = YZ', \tag{2.59}$$

$$YY' = cI + (l-c)J, \tag{2.60}$$

and

$$ZZ' = (l+1)I - J, \tag{2.61}$$

where $c = l - h + 1$.

Then there exists an Hadamard matrix of order lh.

Proof. This corollary is, in fact, a special case of Theorem 2.2.2 for $X = Y$, $a = c$ and $m = 1$.   **Q.E.D.**

Corollary 2.2.3 *If there exists a skew Hadamard matrix of order h, then there exists an Hadamard matrix of order $h(h-1)$.*

Proof. Let H be a skew Hadamard matrix of order h, and $Z - I_{h-1}$ the kernel of H. Then, by Lemma 2.1.3, Z is a (±1)-valued matrix satisfying

$$(Z - I_{h-1})(Z - I_{h-1})' = (h-1)I_{h-1} - J_{h-1}, \tag{2.62}$$

$$(Z - I_{h-1})J_{h-1} = 0, \tag{2.63}$$

$$(Z - I_{h-1})' = -(Z - I_{h-1}). \tag{2.64}$$

Therefore,

$$Z' = -Z + 2I_{h-1} \quad \text{(by Eq. (2.64))}, \tag{2.65}$$

$$J_{h-1}Z' = J_{h-1} = ZJ_{h-1} \quad \text{(by Eq. (2.63))}, \tag{2.66}$$

$$ZZ' = hI_{h-1} - J_{h-1} \quad \text{(by Eq. (2.65) + (2.62))}. \tag{2.67}$$

Note that

$$J_{h-1}J_{h-1} = (h-1)J_{h-1}. \tag{2.68}$$

Thus if we choose $Y = J$, $l = h - 1$, and $c = 0$, then Equations (2.66), (2.68), and (2.67) reduce to Equations (2.59), (2.60), and (2.61), respectively. The proof is finished by Corollary 2.2.2.   **Q.E.D.**

Corollary 2.2.4 *An Hadamard matrix of order $h(h + 3)$ can be constructed from a skew Hadamard matrix of order h and a symmetric Hadamard matrix of order $h + 4$.*

Proof. Let

$$A = \begin{pmatrix} 1 & e \\ e' & C \end{pmatrix}$$

be a symmetric Hadamard matrix of order $h + 4$, where C is the kernel. Thus C is also symmetric and

$$C^2 = (h+4)I_{h+3} - J_{h+3} \quad \text{and} \quad CJ_{h+3} = J_{h+3}C = -J_{h+3}. \tag{2.69}$$

Let $Y = J_{h+3} - 2I_{h+3}$. Then Y is a symmetric (± 1)-valued matrix and

$$YC' = (J_{h+3} - 2I_{h+3})C$$
$$= -J_{h+3} - 2C$$
$$= C(J_{h+3} - 2I_{h+3})$$
$$= CY', \tag{2.70}$$

$$YY' = Y^2 = (h-1)J_{h+3} + 4I_{h+3}. \tag{2.71}$$

If we choose $Z = C$, $c = 4$, and $l = h + 3$, then Equations (2.70), (2.71), and (2.69) reduce to Equations (2.59), (2.60), and (2.61), respectively. The proof is finished by Corollary 2.2.2.   **Q.E.D.**

2.2.2 Amicable Hadamard Matrices

We have known that amicable Hadamard matrices are useful in constructing skew Hadamard matrices. In fact, the truth of the conjecture 'amicable Hadamard matrices exist for every order 2 and $4n$, $n \geq 1$', would imply the two conjectures 'skew Hadamard matrices exist for every order 2 and $4n$, $n \geq 1$' and 'symmetric Hadamard matrices existing for every order 2 and $4n$, $n \geq 1$'. Infinite families of amicable Hadamard matrices can be constructed by the following theorems.

Theorem 2.2.3 Let $t \geq 0$, $l \geq 0$, $e_i \geq 1$, p_i a prime satisfying $(p_i)^{e_i} \equiv 3 \pmod 4$, for $1 \leq i \leq l$. If

$$m = 2^t \prod_{i=1}^{l} ((p_i)^{e_i} + 1), \qquad (2.72)$$

then there exist amicable Hadamard matrices of order m.

Proof. The amicable Hadamard matrices are constructed as follows:

Case 1: $m = 1$, i.e., $t = l = 0$. Then $M = N = (1)$ is a pair of amicable Hadamard matrices of order 1.

Case 2: $m = 2$, i.e., $t = 1$, $l = 0$. Then $M = \begin{pmatrix} 1 & 1 \\ -1 & 1 \end{pmatrix}$ and $N = \begin{pmatrix} 1 & 1 \\ 1 & -1 \end{pmatrix}$ is a pair of amicable Hadamard matrices of order 2.

Case 3: $m = p^e + 1$, i.e., $t = 0$, $l = 1$, $p_1 = p$, and $e_1 = e$. Let $q = p^e$, and $\chi(z)$ be the quadratic characteristic function over $GF(q) - \{0\}$, i.e.,

$$\chi(z) = \begin{cases} 1 & \text{if } z = a^2, \text{ for some non-zero } a \in GF(q), \\ 0 & \text{if } z = 0, \\ -1 & \text{otherwise.} \end{cases}$$

Number the elements of $GF(q)$ by $GF(q) = \{z_1, z_2, \ldots, z_q\}$ and form a matrix $Q = [Q(i,j)]$, $1 \leq i, j \leq q$, by $Q(i,j) = \chi(z_j - z_i)$. Let

$$A = \begin{pmatrix} 0 & e \\ -e' & Q \end{pmatrix}, \qquad (2.73)$$

where e is a row vector of length q. Because p is odd, we can rearrange $GF(q)$ so that $z_1 = 0$, and

$$z_{q+2-i} = -z_i, \quad 2 \leq i \leq q. \qquad (2.74)$$

The equation $p^e \equiv 3 \pmod 4$ implies that the matrix A in Equation (2.73) satisfies $A' = -A$. Thus

$$AA' = \begin{pmatrix} q & eQ' \\ QQ' & J_q + QQ' \end{pmatrix} = \begin{pmatrix} q & 0 \\ 0 & qI_q \end{pmatrix} = qI_m.$$

Let $M := A + I$. Then M is (± 1)-valued and $MM' = AA' + A + A' + I = (q+1)I = mI$. Hence, M is a skew Hadamard matrix of order m.

Let

$$W = \begin{pmatrix} 0 & 0 \cdots 0 & 1 \\ 0 & \cdots & 0 & 1 & 0 \\ 0 & \cdots & 1 & 0 & 0 \\ \vdots & \vdots & \vdots & \vdots & \vdots \\ 1 & 0 \cdots 0 & 0 \end{pmatrix} \quad \text{and} \quad V = \begin{pmatrix} 1 & 0 \\ 0 & W \end{pmatrix}. \tag{2.75}$$

Thus

$$W' = W, \qquad W^2 = I_{q-1}, \qquad V' = V, \qquad \text{and} \qquad V^2 = I_q. \tag{2.76}$$

Let N be a (± 1)-valued matrix of order $q + 1$ defined by

$$N := \begin{pmatrix} 1 & 0 \\ 0 & -V \end{pmatrix} M = \begin{pmatrix} 1 & 0 \\ 0 & -V \end{pmatrix} + \begin{pmatrix} 0 & e \\ e' & -VQ \end{pmatrix},$$

because $M = A + I$. Note that the $(1,j)$-th element of VQ is $\chi(z_j - z_1) = \chi(z_j + z_1)$; moreover the (i,j)-th element of VQ is

$$\chi(z_j - z_{q+2-i}) = \chi(z_j - (-z_i)) = \chi(z_j + z_i),$$

in other words, the matrix VQ is symmetric, therefore the N is.

From Equation (2.76), we have

$$NN' = \begin{pmatrix} 1 & 0 \\ 0 & -V \end{pmatrix} MM' \begin{pmatrix} 1 & 0 \\ 0 & -V \end{pmatrix} = m \begin{pmatrix} 1 & 0 \\ 0 & V^2 \end{pmatrix} = mI_m.$$

Thus N is an Hadamard matrix of order m. Furthermore,

$$MN = MN' = MM' \begin{pmatrix} 1 & 0 \\ 0 & -V \end{pmatrix} = m \begin{pmatrix} 1 & 0 \\ 0 & -V \end{pmatrix},$$

and

$$NM' = \begin{pmatrix} 1 & 0 \\ 0 & -V \end{pmatrix} \quad MM' = m \begin{pmatrix} 1 & 0 \\ 0 & -V \end{pmatrix},$$

i.e., $MN = NM'$. Therefore M and N are indeed a pair of amicable Hadamard matrices of order $q + 1$.

Now it is necessary to prove that if there exist simultaneously pairs of amicable Hadamard matrices of orders m and h, then there exist a pair of amicable Hadamard matrices of order mh.

In fact, if $M_h = S_h + I_h$ and N_h (resp., $M_m = S_m + I_m$ and N_m) is a pair of amicable Hadamard matrices of order h (resp. m), where S_h and S_m are anti-symmetric, while N_m and N_h are symmetric.

Let $M_{hm} := I_h \otimes M_m + S_h \otimes N_m$, and $N_{hm} := N_h \otimes N_m$. Then M_{hm} and N_{hm} are Hadamard matrices with N_{hm} being symmetric.

Because of

$$M_{hm} - I_{hm} = I_h \otimes (S_m + I_m) + S_h \otimes N_m - I_{hm}$$
$$= I_h \otimes S_m + S_h \otimes N_m,$$

we know that $M_{hm} - I_{hm}$ is anti-symmetric. By Lemma 2.1.4 we have

$$M_{hm}N_{hm} = (I_h \otimes M_m + S_h \otimes N_m)(N_h \otimes N_m)$$
$$= N_h \otimes M_m N_m + S_h N_h \otimes N_m^2$$
$$= N_h \otimes N_m M_m' + N_h S_h' \otimes N_m^2$$
$$= (N_h \otimes N_m)(I_h \otimes M_m' + S_h' \otimes N_m)$$
$$= N_{hm} M_{hm}'.$$

Up to now we have proved that M_{hm} and N_{hm} are a pair of amicable Hadamard matrices.   **Q.E.D.**

Theorem 2.2.4 *If A, B, C, and D are four (± 1)-valued matrices satisfying*

$$C = I + U, \quad U' = -U,$$
$$A' = A, \quad B' = B, \quad D' = D,$$
$$AA' + BB' = CC' + DD' = 2(v + 1)I - 2J,$$
$$eA' = eB' = eC' = eD' = e,$$
$$AB = BA, \quad CD = DC',$$

where e is the all ones vector of length v. Let

$$N = \begin{pmatrix} 1 & 1 & e & e \\ 1 & -1 & -e & e \\ e' & -e' & A & -B \\ e' & e' & -B & -A \end{pmatrix}$$

and

$$M = \begin{pmatrix} 1 & 1 & e & e \\ -1 & 1 & e & -e \\ -e' & -e' & C & D \\ -e' & e' & -D & C \end{pmatrix}.$$

Then N is a symmetric Hadamard matrix of order $2(v+1)$, and M is a skew Hadamard matrix of order $2(v+1)$. In addition, if both $AC' - BD$ and $BC' + AD$ are symmetric, them M and N form a pair of amicable Hadamard matrices of order $2(v+1)$.

Proof. It can be proved easily by direct verification.   **Q.E.D.**

The other important result about amicable Hadamard matrices is

Theorem 2.2.5 *Let $t = p^e \equiv 1(\bmod 4)$; $q = g^f = 2t + 1$, where p and g are primes, and e and f positive integers. Then there exist amicable Hadamard matrices of order $2(t+1)$.*

In conclusion, it has been proved that there exist amicable Hadamard matrices of the following orders([1]):

Family 1: The order of 2^t, where $t \geq 0$ an integer;

Family 2: The order of $p^r + 1$, where $p^r \equiv 3(\bmod 4)$;

Family 3: The order of $(p-1)^u + 1$, where p the order of normalized amicable Hadamard matrices, $u > 0$ an odd integer;

Family 4: The order of $2(q+1)$, where $2q + 1$ is a prime power, and q a prime satisfying $q \equiv 1(\bmod 4)$;

Family 5: The order of $(|t|+1)(q+1)$, where q is a prime power satisfying $q \equiv 5 \ (\bmod 8) = s^2 + 4t^2$, $s \equiv 1 \ (\bmod 4)$, and $|t|+1$ is the order of amicable orthogonal designs of type $(1 + |t|; (1, |t|); ((|t| + 1)/2, (|t| + 1)/2))$. (The definition of amicable orthogonal design will be introduced in Definition 2.3.1);

Family 6: The order of $2^r(q+1)$, where q is a prime power satisfying

$$q \equiv 5(\text{mod}8) = s^2 + 4(2^r - 1)^2, \qquad s \equiv 1(\text{mod}4),$$

r a positive integer;

Family 7: The order of $2(q+1)$, where $q \equiv 5(\text{mod}8)$;

Family 8: The order of S, where S is the product of two amicable Hadamard matrices.

2.2.3 Skew Hadamard Matrices

Some of the most powerful methods for constructing Hadamard matrices depend on the existence of skew Hadamard matrices. While it is not an easy task to find skew Hadamard matrices. For example, the Kronecker product of skew Hadamard matrices is not a skew Hadamard matrix, although it has been proved that if h_1 and h_2 are the orders of amicable Hadamard matrices, then there are amicable Hadamard matrices of order h_1h_2. Fortunately, many constructions for amicable Hadamard matrices are also valid for the skew Hadamard matrices.

The following theorems generate some new skew Hadamard matrices by amicable Hadamard matrices and skew Hadamard matrices.

Theorem 2.2.6 *If there exist a skew Hadamard matrix of order h and amicable Hadamard matrices of order m, then there also exist skew Hadamard matrices of order mh.*

Proof. Let $S = I_h + \overline{S}$ be a skew Hadamard matrix of order h. Then

$$\overline{S}' = -\overline{S}, \quad \overline{S}\,\overline{S}' = (h-1)I_h. \tag{2.77}$$

Let $M = I_m + \overline{M}$ and N is a pair of amicable Hadamard matrices of order m. Then

$$MN = NM' \tag{2.78}$$

Let $K = I_h \otimes M + \overline{S} \otimes N$, which is a (± 1)-valued matrix of order mh. From Equations (2.77) and (2.78), we know that

$$\begin{aligned} KK' &= (I_h \otimes M + \overline{S} \otimes N)(I_h \otimes M' - \overline{S} \otimes N) \\ &= I_h \otimes (MM') - \overline{S}^2 \otimes N^2 \\ &= I_h \otimes mI_m + (h-1)I_h \otimes mI_m \\ &= mhI_{mh} \end{aligned}$$

and

$$(K - I_{mk})' = (I_h \otimes \overline{M} + \overline{S} \otimes N)'$$
$$= I_h \otimes M' + \overline{S}' \otimes N$$
$$= -I_h \otimes \overline{M} - \overline{S} \otimes N$$
$$= -(K - I_{mh}).$$

Thus we have proved that K is a skew Hadamard matrix of order mh. **Q.E.D.**

The other important results about skew Hadamard matrices are

Theorem 2.2.7 *Let p be a prime and $p^r = 2m + 1 \equiv 5 \pmod 8$. Then there exists a skew Hadamard matrix of order $4(m + 1)$.*

Theorem 2.2.8 ([6]) *If A, B, C, D are square circulant matrices of order m, R is the back diagonal $(0, 1)$ matrix, then if A is a skew Hadamard matrix and if*

$$AA' + BB' + CC' + DD' = 4mI,$$

then the following matrices

$$
\begin{bmatrix}
A & BR & CR & DR \\
-BR & A & -D'R & C'R \\
-CR & D'R & A & -B'R \\
-DR & -C'R & B'R & A
\end{bmatrix}
\quad or \quad
\begin{bmatrix}
A & BR & CR & DR \\
-BR & A & D'R & -C'R \\
-CR & -D'R & A & B'R \\
-DR & C'R & -B'R & A
\end{bmatrix}
$$

are skew Hadamard matrices of order $4m$.

In conclusion, it has been proved that there exist skew Hadamard matrices of the following orders([1]):

Family 1: The order of $2^t \prod k_i$, where $k_i = p_i^{r_i} + 1 \equiv 0 \pmod 4$, p_i prime, r_i and t positive integers;

Family 2: The order of $(p - 1)^u + 1$, where u a positive odd integer, and p the order of another skew Hadamard matrix;

Family 3: The order of $2(q+1)$, where $q \equiv 5 \pmod 8$ is a prime power;

Family 4: The order of $2(q + 1)$, where $q = p^t$ is a prime power with $p \equiv 5 \pmod 8$ and $t \equiv 2 \pmod 4$;

Family 5: The order of $2^s(q+1)$, where $q = p^t$ is a prime power, $p \equiv 5 \pmod 8$, $t \equiv 2 \pmod 4$, and $s \geq 1$ an integer;

Family 6: The order of $4m$, where m is odd integer in $[3, 31]$ or $m \in \{37, 39, 43, 49, 65, 67, 93, 113, 121, 127, 129, 133, 157, 163, 181, 217, 219, 241, 267\}$;

Family 7: The order of $n(n-1)(m-1)$, where m and n are the orders of amicable Hadamard matrices such that $(m-1)n/m$ is the order of another skew Hadamard matrix;

Family 8: The order of $mn(n-1)$, where n is the order of amicable orthogonal designs of types $((1, n-1); (n))$ and mn the order of an orthogonal design of type $(1, m, mn-m-1)$ (see Definition 2.3.1);

Family 9: The order of $4(q+1)$, where $q \equiv 9 \pmod{16}$ a is prime power;

Family 10: The order of $(\mid t \mid +1)(q+1)$, where $q = s^2 + 4t^2 \equiv 5 \pmod 8$ is a prime power, and $\mid t \mid +1$ the order of another skew Hadamard matrix;

Family 11: The order of $4(1+q+q^2)$, where q is a prime power such that (1). $1+q+q^2 \equiv 3, 5,$ or $7 \pmod 8$ is a prime power; or (2). $3+2q+2q^2$ is a prime power;

Family 12: The order of $2^t q$, where $q = s^2 + 4r^2 \equiv 5 \pmod 8$ is a prime power, and an orthogonal design of type $(2^t; 1, a, b, c, c+ \mid r \mid)$ exists where $1 + a + b + 2c+ \mid r \mid = 2^t$ and $a(q+1) + b(q-4) = 2^t$;

Family 13: The order of hm, where h is the order of another skew Hadamard matrix, and m the order of a pair of amicable Hadamard matrices.

2.2.4 Symmetric Hadamard Matrices

One of the two matrices in a pair of amicable Hadamard matrices is symmetric. Thus Theorems 2.2.3 and 2.2.5, respectively imply the following:

Theorem 2.2.9 Let $t \geq 0$, $l \geq 0$, $e_i \geq 1$, p_i a prime satisfying $(p_i)^{e_i} \equiv 3 \pmod 4$, for $1 \leq i \leq l$. If

$$m = 2^t \prod_{i=1}^{l}((p_i)^{e_i} + 1), \qquad (2.79)$$

then there exist symmetric Hadamard matrix of order m.

and:

Theorem 2.2.10 *Let $t = p^e \equiv 1(\mathrm{mod}4)$; $q = g^f = 2t + 1$, where p and g are primes, e and f positive integers. Then there exists symmetric Hadamard matrix of order $2(t+1)$.*

Theorem 2.2.11 *Let p be a prime, l a positive integer, and $q = p^l \equiv 1(\mathrm{mod}4)$. Then there exists a symmetric Hadamard matrix of order $2(q+1)$.*

Proof. Let G, q, and Q are the same as those in the proof of Theorem 2.2.3. Let

$$P = \begin{pmatrix} 0 & e \\ e' & Q \end{pmatrix}, \quad N = P + I_{q+1}. \tag{2.80}$$

Then the diagonal of Q is all-zero, $Q' = Q$, $QQ' = qI - J$, $QJ = JQ = 0$. Thus N is a symmetric (± 1)-valued matrix, and

$$P^2 = \begin{pmatrix} q & eQ \\ Qe' & J + Q^2 \end{pmatrix} = qI_{q+1}. \tag{2.81}$$

Let

$$H = \begin{pmatrix} -N & P - I \\ P - I & N \end{pmatrix}, \tag{2.82}$$

which is a symmetric (± 1)-valued matrix of order $2(q+1)$. From Equation (2.81), we have

$$HH' = \begin{pmatrix} (P+I)^2 + (P-I)^2 & 0 \\ 0 & (P+I)^2 + (P-I)^2 \end{pmatrix}$$

$$= 2 \begin{pmatrix} (P^2 + I) & 0 \\ 0 & (P^2 + I) \end{pmatrix}$$

$$= 2(q+1)I_{2(q+1)}.$$

In other words, the matrix H is the symmetric Hadamard matrix of order $2(q+1)$.   **Q.E.D.**

Lemma 2.2.1 *Let H_1 and H_2 are symmetric Hadamard matrices of orders m_1 and m_2, respectively. Then $H_1 \otimes H_2$ is a symmetric Hadamard matrix of order $m_1 m_2$.*

This lemma will be used in the proofs of the following theorems.

Theorem 2.2.12 *Let $t \geq 0$, $l \geq 0$, $e_i \geq 1$, p_i an odd prime and*

$$m = 2^t \prod_{i=1}^{l} ((p_i)^{e_i} + 1). \tag{2.83}$$

If the number of p_i satisfying $(p_i)^{e_i} \equiv 1 \pmod 4$ is upper bounded by t, then there exists a symmetric Hadamard matrix of order m.

Proof. This theorem follows the direct multiplication of the Hadamard matrices presented in Equation (2.82), in Theorem 2.2.3 and some $\begin{pmatrix} 1 & 1 \\ 1 & -1 \end{pmatrix}$.

Q.E.D.

Theorem 2.2.13 *Let p be a prime, e positive integer, and $q = p^e \equiv 1 \pmod 4$. Then there exists a symmetric Hadamard matrix of order $2q(q + 1)$.*

Proof. Let P and N be those matrices of order $q+1$ defined by Equation (2.80). Let

$$X = \begin{pmatrix} 1 & 1 \\ 1 & -1 \end{pmatrix} \otimes J_q,$$

and

$$
\begin{aligned}
Y &= \begin{pmatrix} Q + I & Q - I \\ Q - I & -Q - I \end{pmatrix} \\
&= \begin{pmatrix} 1 & 1 \\ 1 & -1 \end{pmatrix} \otimes Q + \begin{pmatrix} 1 & -1 \\ -1 & -1 \end{pmatrix} \otimes I_q,
\end{aligned}
$$

where Q is the matrix which appears in the matrix P. Then

$$X' = X, \quad Y' = Y, \quad X^2 = 2qI_2 \otimes J_q, \tag{2.84}$$

$$
\begin{aligned}
Y^2 &= 2I_2 \otimes Q^2 + 2I_2 \otimes I_q \\
&= 2I_2 \otimes ((q+1)I_q - J_q), \tag{2.85}
\end{aligned}
$$

$$XY + YX = 2I_2 \otimes JQ + 2\begin{pmatrix} 0 & -1 \\ 1 & 0 \end{pmatrix} \otimes J_q + 2I_2 \otimes QJ$$

$$+2\begin{pmatrix} 0 & 1 \\ -1 & 0 \end{pmatrix} = 0 \tag{2.86}$$

Let $H = I_{q+1} \otimes X + P \otimes Y$. Then H is a (± 1)-valued symmetric matrix. From Equations (2.84)–(2.86), we have

$$HH' = I_{q+1} \otimes X^2 + P^2 \otimes Y^2$$
$$= 2q(q+1)I_{2q(q+1)}.$$

In other words, this H is a symmetric Hadamard matrix of order $2q(q+1)$. **Q.E.D.**

Theorem 2.2.14 Let $t \geq l \geq 0$, p_i a prime, and $p_i^{e_i} \equiv 1 \pmod 4$, $1 \leq i \leq l$. Let

$$m = 2^t \prod_{i=1}^{l} (p_i^{e_i}(p_i^{e_i} + 1)).$$

Then there exists a symmetric Hadamard matrix of order m.

Proof. This theorem is the direct consequence of Lemma 2.2.1 and , Theorem 2.2.13 and the known symmetric Hadamard matrix $\begin{pmatrix} 1 & 1 \\ 1 & -1 \end{pmatrix}$. **Q.E.D.**

2.3 Existence

2.3.1 Orthogonal Designs and Hadamard Matrices

A $(0, \pm x_1, \ldots, \pm x_l)$-valued matrix, X, of order n is called an orthogonal design of order n and type $(s_1, \ldots, s_l)$, s_i positive integer, if and only if

$$XX' = (\sum_{i=1}^{l} s_i x_i^2)I_n.$$

Alternatively, X is an orthogonal matrix with its each row containing exactly s_i entries of the type $\pm x_i$, $1 \leq i \leq l$.

It is obvious that Hadamard matrices of order n are (± 1)-valued orthogonal designs of type (n). Some other orthogonal designs of special interest are ([7]):

(1) The following matrix X is an orthogonal design of order 4 and type $(1, 1, 1, 1)$:

$$X = \begin{bmatrix} A & B & C & D \\ -B & A & -D & C \\ -C & D & A & -B \\ -D & -C & B & A \end{bmatrix} \quad \text{or} \quad \begin{bmatrix} A & B & C & D \\ -B & A & D & -C \\ -C & -D & A & B \\ -D & C & -B & A \end{bmatrix}.$$

(2) The following matrix X is an orthogonal design of order 8 and type $(1, 1, 1, 1, 1, 1, 1, 1)$:

$$X = \begin{bmatrix} A & B & C & D & E & F & G & H \\ -B & A & D & -C & F & -E & -H & G \\ -C & -D & A & B & G & H & -E & -F \\ -D & C & -B & A & H & -G & F & -E \\ -E & -F & -G & -H & A & B & C & D \\ -F & E & -H & G & -B & A & -D & C \\ -G & H & E & -F & -C & D & A & -B \\ -H & -G & F & E & -D & -C & B & A \end{bmatrix}.$$

In addition orthogonal designs of order 12 and type $(3, 3, 3, 3)$, order 24 and type $(3, 3, 3, 3, 3, 3, 3, 3)$, order 20 and type $(5, 5, 5, 5)$, and order 36 and type $(9, 9, 9, 9)$ can also be constructed.

The following existence result about orthogonal designs will be used in the construction of Hadamard matrices.

Lemma 2.3.1 ([6], [7]) *There exist orthogonal designs of type $(1, m - 1, nm - n - m)$ and order $2^t = (m - 1)n$.*

Definition 2.3.1 ([6], [7]) *Let $(0, \pm x_1, \ldots, \pm x_s)$-valued X (resp. $(0, \pm y_1, \ldots, \pm y_t)$-valued Y) be an orthogonal design of order n and type $(u_1, \ldots, u_s)$ (resp., $(v_1, \ldots, v_t)$). If $XY' = YX'$ is satisfied, the pair X and Y are called amicable orthogonal designs. Sometimes this pair are also called amicable orthogonal designs of type $[(u_1, \ldots, u_s); (v_1, \ldots, v_t)]$ and order n.*

For example([6]),

$$X = \begin{bmatrix} x_1 & x_2 \\ x_2 & -x_1 \end{bmatrix} \qquad \text{and} \qquad Y = \begin{bmatrix} y_1 & y_2 \\ -y_2 & y_1 \end{bmatrix} \qquad (2.87)$$

are amicable orthogonal designs of type $[(1,1);(1,1)]$ and order 2.

$$X = \begin{bmatrix} x_1 & x_2 & x_3 & x_3 \\ -x_2 & x_1 & x_3 & -x_3 \\ x_3 & x_3 & -x_1 & -x_2 \\ x_3 & -x_3 & x_2 & -x_1 \end{bmatrix} \qquad \text{and} \qquad Y = \begin{bmatrix} y_1 & y_2 & y_3 & y_3 \\ y_2 & -y_1 & y_3 & -y_3 \\ -y_3 & -y_3 & y_2 & y_1 \\ -y_3 & y_3 & y_1 & -y_2 \end{bmatrix} \qquad (2.88)$$

are amicable orthogonal designs of the type $[(1,1,2);(1,1,2)]$ and order 2 ([6]).

Comparing the definition of amicable Hadamard matrices with Definition 2.3.1, the following lemma is obtained.

Lemma 2.3.2 ([6], [7]) *Let* $M = I + S$ *and* N *be amicable Hadamard matrices of order* n. *Then* (1) M *and* N *are amicable orthogonal designs of type* $[(n);(n)]$ *and order* n; (2) S *and* N *are amicable orthogonal designs of type* $[(1,n-1);(n)]$ *and order* n.

Proof. The first part is directly owed to Definitions 2.3.1 and 2.1.5. The second part is obtained by Lemmas 2.1.4 and 2.1.2.   **Q.E.D.**

It has been proved that:

Lemma 2.3.3 ([6], [7]) *For each integer* t, *there exist amicable orthogonal designs of types*

$$((1,1,2,4,\ldots,2^t);(2^t,2^t)) \qquad \text{and} \qquad ((1,2^{t+1}-1);(2^t,2^t)).$$

Let $G = \{z_1, z_2, \ldots, z_t\}$ be an additive Abelian group of order t. Let ψ and ϕ be two mappings from G to a commutative ring. Define two matrices $M = [m(i,j)]$ and $N = [n(i,j)]$ by

$$m(i,j) = \psi(z_j - z_i) \qquad \text{and} \qquad n(i,j) = \phi(z_j + z_i), \quad 1 \le i,j \le t.$$

These matrices M and N are called type 1 and type 2 matrices, respectively. Here the words 'type i' illustrate the way the elements of G are ordered and which one of the following two functions ψ or ϕ is used.

Let X be a subset of G and $0 \notin X$. If ψ and ϕ are defined by

$$\psi(x) = \begin{cases} a & x = 0, \\ b & x \in X, \\ c & \text{otherwise,} \end{cases} \quad \text{and} \quad \phi(x) = \begin{cases} d & x = 0, \\ e & x \in X, \\ f & \text{otherwise.} \end{cases}$$

Then the M (resp., N) is called type 1 (a, b, c) (resp., type 2 (d, e, f)) incidence matrix generated by X.

If the restriction $0 \notin X$ is dropped, and

$$\psi(x) = \phi(x) = \begin{cases} 1 & x \in X, \\ -1 & \text{otherwise,} \end{cases}$$

then the M (resp., N) is called type 1 (resp., type 2) ± 1 incidence matrix generated by X, and if

$$\psi(x) = \phi(x) = \begin{cases} 1, & x \in X \\ 0, & \text{otherwise,} \end{cases}$$

then the M (resp., N) is called type 1 (resp., type 2) $(0,1)$ incidence matrix generated by X.

Definition 2.3.2 ([7], [8]) *Let α be a primitive element of $GF(q)$, where $q = p^k = ef + 1$ is a prime power. Let $G = \langle x \rangle$, the Abelian group generated by x. The following subsets, C_i, $0 \le i \le e - 1$, of $GF(q)$ are called the cyclotomic classes, where*

$$C_i := \{\alpha^{es+i} : 0 \le s \le f - 1\}, \quad 0 \le i \le e - 1.$$

Hence these cyclotomic classes are disjoint with each other and their union is exactly the group G itself.

Theorem 2.3.1 ([6], [7]) *Let q be a prime power and $q \equiv 5 \pmod 8$, $q = s^2 + 4t^2$, $s \equiv 1 \pmod 4$. If there are amicable orthogonal designs of type $[(1, 2r - 1); (r, r)]$ and order $2r = |t| + 1$, then there exist amicable Hadamard matrices of order $(|t| + 1)(q + 1)$.*

Proof. Let C_i be the cyclotomic classes in Definition 2.3.2. Then the sets

$$C_0 \cap C_1, \overbrace{C_0 \cap C_2, C_0 \cap C_2, \ldots, C_0 \cap C_2}^{|t|}$$

are $(|t| + 1)$-$\{q; (q - 1)/2; (|t| + 1)(q - 3)/4\}$ supplementary difference sets(SDS) satisfying

$$x \in C_0 \cap C_1 \implies -x \notin C_0 \cap C_1$$

and

$$y \in C_0 \cap C_2 \implies -y \in C_0 \cap C_2.$$

In addition, the sets

$$\overbrace{C_0 \cap C_2, \ldots, C_0 \cap C_2}^{(|t|)/2} \overbrace{C_1 \cap C_3, \ldots, C_1 \cap C_3}^{(|t|)/2}$$

are $(|t| + 1)$-$\{q; (q - 1)/2; (|t| + 1)(q - 3)/4\}$ supplementary difference sets(SDS) satisfying

$$y \in C_0 \cap C_2 \implies -y \in C_0 \cap C_2$$

and

$$z \in C_1 \cap C_3 \implies -z \in C_1 \cap C_3.$$

Let A be the type 1 (±1) incidence matrix generated by $C_0 \cap C_1$, and B (resp. C) the type 2 (±1) incidence matrix generated by $C_0 \cap C_2$ (resp. $C_1 \cap C_3$). Then

$$AJ = BJ = CJ = -J, \quad (A + I)' = -(A + I), \quad B' = B, \quad C' = C,$$

and

$$AA' + |t| BB' = (|t| + 1)/2(BB' + CC')$$
$$= (|t| + 1)(q + 1)I - (|t| + 1)J.$$

Let $P = x_0 U + x_1 V$ and $Q = x_3 X + x_4 Y$ be the amicable orthogonal designs of order $2r$ and type $((1, 2r - 1); (r, r))$. Without loss of generality it can be assumed that $U = I$, $V' = -V$, $X' = X$, and $Y' = Y$, in fact, if otherwise simultaneously multiplying P and Q by some matrix W. Let e be the all ones vector of length q. Then the required amicable Hadamard matrices are

$$E = \begin{bmatrix} U + V & (U + V) \otimes e \\ (-U + V) \otimes e' & U \otimes (-A) + V \otimes B \end{bmatrix}$$

and

$$F = \begin{bmatrix} X+Y & (X+Y)\otimes e \\ (X+Y)\otimes e' & X\otimes C+Y\otimes D \end{bmatrix}.$$

Q.E.D.

Corollary 2.3.1 ([6], [7]) *Let q be a prime power, $q \equiv 5 \pmod 8$, and $q = s^2 + 4t^2$, $s \equiv 1 \pmod 4$. If $\mid t \mid = 2^r - 1$, r a positive integer, then there exist amicable Hadamard matrices of order $2^r(q+1)$.*

Proof. This is a simple consequence of Theorem 2.3.1 and Lemma 2.3.3.
Q.E.D.

Corollary 2.3.2 ([7], [9]) *Let q be a prime power, $q \equiv 5 \pmod 8$, and $q = s^2 + 4$, or $q = s^2 + 36$, $s \equiv 1 \pmod 4$. Then there exist amicable Hadamard matrices of orders $2(s^2 + 5)$ or $4(s^2 + 37)$.*

Definition 2.3.3 ([7], [10]) *A set of m (± 1)-valued matrices $A_1, A_2, \ldots, A_m$ of order n are called suitable matrices for the orthogonal design of type $(s_1, s_2, \ldots, s_m)$ if the following two equations are satisfied:*

1. $A_i A_j' = A_j A_i$, $1 \le i, j \le m$;

and

2. $\sum_{i=1}^m s_i A_i A_i' = (\sum s_i) n I_n$.

A simple observation of suitable matrices and orthogonal designs is:

Theorem 2.3.2 ([7], [11]) *Let X be a $(0, \pm x_1, \pm x_2)$-valued orthogonal design of type $(1, m-1, m(h-1))$ and order mh. Let A_1, A_2 are suitable matrices of order n. Then replacing the variable x_i in X by the matrix A_i, $1 \le i \le 2$, an Hadamard matrix of order mhn is generated.*

Furthermore, the following results are also true:

Theorem 2.3.3 ([7], [11]) *Suppose that there exist m suitable matrices of order n and an orthogonal design of type $(s_1, s_2, \ldots, s_m)$ and order $\sum s_i$. Then there exists an Hadamard matrix of order $(\sum s_i)m$.*

Theorem 2.3.4 ([7], [12]) *Suppose that there exists an orthogonal design of order $4t$ and type (t, t, t, t) and that there exist 4 suitable matrices A, B, C, D of order m satisfying*

$$AA' + BB' + CC' + DD' = 4mI_m.$$

Then there exists an Hadamard matrix of order $4m$.

Corollary 2.3.3 ([7], [13]) *Let n be the order of Hadamard matrix H. If there exists an orthogonal design, say D, of type $(1, m - 1, nm - n - m)$ and order $n(m - 1)$, then there exists an Hadamard matrix of order $n(n - 1)(m - 1)$.*

Proof. Without loss of generality, we assume

$$H = \begin{bmatrix} 1 & e \\ -e' & P \end{bmatrix},$$

where e is the all one row of length $n - 1$. Then the matrix P satisfies

$$PJ = J \quad and \quad PP' = nI - J.$$

The proof is finished by replacing the variables of the orthogonal design D by the matrices P, J, P, respectively.   **Q.E.D.**

Corollary 2.3.4 ([7], [8]) *There exists an Hadamard matrix of order $2^s(2^s - 1)(2^t - 1)$ for each pair of nonnegative integers s and t.*

Proof. It can be proved by an obvious consequence of Lemma 2.3.1 and Corollary 2.3.3.   **Q.E.D.**

Definition 2.3.4 *A weighting matrix $W = W(p, p - 1)$ is called a symmetric conference matrix if $W' = W$ and $p \equiv 2 \pmod 4$.*

Theorem 2.3.5 ([7], [10]) *If there exists an Hadamard matrix, H, of order $k > 1$, a symmetric conference matrix, C, of order n, a pair of amicable orthogonal designs, M, N, of order m and type $((1, m-1); (m/2, m/2))$, and suitable matrices of order p, then there exists an Hadamard matrix of order $nmkp$.*

Proof. Let P be a matrix defined by

$$P = \begin{bmatrix} 0 & -1 \\ 1 & 0 \end{bmatrix} \otimes I_{k/2}.$$

Then the matrix $R := C \otimes H \otimes N + I \otimes PH \otimes M$ is an orthogonal design of type $(k, (m-1)k, (n-1)mk/2, (n-1)nk/2)$ and order nkm. The proof is finished by replacing the variables in this orthogonal design by the suitable matrices.   **Q.E.D.**

2.3.2   Existence Results

Most of the following existence results are based on the orthogonal designs. First we restate a well known theorem in number theory.

Theorem 2.3.6 ([7], [11]) *Let x and y be two positive co-prime integers, and N an integer satisfying $N > (x-1)(y-1)$. Then there exist non-negative integers a and b such that $N = ax + by$.*

A direct consequence of this theorem is:

Corollary 2.3.5 ([7], [11]) *Let $v \geq 9$ be an odd integer, $x = v + 1$, and $y = v - 3$. Then there exist non-negative integers a, b, and t such that $a(v+1) + b(v-3) = 2^t$.*

Proof. Let $g = gcd(v+1, v-3)$. Then $g = 1, 2$, or 4. Let m be the smallest number of the form $m = 2^k$ such that

$$m > [(v+1)/g - 1][(v-3)/g - 1].$$

Then by Theorem 2.3.6, there exist integers a and b such that

$$a(v+1)/g + b(v-3)/g = m.$$

The corollary follows.   **Q.E.D.**

The first existence result is:

Lemma 2.3.4 ([7], [6]) *Let $p \geq 9$ be a prime and $p \equiv 3 \pmod 4$. Then there exists an integer t such that an Hadamard matrix of order $2^s p$ can always be obtained if $s \geq t$.*

Proof. Let $x = p+1$ and $y = p-3$. Then, by Corollary 2.3.5, there exist a and b satisfying $ax + by = 2^t := n$ for some t. Let D be an orthogonal design of the variables x_1, x_2 and x_3, of order 2^t and type $(a, b, n-a-b)$. The required Hadamard matrix is obtained by replacing each variable x_1 by the matrix J, each variable x_2 by $J - 2I$ and each variable x_3 by the back circulant $(\pm)$-valued matrix $B = (Q+1)R$, where R is the back diagonal matrix and $Q = [Q_{ij}]$ is defined by

$$Q_{ij} = \begin{cases} 0 \text{ if } i = j \\ 1 \text{ if } j - i = y^2 \text{ for some } y \in GF(p) \\ -1 \text{ otherwise.} \end{cases}$$

Q.E.D.

Lemma 2.3.5 ([7], [6]) *Let $p \geq 9$ be a prime and $p \equiv 1 \pmod 4$. Then there exists an integer t such that an Hadamard matrix of order $2^s p$ can always be obtained if $s \geq t + 1$.*

This lemma can be proved in the same way as that of Lemma 2.3.4. In fact, let F be an orthogonal design of the variables x_1, x_2, x_3 and x_4, of order $2^{t+1} := n$ and type $(2a, 2b, n - a - b, n - a - b)$. Then the required matrix is obtained by replacing each variable x_1 by J, each variable x_2 by $J - 2I$, and the variables x_3 and x_4 by $X = I + Q$ and $Y = I - Q$, respectively.

The previous two lemmas complete the proof of the existence of Hadamard matrices of order $2^t.p$ for all prime p except $p = 2, 3, 5$, and 7.

The existence of Hadamard matrices of order $4m$ and $4n$ implies the existence of that of order $4 \times 4mn = 16mn$ by the direct multiplication. The following result reduces the order to 8mn in the resulting Hadamard matrix.

Lemma 2.3.6 ([7], [10]) *Let H and E be Hadamard matrices of orders $4m$ and $4n$, respectively. Then there is an Hadamard matrix of order $8mn$.*

Proof. Write the H and E as

$$H = \begin{bmatrix} P & Q \\ R & S \end{bmatrix} \quad \text{and} \quad E = \begin{bmatrix} K & L \\ M & N \end{bmatrix}.$$

Because of $HH' = 4mI$ and $EE' = 4nI$, we have

$$PP' + QQ' = RR' + SS' = 2mI,$$
$$PR' + QS' = 0 = RP' + SQ',$$
$$KK' + LL' = MM' + NN' = 2nI,$$
$$KM' + LN' = 0 = MK' + NL'.$$

The resulting Hadamard matrix of order $8mn$ is

$$\begin{bmatrix} \frac{1}{2}(P+Q) \otimes K + \frac{1}{2}(P-Q) \otimes M & \frac{1}{2}(P+Q) \otimes L + \frac{1}{2}(P-Q) \otimes N \\ \frac{1}{2}(R+S) \otimes K + \frac{1}{2}(R-S) \otimes M & \frac{1}{2}(R+S) \otimes L + \frac{1}{2}(R-S) \otimes N \end{bmatrix}.$$

Q.E.D.

For example, we have known that there are Hadamard matrices of orders 20 and 12. Thus Lemma 2.3.6 guarantees the existence of an Hadamard matrix of order $8 \times 5 \times 3 = 120$.

Lemma 2.3.7 ([7], [6]) *There exist Hadamard matrices of order $2^t p$ for $p = 3, 5, 7$.*

Proof. The required Hadamard matrices can be formed by applying Lemma 2.3.6 to the known Hadamard matrices of orders 12, 20, 28, and 2^t.   **Q.E.D.**

Theorem 2.3.7 ([7], [8]) *Let q be a positive integer. Then there exists $t = t(q)$ such that an Hadamard matrix of order $2^s q$ exists for every $s \geq t$.*

Proof. Applying Lemmas 2.3.4, 2.3.5 and/or 2.3.7 to each prime factor of q. The proof of the theorem is finished by the direct multiplication of Hadamard matrices.   **Q.E.D.**

A strengthened form of Lemma 2.3.6 can be stated as:

Theorem 2.3.8 ([7], [10]) *Suppose that there are Hadamard matrices of orders $4m$, $4n$, $4p$, and $4q$. Then there exists an Hadamard matrices of order $16mnpq$.*

A natural generalization of Hadamard matrix is the so-called complex Hadamard matrix, which is defined by:

Definition 2.3.5 ([7], [13]) *A matrix C of order $2n$ with elements ± 1, $\pm i$ that satisfies $CC^* = 2nI$ is called a complex Hadamard matrix, where C^* be the transpose of the conjugate matrix of C.*

The following result is similar to Lemma 2.3.6.

Theorem 2.3.9 ([7], [13]) *Suppose that there are a complex Hadamard matrix of order $2n$ and an Hadamard matrix of order $4m$. Then there exists an Hadamard matrix of order $8mn$.*

Some of the other strongest existence results are listed in the following ([1], [5], [7]):

Class 1: Let $p \equiv 3 \pmod 4$ be a prime power. Then there exist a Hadamard matrix of order $p + 1$;

Class 2: Let $p \equiv 1 \pmod 4$ be a prime power. Then there exist a Hadamard matrix of order $2(p + 1)$;

Class 3: Suppose that there is an Hadamard matrix of order n. Then there is a symmetric regular Hadamard matrix with constant diagonal of order n^2;

Class 4: Let q be any positive integer. Then there exists an Hadamard matrix of order $2^t q$ for every $t \geq [2log_2(q-3)] + 1$;

Class 5: Let p and $p + 2$ be twin prime powers. Then there exists a $t \leq [log_2(p+3)(p-1)(p^2+2p-7)] - 2$ so that there is an Hadamard matrix of order $2^t p(p+2)$;

Class 6: Let $p + 1$ be the order of a symmetric Hadamard matrix. Then there exists a $t \leq [log_2(p-3)(p-7)] - 2$ so that there is an Hadamard matrix of order $2^t p$;

Class 7: Let pq be an odd natural number. Suppose that all orthogonal design of order $2^s p$ and type $(2^r a, 2^r b, 2^r c)$ exist, $s \geq s_0$, $2^{s-r} p = a + b + c$. Then there exists an Hadamard matrix of order $2^t . p.q$;

Class 8: Let q be a positive integer. Then there exists a $t = t(q)$ so that there is an Hadamard matrix of order $2^s 3q$ for all $s \geq t$;

Class 9: There is an Hadamard matrix of order $8 \cdot 49 \cdot 3^t = 342 \cdot 3^t$ for all $t \geq 0$;

Class 10: If $n \equiv 3(\text{mod}4)$ is a prime power, there is an Hadamard matrix of order $n^2(n+1)9^t$ for all $t \geq 0$;

Class 11: There exist Hadamard matrices of orders $4 \cdot 3^t$, $4 \cdot 5^t$, $4 \cdot 13^t$, $4 \cdot 17^t$, $4 \cdot 29^t$, $4 \cdot 37^t$, $4 \cdot 41^t$, $4 \cdot 53^t$, $4 \cdot 61^t$, $4 \cdot 101^t$, $t \geq 0$; $4 \cdot g^{4i}$, $4 \cdot g^{4i+1}$, $4 \cdot g^{4i+2}$, $8 \cdot g^{4i+3}$, $i \geq 0$, $g = 7, 11$; and $4.p^k$ whenever $p = 1 + 2^a \cdot 10^b \cdot 26^c$ is prime, $a, b, c \geq 0$.

Up to now, many new progresses on Hadamard matrices have been achieved. For the updated papers in this area the readers are recommend to the references [14-71].

Bibliography

[1] R.Craigen, W.Wallis, *Hadamard Matrices:1893-1993*, Proc. of the 24th Southeastern Int. Conf. On Combinatorics, Graph Theory, and Computing, Boca Raton, FL, 1993, Congr. Numer, 97(1993)99-129.

[2] S.S. Agaian, *Hadamard Matrices and Their Applications*, Lecture Notes in Mathematics 1168, Springer-Verlag, Berlin, 1985.

[3] E.Assmus, *Hadamard Matrices and Their Designs: A Coding-Theoretic Approach*, Trans. Amer. Math. Soc. 330(1990), No.1, 269-293.

[4] K.Balasubramanian, *Computer Generation of Hadamard Matrices*, J. Comput. Chem., 14(1993), No.5, 603-619.

[5] R.Craigen and H.Kharaghani, *On the Existence of Regular Hadamard Matrices* , 21th Manitoba Conf. On Numerical Math. and Comput., Congr., Numer., 99(1994), 277-283.

[6] J. Seberry and M.Yamada, *Amicable Hadamard Matrices and Amicable Orthogonal Designs*, Utilitas Math., 40(1991), 179-192.

[7] A. Geramita and J. Seberry, *Orthogonal Designs* , Marcel Dekker, Inc. New York, 1979.

[8] J. Seberry, *Existence of $SBIBD(4k^2, 2k^2 \pm k, k^2 \pm k)$ and Hadamard Matrices With Maximal Excess*, Australas J. Combin., 4(1991), 87-91.

[9] J.Seberry, *SBIBD($4k^2, 2k^2 + k, k^2 + k$) and Hadamard Matrices of Order $4k^2$ With Maximum Excess Are Equivalent*, Graphs Combin. 5(1989), No.4, 373-383.

[10] J.Seberry, A.L.Whiteman, *New Hadamard Matrices and Conference Matrices Obtained Via Mathon's Construction*, Graphs Combin, 4(1988), No.4, 355-377.

[11] J.Seberry, and M.Yamada, *Hadamard Matrices, Sequences, and Block Designs*, In: Contemporary Design Theory: A Collection of Surveys, Edited by J.H.Dinitz and Douglas R. Stinson, John Wiley and Sons, Inc., pp431-560, 1992.

[12] J.Seberry, *A New Construction for Williamson-Type Matrices*, Graphs Combin. 2(1986), No.1, 81-87.

[13] J.Seberry and M. Yamada, *On the Multiplication Theorem of Hadamard Matrices of Generalized Quaternion Type Using M-Structures*, J. Combin. Math. Combin. Comput., 13(1993), 97-106.

[14] K.Balasubramanian, *Characterization of Hadamard Matrices*, Molecular Phys., 78 (1993), No.5, 1309-1329.

[15] F.Bussemaker and V.D.Tonchev, *New Extremal Double-Even Codes of Length 56 Derived From Hadamard Matrices of Order 28*, Discrete Math., 76(1989), 45-59.

[16] F.Bussemaker and V.D.Tonchev, *Extremal Double-Even Codes of Length 40 Derived From Hadamard Matrices of Order 20*, Discrete Math., 82(1990), 317-321.

[17] L. Cantian, etal., *Codes From Hadamard Matrices and Profiles of Hadamard Matrices*, J. Combin. Math. Combin. Comput., 12(1992), 57-64.

[18] W.Chan, *Necessary Conditions for Menon Difference Sets*, Des. Codes Cryptogr. 3(1993), No.2, 147-154.

[19] G.Cohen, etal., *A Survey of Base Sequences, Disjoint Complementary Sequences and OD(4t; t,t,t,t)*, J. Combin. Math. Combin. Comput., 5(1989), 69-103.

[20] R.Craigen, *Product of Four Hadamard Matrices*, J. of Combin., Ser. A, 59, 318-320, 1992.

[21] R.Craigen *Constructing Hadamard Matrices With Orthogonal Pairs*, Ars. Combin., 33(1992), 57-64.

[22] R.Craigen, *Equivalence Classes of Inverse Orthogonal and Unit Hadamard Matrices* , Bull. Austral., Math., Soc., 44(1991), No.1, 109-115.

[23] R.Craigen, *Embedding Rectangular Matrices in Hadamard Matrices,* Linear and Multilinear Algebra 29(1991), No.2, 91-92.

[24] J.Davis and J.Jedwab, *A Summary of Menon Difference Sets,* Congr. Numer., 93(1993), 203-207.

[25] D.Dokvic, *Williamson Matrices of Order 4n for n=33, 35, 39,* Discrete Math. 115(1993), 267-271.

[26] D.Dokvoick, *Williamson Matrices of Order* 4 × 29 *and* 4 × 31, J. of Combin. Ser. A, 59, 309-311 (1992).

[27] D.Dokvic, *Ten Hadamard Matrices of Order 1852 of Goethals-Seidel Type,* European J. Combin. 13(1992), No.4, 245-248.

[28] D.Dokvic, *Construction of Some New Hadamard Matrices,* Bull. Austral., Math., Soc., 45(1992), No.2, 327-332.

[29] D.Dragomir, *Ten New Orders for Hadamard Matrices of Skew Type,* Univ. Beograd., Publ. Elektrotehn, Fak., Ser., Mat., 3(1992), 47-59.

[30] D.Gluck, *Hadamard Difference Sets in Groups of Order 64,* J. of Combin., Ser. A, 51, 138-140 , 1989.

[31] W.H.Holzmann and H.Kharaghani, *On the Access of Hadamard Matrices,* Congr. Numer, 92(1993), 257-260.

[32] H.Kharaghani, *A Construction for Hadamard Matrices,* Discrete Math., 120(1993), 115-120.

[33] H.Kimura, *New Hadamard Matrix of Order 24,* Graphs Combin., 5(1989), No.3, 235-242.

[34] C.Koukouvinos, and S. Kounias, *Construction of Some Hadamard Matrices With Maximum Excess,* Discrete Math, 85(1990), 295-300.

[35] C.Koukouvinos, J.Seberry, *Hadamard Matrices of Order 8(mod16) With Maximum Excess,* Discrete Math., 92(1991), 173-176.

[36] C.Koukouvinos, and S. Kounias, *Hadamard Matrices of the Williamson Type of Order 4m, m=pq, An Exhaustive Search for m=33,* Discrete Math., 68(1988), 45-57.

[37] S.Kounias and N.Farmakis, *On the Excess of Hadamard Matrices,* Discrete Math., 68(1988), 59-69.

[38] C.Koukouvinos, *On the Excess of Hadamard Matrices,* Utilitas Math., 45(1994), 97-101.

[39] C.Koukouvinos and J.Seberry, *Constructing Hadamard Matrices from orthogonal Designs,* Australas. J. Combin, 6(1992), 267-278.

[40] C.Koukouvinos and J.Seberry, *Construction of New Hadamard Matrices With Maximal Excess and Infinity Many New* $SBIBD(4k^2, 2k^2 + k, k^2 + k)$, Graphs, Matrices, and Designs, 255-267, Lecture Notes in Pure and Appl. Math., 139, Dekker, New York, 1993.

[41] C.Koukouvinos, S.Kounias, and J.Seberry, *Further Results On Base Sequences, Disjoint Complementary Sequences,* $OD(4t; t, t, t, t)$ *and the Excess of Hadamard Matrices,* Ars. Combin., 30(1990), 241-255.

[42] C.Koukouvinos, S.Kounias, and J.Seberry, *Further Hadamard Matrices With Maximal Excess and New SBIBD($4k^2, 2k^2 + k, k^2 + k$)* , Utilitas Math., 36(1989), 135-150.

[43] R.Kraemer, *Proof of a Conjecture on Hadamard 2-Groups,* J. of Combin., Ser. A, 63, 1-10, (1993).

[44] W.Launey, *A Product of Twelve Hadamard Matrices,* Australas., J. Combin., 7(1993), 123-127.

[45] C.Lin, W.Wallis, *Symmetric and Skew Equivalence of Hadamard Matrices,* Congr. Numer. 85(1991), 73-79.

[46] C.Lin, W.Wallis, *Profiles of Hadamard Matrices of Order 24,* Congr. Numer, 66(1988), 93-102.

[47] R.L. Mcfarland, *Necessary Conditions for Hadamard Difference Sets,* In: D.Ray-Chaudhuri, ed., Coding Theory and Design Theory, Part II (Springer, Berlin, 1990), pp257-272.

[48] R.L. Mcfarland, *Subdifference Sets of Hadamard Difference Sets,* J. Combin. Theory, Ser. A 54(1990) 112-122.

[49] R.L. Mcfarland, *Necessary Conditions for Hadamard Difference Sets,* Coding Theory and Design Theory, Part II, 257-272, IMA Vol. Math. Appl. 21, Springer, New York, 1990.

[50] D.Meisner, *Families of Menon Difference Sets,* Combinatorics'90 (Gaeta, 1990), 365-380, Ann., Discrete Math., 52, North-Holland, Amsterdam, 1992.

[51] D.Meisner, *On a Construction of Regular Hadamard Matrices,* Math., Appl. 3(1992), No.4, 233-240.

[52] M.Miyamoto, *A Construction of Hadamard Matrices,* J. of Combin., Ser. A, 57, 86-108 (1991).

[53] K.Momura, *Spin Models Constructed From Hadamard Matrices* , J. Combin. Theory, Ser. Vol 68, No.2 (1994), 251-261.

[54] R.C.Mullin and D.Wevrick, *Singular Points in Pair Covers and Their Relation to Hadamard Designs,* Discrete Math., 92(1991), 221-225.

[55] I.Noboru, *Note on Hadamard Groups of Quadratic Residue Type,* Hokkaido Math. J., 22(1993), No.3, 373-378.

[56] I.Noboru, *Nearly Triply Regular Hadamard Designs and Tournaments,* Math. J. Okayama. Univ., 32(1990), 1-5.

[57] K.W. Smith, *Non-Abelian Hadamard Difference Sets,* J. of Combinatorial Theory, Ser. A, Vol 70, No.1 (1995), 144-156.

[58] P.Sole, A. Ghafoor and S.A.Sheikh, *The Covering Radius of Hadamard Codes in Odd Graphs,* Discrete Applied Math. 37/38 (1992) 501-510.

[59] E.Spence, *Classification of Hadamard Matrices of Order 24 and 28,* Discrete Mathematics, Vol.140, Nos.1-3, pp185-243, 1995.

[60] V.D.Tonchev, *Self-Dual Codes and Hadamard Matrices,* Discrete Applied Math. 33(1991), Nos1-3, pp235-240.

[61] W.Wallis, *Hadamard Matrices,* Combinatorial and Graph-Theoretical Problems in Linear Algebra (Minneapolis, MN, 1991), 235-243, IMA Vol. Math. Appl., 50, Springer, New York, 1993.

[62] M. Xia, *Some Infinite Classes of Special Williamson Matrices and Difference Sets*, J. of Combinatorial Theory, Ser. A, Vol 61(1992), 230-242.

[63] M. Xia and T. Xia, *Hadamard Matrices Constructed From Supplementary Difference Sets in the Class F_1*, J. of Combin. Des., 2(1994), No.5, 325-339.

[64] M.Xia, *An Infinite Classes of Supplementary Difference Sets and Williamson Matrices*, J. of Combin., Ser.A, 58, 310-317, (1991).

[65] M.Xia, *Some Infinite Classes of Williamson Matrices and Weighting Matrices*, Australas J. Combin., 6(1992), 107-110.

[66] M.Xia, *Hadamard Matrices, Combinatorial Designs and Applications*, (Hungshan, 1988), 179-181, Lecture Notes in Pure and Appl. Math., 126, Dekker, New York, 1990.

[67] M.Yamada, *Some New Series of Hadamard Matrices*, J. Austral. Math. Soc. Ser. A 46(1989), No.3, 371-383.

[68] Y.X. Yang, and X.D.Lin, *Coding and Cryptography*, PPT Press, Beijing, 1992.

[69] M.Yamada, *Hadamard Matrices of Generalized Quaternion Type*, Discrete Math., 87(1991), 187-196.

[70] L.Zhu, *Equivalence Classes of Hadamard Matrices of Order 32*, Congr. Numer, 95 (1993), 179-182.

[71] L.Zhu, *An Infinite Family of Complex Amicable Hadamard Matrices*, Bull. Inst. Combin. Appl. 1(1991), 37-40.

Part II
Lower-Dimensional Cases

Chapter 3
Three-Dimensional Hadamard Matrices

3.1 Definitions and Constructions

3.1.1 Definitions (see [2])

It is known that a 2-dimensional Hadamard matrix $H = [H(i,j)]$, $0 \leq i, j \leq m-1$, of order $m \times m$ is in fact a binary (± 1)-valued orthogonal matrix satisfying $HH' = mI_m$. In other words, a 2-dimensional Hadamard matrix is a matrix such that its $(2-1)$-dimensional layers, in each normal orientation of axes, are orthogonal to each other, i.e., the $(2-1)$-dimensional layers $(H(0,a), \ldots, H(m-1,a); (H(0,b), \ldots, H(m-1,b)$ in y-axis orientation and the $(2-1)$-dimensional layers $(H(a,0), \ldots, H(a,m-1); (H(b,0), \ldots, H(b,m-1)$ in x-axis orientation satisfy $H(i,j) = -1$ or 1 and

$$\sum_{i=0}^{m-1} H(i,a)H(i,b) = \sum_{j=0}^{m-1} H(a,j)H(b,j)$$
$$= m\delta_{ab}, \tag{3.1}$$

where $\delta_{ab} = 1$ iff $a = b$, otherwise $\delta_{ab} = 0$.

Similarly to the definition of 2-dimensional Hadamard matrices, a 3-dimensional Hadamard matrix of order $m \times m$ can be defined as a (± 1)-valued matrix $H = [H(i,j,k)]$, $0 \leq i, j, k \leq m-1$, such that its $(3-1)$-dimensional layers, in each normal orientation of axes, are orthogonal to each other, i.e., the $(3-1)$-dimensional layers $[H(i,j,a) : 0 \leq i, j \leq m-1]$ and $[H(i,j,b) : 0 \leq i, j \leq m-1]$ in z-axis orientation are orthogonal to each other; the $(3-1)$-dimensional layers $[H(i,a,k) : 0 \leq i, k \leq m-1]$ and

$[H(i, b, k) : 0 \leq i, k \leq m - 1]$ in y-axis orientation are orthogonal to each other; and the $(3 - 1)$-dimensional layers $[H(a, j, k) : 0 \leq j, k \leq m - 1]$ and $[H(b, j, k) : 0 \leq j, k \leq m - 1]$ in x-axis orientation are also orthogonal to each other. In other words, a three-dimensional Hadamard matrix is one in which all parallel two-dimensional layers, in normal orientations of all axes, are orthogonal to each other. That is, the three-dimensional matrix, $H = [H(i, j, k)]$, $0 \leq i, j, k \leq m - 1$, of order m is Hadamardian if all $H(i, j, k) = -1$ or 1 and if the following are satisfied ([1] , [3]):

$$\sum_{i=0}^{m-1} \sum_{j=0}^{m-1} H(i, j, a) H(i, j, b)$$

$$= \sum_{j=0}^{m-1} \sum_{k=0}^{m-1} H(a, j, k) H(b, j, k)$$

$$= \sum_{i=0}^{m-1} \sum_{k=0}^{m-1} H(i, a, k) H(i, b, k)$$

$$= m^2 \delta_{ab}. \tag{3.2}$$

A much stronger definition of a three-dimensional Hadamard matrix is defined by those matrices in which all of the 2-dimensional layers, in normal orientations of all axes, are in themselves (two-dimensional) Hadamard matrices ([1], [3]). In other words,

$$\sum_{i=0}^{m-1} H(i, a, r) H(i, b, r) = \sum_{j=0}^{m-1} H(a, j, r) H(b, j, r)$$

$$= m \delta_{ab}, \quad r = \text{any value of } k , \tag{3.3}$$

i.e., the 2-dimensional layers in z-normal orientation are 2-dimensional Hadamard matrices.

$$\sum_{i=0}^{m-1} H(i, q, a) H(i, q, b) = \sum_{k=0}^{m-1} H(a, q, k) H(b, q, k)$$

$$= m \delta_{ab}, \quad q = \text{any value of } j, \tag{3.4}$$

i.e., the 2-dimensional layers in normal orientation of y-axis are 2-dimensional Hadamard matrices.

$$\sum_{j=0}^{m-1} H(p, j, a) H(p, j, b) = \sum_{k=0}^{m-1} H(p, a, k) H(p, b, k)$$

$$= m \delta_{ab}, \quad p = \text{any value of } i, \tag{3.5}$$

i.e., the 2-dimensional layers in normal orientation of x-axis are 2-dimensional Hadamard matrices.

Matrices satisfying Equations (3.3), (3.4), and (3.5) are precisely called 'absolutely proper' three-dimensional Hadamard matrices. Matrices satisfying Equation (3.2) but not all of Equations (3.3)–(3.5) are called 'improper'. Thus, a matrix satisfying Equations (3.2), (3.3), and (3.4) but not (3.5) are called 'improper in the x-direction.' Matrices satisfying Equations (3.2), (3.3) and (3.4) are called 'proper' in the z-direction. Three-dimensional Hadamard matrices which are improper in every axial direction are called absolutely improper.

Each three-dimensional matrix $A = [A(i,j,k)]$, $0 \leq i,j,k \leq m-1$, can be described by its two-dimensional layers in each direction, e.g. the m layers $[A(i,j,0)]$, $[A(i,j,1)]$, $[A(i,j,2)]$, ..., $[A(i,j,m-1)]$ in the z-direction.

Thus it is easy to verify that the 3-dimensional matrix $A = [A(i,j,k)]$, $0 \leq i,j,k \leq 1$, described by

$$[A(i,j,0)] = \begin{bmatrix} 1 & 1 \\ 1 & -1 \end{bmatrix} \quad \text{and} \quad [A(i,j,1)] = \begin{bmatrix} -1 & 1 \\ 1 & 1 \end{bmatrix}$$

is an absolutely proper three-dimensional Hadamard matrix ([1] , [3]).

The 3-dimensional matrix $B = [B(i,j,k)]$, $0 \leq i,j,k \leq 1$, described by

$$[B(i,j,0)] = \begin{bmatrix} -1 & 1 \\ 1 & 1 \end{bmatrix} \quad \text{and} \quad [B(i,j,1)] = \begin{bmatrix} -1 & 1 \\ -1 & -1 \end{bmatrix}$$

is improper in x-direction. In fact, its x-direction layers are

$$[B(0,j,k)] = \begin{bmatrix} -1 & -1 \\ 1 & 1 \end{bmatrix} \quad \text{and} \quad [B(1,j,k)] = \begin{bmatrix} 1 & 1 \\ 1 & -1 \end{bmatrix}.$$

One of them, $[B(0,j,k)]$, is not a 2-dimensional Hadamard matrix ([1], [3]).

The 3-dimensional matrix $C = [C(i,j,k)]$, $0 \leq i,j,k \leq 1$, described by

$$[C(i,j,0)] = \begin{bmatrix} 1 & 1 \\ 1 & -1 \end{bmatrix} \quad \text{and} \quad [C(i,j,1)] = \begin{bmatrix} 1 & -1 \\ -1 & -1 \end{bmatrix}$$

is an absolutely proper three-dimensional Hadamard matrix ([1] , [3]).

The 3-dimensional matrix $D = [D(i, j, k)]$, $0 \le i, j, k \le 1$, described by

$$[D(i, j, 0)] = \begin{bmatrix} -1 & 1 \\ 1 & 1 \end{bmatrix} \qquad \text{and} \qquad [D(i, j, 1)] = \begin{bmatrix} 1 & -1 \\ 1 & 1 \end{bmatrix}$$

is improper in the x-direction. In fact, its x-direction layers are

$$[D(0, j, k)] = \begin{bmatrix} -1 & 1 \\ 1 & -1 \end{bmatrix} \qquad \text{and} \qquad [D(1, j, k)] = \begin{bmatrix} 1 & 1 \\ 1 & 1 \end{bmatrix},$$

which are not 2-dimensional Hadamard matrices ([1] , [3]).

3.1.2   Constructions Based on Direct Multiplications

For the details of the contents of this subsection the readers are recommended to [1].

Direct multiplication (or Kronecker production) of two matrices A and B is the matrix C constructed from A and B by replacing each element $A(i, j)$ by the submatrix $A(i, j)B$. Exactly, if the given two matrices are

$$A = \begin{bmatrix} A(0,0) & A(0,1) & \cdots & A(0, m-1) \\ A(1,0) & A(1,1) & \cdots & A(1, m-1) \\ \vdots & \vdots & \vdots & \vdots \\ A(n-1, 0) & A(n-1, 1) & \cdots & A(n-1, m-1) \end{bmatrix}$$

and

$$B = \begin{bmatrix} B(0,0) & B(0,1) & \cdots & B(0, M-1) \\ B(1,0) & B(1,1) & \cdots & B(1, M-1) \\ \vdots & \vdots & \vdots & \vdots \\ B(N-1, 0) & B(N-1, 1) & \cdots & B(N-1, M-1) \end{bmatrix},$$

then the direct multiplication of A and B is another matrix $C = [C(i, j)]$ of order $(nN) \times (mM)$ defined by

$$C = \begin{bmatrix} A(0,0)B & A(0,1)B & \cdots & A(0, m-1)B \\ A(1,0)B & A(1,1)B & \cdots & A(1, m-1)B \\ \vdots & \vdots & \vdots & \vdots \\ A(n-1, 0)B & A(n-1, 1)B & \cdots & A(n-1, m-1)B \end{bmatrix}.$$

This definition is denoted by $C = A \otimes B$, or equivalently, the general elements of the direct multiplication matrix C are

$$C(i,j) = A\left(\left\lfloor \frac{i}{N} \right\rfloor, \left\lfloor \frac{j}{M} \right\rfloor\right) B([i]_N, [j]_M),$$

$$\text{for } 0 \le i \le (nN-1),\ 0 \le j \le (mM-1), \qquad (3.6)$$

where $\lfloor x \rfloor$, the floor function, denotes the largest integer not larger than x, and $[i]_K \equiv i \bmod K$, the remainder of i divided by K. Thus $i = \lfloor \frac{i}{N} \rfloor N + [i]_N$ and $j = \lfloor \frac{j}{M} \rfloor M + [j]_M$.

The concept of direct multiplication defined by Equation (3.6) can be generalized to three-dimensional cases as follows ([21]):

Let

$$A = [A(i,j,k)], \quad 0 \le i \le n-1,\ 0 \le j \le m-1,\ 0 \le k \le p-1$$

and

$$B = [B(i,j,k)], \quad 0 \le i \le N-1,\ 0 \le j \le M-1,\ 0 \le k \le P-1$$

are two three-dimensional matrices of orders $n \times m \times p$ and $N \times M \times P$, respectively. Then the direct product of A and B is a three-dimensional matrix

$$C = A \otimes B = [C(i,j,k)], 0 \le i \le Nn-1, \quad 0 \le j \le Mm-1,\ 0 \le k \le pP-1$$

of order $(nN) \times (mM) \times (pP)$ defined by

$$C(i,j,k) = A\left(\left\lfloor \frac{i}{N} \right\rfloor, \left\lfloor \frac{j}{M} \right\rfloor, \left\lfloor \frac{k}{P} \right\rfloor\right) B([i]_N, [j]_M, [k]_P), \qquad (3.7)$$

for $0 \le i \le (nN-1),\ 0 \le j \le (mM-1), 0 \le k \le pP-1$. Thus the matrix C is constructed from its mother matrices A and B by replacing each element $A(i,j,k)$ by a 3-dimensional sub-matrix $A(i,j,k)B$.

Direct multiplication is an important construction approach for 2-dimensional Hadamard matrices. In fact, we have known that if A and B are two 2-dimensional Hadamard matrices, then so is their direct multiplication matrix $C = A \otimes B$. This kind of construction is still valid for 3-dimensional cases. Precisely, we have the following theorems.

Theorem 3.1.1 ([1] , [3], [21]) *The direct multiplication of two three-dimensional Hadamard matrices is also a three-dimensional Hadamard matrix. Furthermore, the product matrix is proper in those directions in which both the parent matrices are proper.*

Proof. Let $A = [A(i, j, k)]$ and $B = [B(i, j, k)]$ be two 3-dimensional Hadamard matrices of order n and m, respectively. Their direct production $C = A \otimes B$ is of order (nm). Thus, by Equation (3.7), we have

$$\sum_{i=0}^{nm-1} \sum_{j=0}^{nm-1} C(i, j, a) C(i, j, b)$$

$$= A(\lfloor \frac{i}{m} \rfloor, \lfloor \frac{j}{m} \rfloor, \lfloor \frac{a}{m} \rfloor) . B([i]_m, [j]_m, [a]_m)$$

$$\cdot A(\lfloor \frac{i}{m} \rfloor, \lfloor \frac{j}{m} \rfloor, \lfloor \frac{b}{m} \rfloor) . B([i]_m, [j]_m, [b]_m)$$

$$= \sum_{u=0}^{n-1} \sum_{v=0}^{n-1} \sum_{p=0}^{m-1} \sum_{q=0}^{m-1} A(u, v, \lfloor \frac{a}{m} \rfloor)$$

$$\cdot A(u, v, \lfloor \frac{b}{m} \rfloor) . B(p, q, [a]_m) . B(p, q, [b]_m)$$

$$= \left[\sum_{u=0}^{n-1} \sum_{v=0}^{n-1} A(u, v, \lfloor \frac{a}{m} \rfloor) . A(u, v, \lfloor \frac{b}{m} \rfloor) . \right]$$

$$\left[\sum_{p=0}^{m-1} \sum_{q=0}^{m-1} B(p, q, [a]_m) . B(p, q, [b]_m) \right]$$

$$= (nm)^2 \delta_{a,b},$$

where the last equation is owed to the properties that (1) A and B are themselves 3-dimensional Hadamard matrices; and (2) $a \neq b$ implies that $\lfloor \frac{a}{m} \rfloor \neq \lfloor \frac{b}{m} \rfloor$ or $[a]_m \neq [b]_m$.

Similarly, it can be proved that

$$\sum_{j=0}^{nm-1} \sum_{k=0}^{nm-1} C(a, j, k) C(b, j, k) = \sum_{i=0}^{nm-1} \sum_{k=0}^{nm-1} C(i, a, k) C(i, b, k)$$

$$= (nm)^2 \delta_{a,b}$$

Therefore Equation (3.2) is satisfied by C, i.e., C is indeed a 3-dimensional Hadamard matrix.

Every layer of C in x- (resp., y- or z-) direction is the direct multiplication of two layers of the parents in the x- (resp., y- or z-) direction. Thus C is proper in those directions in which both the parent matrices are proper.   **Q.E.D.**

Corollary 3.1.1 ([1] , [3], [21]) *Three-dimensional Hadamard matrices of order 2^t can be generated from $t-1$ successive direct multiplication among three-dimensional Hadamard matrices of order 2.*

Three-dimensional Hadamard matrices of order m^2 can also be generated by the direct multiplication of three two-dimensional Hadamard matrices of order m in different orientations.

For example, the 2-dimensional Hadamard matrix

$$A = \begin{bmatrix} 1 & 1 \\ 1 & -1 \end{bmatrix}$$

can be respectively treated as a 3-dimensional matrix $A_1 = [A_1(i,j,k)]$, $k = 0$, $0 \leq i,j \leq 1$, of order $2 \times 2 \times 1$ such that $[A_1(i,j,0)] = [A(i,j)]$, a 3-dimensional matrix $A_2 = [A_2(i,j,k)]$, $j = 0$, $0 \leq i,k \leq 1$, of order $2 \times 1 \times 2$ such that $[A_2(i,0,k)] = [A(i,k)]$, and a 3-dimensional matrix $A_3 = [A_3(i,j,k)]$, $i = 0$, $0 \leq j,k \leq 1$, of order $1 \times 2 \times 2$ such that $[A_3(0,j,k)] = [A(j,k)]$. The direct multiplication of these three matrices is

$$C = [C(i,j,k)] = (A_1 \otimes A_2) \otimes A_3, \ 0 \leq i,j,k \leq 3.$$

Its layers in z-direction are:

$$C(i,j,0) = \begin{bmatrix} 1 & 1 & 1 & 1 \\ 1 & 1 & 1 & 1 \\ 1 & 1 & -1 & -1 \\ 1 & 1 & -1 & -1 \end{bmatrix}, \quad C(i,j,1) = \begin{bmatrix} 1 & -1 & 1 & -1 \\ 1 & -1 & 1 & -1 \\ 1 & -1 & -1 & 1 \\ 1 & -1 & -1 & 1 \end{bmatrix}$$

and

$$C(i,j,2) = \begin{bmatrix} 1 & 1 & 1 & 1 \\ -1 & -1 & -1 & -1 \\ 1 & 1 & -1 & -1 \\ -1 & -1 & 1 & 1 \end{bmatrix}, \quad C(i,j,3) = \begin{bmatrix} 1 & -1 & 1 & -1 \\ -1 & 1 & -1 & 1 \\ 1 & -1 & -1 & 1 \\ -1 & 1 & 1 & -1 \end{bmatrix}.$$

It can be verified that the above direct multiplication matrix $C = (A_1 \otimes A_2) \otimes A_3$ is a 3-dimensional Hadamard matrix of order $4 \times 4 \times 4$.

In general, we have the following theorem.

Theorem 3.1.2 ([1], [3], [21]) *Let $A = [A(i, j)]$ be a 2-dimensional Hadamard matrix of order $m \times m$. And let $A_1 = [A_1(i, j, k)]$ (resp., $A_2 = [A_2(i, j, k)]$, and $A_3 = [A_3(i, j, k)]$) be the 3-dimensional matrix of order $m \times m \times 1$ (resp., $m \times 1 \times m$, and $1 \times m \times m$) that produced from the matrix $A = [A(i, j)]$ by $A_1(i, j, 0) = A(i, j)$ (resp., $A_2(i, 0, k) = A(i, k)$, and $A_3(0, j, k) = A(j, k)$). Then the direct multiplication $C = (A_1 \otimes A_2) \otimes A_3$ is a 3-dimensional Hadamard matrix of order $m^2 \times m^2 \times m^2$.*

Proof. By the definition of direct multiplication, the general formula of $C(i, j, k)$, $0 \le i, j, k \le m^2 - 1$, is

$$C(i, j, k) = A\left(\left\lfloor \frac{i}{m} \right\rfloor, \left\lfloor \frac{j}{m} \right\rfloor\right) A\left([i]_m, \left\lfloor \frac{k}{m} \right\rfloor\right) A([j]_m, [k]_m).$$

Therefore

$$\sum_{i,j=0}^{m^2-1} C(i, j, a) C(i, j, b)$$

$$= \sum_{i,j=0}^{m^2-1} A\left([i]_m, \left\lfloor \frac{a}{m} \right\rfloor\right) A([j]_m, [a]_m)$$

$$\cdot A\left([i]_m, \left\lfloor \frac{b}{m} \right\rfloor\right) A([j]_m, [b]_m), \quad \text{for } (A(p, q))^2 = 1$$

$$= m^2 \sum_{i=0}^{m-1} A\left(i, \left\lfloor \frac{a}{m} \right\rfloor\right) \cdot A\left(i, \left\lfloor \frac{b}{m} \right\rfloor\right) \sum_{j=0}^{m-1} A(j, [a]_m) A(j, [b]_m)$$

$$= m^4 \delta_{a,b}.$$

The last equation is due to the facts that: (1) The matrix A is itself a Hadamard matrix; and (2) $a \ne b$ implies $[a]_m \ne [b]_m$ or $\left\lfloor \frac{a}{m} \right\rfloor \ne \left\lfloor \frac{b}{m} \right\rfloor$.

By the same way, it can be proved that

$$\sum_{j,k=0}^{m^2-1} C(a, j, k) C(b, j, k) = \sum_{i,k=0}^{m^2-1} C(i, a, k) C(i, b, k)$$

$$= m^4 \delta_{a,b}.$$

Therefore Equation (3.2) is satisfied by C, i.e., C is indeed a 3-dimensional Hadamard matrix. **Q.E.D.**

The 3-dimensional Hadamard matrices constructed by Theorem 3.1.2 are absolutely improper in every of the possible directions. A more general form of Theorem 3.1.2 will be presented in the third part of the book.

3.1.3 Constructions Based on 2-Dimensional Hadamard Matrices

Besides the direct multiplications, other methods for generating 3-dimensional Hadamard matrices are at present restricted to special cases involving propriety or at least some degree of correlation within the two-dimensional layers ([1] , [3], [21]) .

For example, if the 2-dimensional layers in y-direction are themselves 2-dimensional Hadamard matrices, or if the 2-dimensional layers in x-direction are themselves 2-dimensional Hadamard matrices, then the 2-dimensional layers in z-direction are orthogonal to each other.

Thus the equation

$$\sum_{i=0}^{m-1}\sum_{j=0}^{m-1} H(i,j,a)H(i,j,b) = m^2\delta_{ab}$$

is satisfied if either

$$\sum_{i=0}^{m-1} H(i,j,a)H(i,j,b) = m\delta_{ab} \tag{3.8}$$

or

$$\sum_{j=0}^{m-1} H(i,j,a)H(i,j,b) = m\delta_{ab}. \tag{3.9}$$

Similarly, if the 2-dimensional layers in z-direction are themselves 2-dimensional Hadamard matrices, or if the 2-dimensional layers in y-direction are themselves 2-dimensional Hadamard matrices, then the 2-dimensional layers in x-direction are orthogonal to each other.

Thus the equation

$$\sum_{j=0}^{m-1}\sum_{k=0}^{m-1} H(a,j,k)H(b,j,k) = m^2\delta_{ab}$$

is satisfied if either

$$\sum_{j=0}^{m-1} H(a,j,k)H(b,j,k) = m\delta_{ab} \tag{3.10}$$

or

$$\sum_{k=0}^{m-1} H(a,j,k)H(b,j,k) = m\delta_{ab}. \tag{3.11}$$

If the 2-dimensional layers in x-direction are themselves 2-dimensional Hadamard matrices, or if the 2-dimensional layers in the z-direction are themselves 2-dimensional Hadamard matrices, then the 2-dimensional layers in the y-direction are orthogonal to each other.

Thus the equation

$$\sum_{i=0}^{m-1} \sum_{k=0}^{m-1} H(i,a,k)H(i,b,k) = m^2 \delta_{ab}$$

is satisfied if either

$$\sum_{i=0}^{m-1} H(i,a,k)H(i,b,k) = m\delta_{ab} \tag{3.12}$$

or

$$\sum_{k=0}^{m-1} H(i,a,k)H(i,b,k) = m\delta_{ab}. \tag{3.13}$$

Each of these equations specifies a possible orientation for orthogonality of either rows or columns within the two-dimensional layers in one direction. A three-dimensional Hadamard matrix exists whenever there are orthogonalities having at least one correlation vector in each axial direction, e.g. ([1] , [3], [21]) ,

> Equations(3.8) + (3.10) + (3.12);
> Equations(3.8) + (3.10) + (3.13);
> Equations(3.8) + (3.11) + (3.12);
> Equations(3.8) + (3.11) + (3.13);
> Equations(3.9) + (3.10) + (3.12);
> Equations(3.9) + (3.10) + (3.13);
> Equations(3.9) + (3.11) + (3.12);
> Equations(3.9) + (3.11) + (3.13).

Note that certain pairs of these equations also specify that all layers in one direction are two-dimensional Hadamard matrices, e.g., Equations (3.10) +(3.12) is equivalent to Equation (3.3).

From these relations, it follows that a three-dimensional Hadamard matrix is specified ([1] , [3], [21]) , if

a) all layers in one direction are two-dimensional Hadamard matrices which are orthogonal to each other (e.g., Equations (3.8)+(3.10)+(3.12));

b) all layers in two directions are Hadamard matrices (e.g., Equations (3.9)+ (3.13) + (3.10) + (3.12)) , or

c) in any direction, all layers are orthogonal in at least one layer direction so that collectively there is at least one correlation vector in each axial direction (e.g., Equations (3.8) +(3.10)+ (3.13)) .

By making use of these rules, we can construct three-dimensional Hadamard matrices proper in at least two directions, if we have been given a two-dimensional Hadamard matrix. In fact, if the rows (or columns) of a two-dimensional Hadamard matrix are cyclically permuted, then the resultant Hadamard matrix is orthogonal to the mother matrix. Therefore the set of successive cyclic row-permutations of a given two-dimensional Hadamard matrix $[h(i,j)]$ of order $n \times n$, i.e.,

$$H(i,j,k) = h(i,(j+k) \bmod n) \tag{3.14}$$

form the successive layers of a three-dimensional matrix which satisfies both rules a) and b). This matrix is therefore a three-dimensional Hadamard matrix proper in at least two directions.

To sum up, it has been found possible to construct the following three-dimensional Hadamard matrices ([1] , [3], [21]) :

- Absolutely proper 3-dimensional Hadamard matrices of order 2^t, plus a variety of partially proper and improper ones, by direct multiplication of 2^3 matrices;

- Absolutely improper 3-dimensional Hadamard matrices of order m^2 by successive direct multiplication of two-dimensional m^2 matrices in different orientations;

- Two-directional proper 3-dimensional Hadamard matrices of order m by cyclic permutation of the rows of any m^2 two-dimensional matrix;

- In some cases, absolutely proper matrices of order m from mirror-symmetrical m^2 matrices by assuming threefold symmetry.

3.2   Three-Dimensional Hadamard Matrices of Order $4k + 2$

One of the most important necessary conditions for a 2-dimensional Hadamard matrix is that its order n satisfies $n = 4k$ or $n = 2$. In the following section, we will show, by example, that the order n of some 3-dimensional Hadamard matrix can take the form or $n = 4k + 2$, $k \geq 1$.

Theorem 3.2.1 *If $H = [H(i, j, k)]$, $0 \leq i, j, k \leq n-1$, is a 3-dimensional Hadamard matrix of order n, then this n must be an even integer.*

Proof. $H(i, j, k) = \pm 1$ implies $(H(i, j, k))^2 = 1$. The orthogonality of 3-dimensional Hadamard matrix implies the following equations

$$\sum_{i=0}^{n-1} \sum_{j=0}^{n-1} H(i, j, 0) H(i, j, 1) = 0 \tag{3.15}$$

and

$$\sum_{i=0}^{n-1} \sum_{j=0}^{n-1} H(i, j, 0) H(i, j, 0) = n^2. \tag{3.16}$$

Therefore

$$\sum_{i=0}^{n-1} \sum_{j=0}^{n-1} H(i, j, 0)[H(i, j, 0) + H(i, j, 1)]$$

$$= \sum_{i=0}^{n-1} \sum_{j=0}^{n-1} H(i, j, 0) H(i, j, 0) + \sum_{i=0}^{n-1} \sum_{j=0}^{n-1} H(i, j, 0) H(i, j, 1)$$

$$= n^2 + 0$$

$$= n^2. \tag{3.17}$$

On the other hand,

$$H(i, j, 0) + H(i, j, 1) = (\pm 1) + (\pm 1) = \text{even}.$$

Thus the number n^2 must be even, and so is n itself.   **Q.E.D.**

The constructions stated in the last section produce 3-dimensional Hadamard matrices of order $4k$ only. While Theorem 3.2.1 motivates us trying to find those of order $4k + 2$, $k \geq 1$. The smallest such order is $4 \times 1 + 2 = 6$, i.e. the case of $k = 1$. This subsection will show an example of 3-dimensional Hadamard matrix of order 6 which is based on the concept of perfect binary array, defined by

Definition 3.2.1 ([4], [23]) *A* (± 1)-*valued matrix* $S = [A(i,j)]$, $0 \le i \le n-1$, $0 \le j \le m-1$, *is called a* (2-*dimensional*) *perfect binary array of order* $n \times m$, *abbreviated as* $\mathrm{PBA}(n,m)$, *if and only if its* 2-*dimensional cyclic auto-correlation is a* δ-*function, i.e.*,

$$R_A(s,t) = \sum_{i=0}^{n-1} \sum_{j=0}^{m-1} A(i,j)A(i+s, j+t) = 0 \quad \text{if} \quad (s,t) \ne (0,0), \quad (3.18)$$

where $(i+s)$ *and* $(j+t)$ *refer to* $(i+s)$modn *and* $(j+t)$ mod*m*, *respectively.*

For example ([14], [25], [9]), the matrix

$$A = \begin{bmatrix} - & + & + & + & + & - \\ + & - & + & + & + & - \\ + & + & - & + & + & - \\ + & + & + & - & + & - \\ + & + & + & + & - & - \\ - & - & - & - & - & + \end{bmatrix} \tag{3.19}$$

is a $\mathrm{PBA}(6,6)$(see [27]). Where '$-$', and '$+$' refer to -1, and $+1$, respectively.

Theorem 3.2.2 *If* $A = [A(i,j)]$, $0 \le i,j \le m-1$, *is a* $\mathrm{PBA}(m,m)$, *then the matrix A produces a* 3-*dimensional Hadamard matrix, say* $B = [B(i,j,k)]$, $0 \le i,j,k \le m-1$, *of order m, by*

$$B(i,j,k) = A(k+i, k+j), \quad 0 \le i,j,k \le m-1, \tag{3.20}$$

where $i+k$ *and* $j+k$ *refer to* $(i+k)$mod*m* *and* $(j+k)$mod*m*, *respectively.*
Proof. Let $0 \le a \ne b \le m-1$. Then

$$\sum_{i=0}^{m-1} \sum_{j=0}^{m-1} B(i,j,a)B(i,j,b) = \sum_{i=0}^{m-1} \sum_{j=0}^{m-1} A(i+a, j+a)A(i+b, j+b)$$

$$= \sum_{i=0}^{m-1} \sum_{j=0}^{m-1} A(i,j)A(i+(b-a), j+(b-a))$$

$$= 0.$$

The last equation is owed to the property that the matrix $A = [A(i,j)]$ is itself a PBA.

$$\sum_{i=0}^{m-1} \sum_{j=0}^{m-1} B(i,a,k)B(i,b,k) = \sum_{i=0}^{m-1} \sum_{j=0}^{m-1} A(i+k, a+k)A(i+k, b+k)$$

$$= \sum_{i=0}^{m-1} \sum_{j=0}^{m-1} A(i,j)A(i,j+(b-a))$$

$$= 0.$$

Similarly, it can be proved that

$$\sum_{i=0}^{m-1} \sum_{j=0}^{m-1} B(a,j,k)B(b,j,k) = 0.$$

Thus the matrix $B = [B(i,j,k)]$ satisfies Equation (3.2).   **Q.E.D.**

Applying Theorem 3.2.2 to the PBA$(6,6)$ in Equation (3.19), we obtain a 3-dimensional Hadamard matrix of order 6 which is described by its layers in z-direction as follows

$$B(i,j,0) = \begin{bmatrix} - & + & + & + & + & - \\ + & - & + & + & + & - \\ + & + & - & + & + & - \\ + & + & + & - & + & - \\ + & + & + & + & - & - \\ - & - & - & - & - & + \end{bmatrix}, \quad B(i,j,1) = \begin{bmatrix} - & + & + & + & - & + \\ + & - & + & + & - & + \\ + & + & - & + & - & + \\ + & + & + & - & - & + \\ - & - & - & - & + & - \\ + & + & + & + & - & - \end{bmatrix},$$

$$B(i,j,2) = \begin{bmatrix} - & + & + & - & + & + \\ + & - & + & - & + & + \\ + & + & - & - & + & + \\ - & - & - & + & - & - \\ + & + & + & - & - & + \\ + & + & + & - & + & - \end{bmatrix}, \quad B(i,j,3) = \begin{bmatrix} - & + & - & + & + & + \\ + & - & - & + & + & + \\ - & - & + & - & - & - \\ + & + & - & - & + & + \\ + & + & - & + & - & + \\ + & + & - & + & + & - \end{bmatrix},$$

$$B(i,j,4) = \begin{bmatrix} - & - & + & + & + & + \\ - & + & - & - & - & - \\ + & - & - & + & + & + \\ + & - & + & - & + & + \\ + & - & + & + & - & + \\ + & - & + & + & + & - \end{bmatrix}, \quad B(i,j,5) = \begin{bmatrix} + & - & - & - & - & - \\ - & - & + & + & + & + \\ - & + & - & + & + & + \\ - & + & + & - & + & + \\ - & + & + & + & - & + \\ - & + & + & + & + & - \end{bmatrix}.$$

Lemma 3.2.1 ([23]) *For each positive integer b there exists at least one* PBA$(2.3^b, 2.3^b)$.

This lemma is, in fact, a direct corollary of a theorem in the Chapter 6 on General Higher-Dimensional Hadamard Matrices. Thus we omit here the proof.

Lemma 3.2.1 together with Theorem 3.2.2 reduces the following result.

Theorem 3.2.3 *For each positive integer b there exists at least one 3-dimensional Hadamard matrix of order $2 \cdot 3^b$.*

The above 3-dimensional Hadamard matrix of order 6 is clearly an example of a matrix of Theorem 3.2.3. It is very strange that except for the above 3-dimensional Hadamard matrices, no other 3-dimensional Hadamard matrix of order $4k + 2 \neq 2 \cdot 3^b$, $k > 1$, $b \geq 0$, has been found. Thus it is an open problem to find such 3-dimensional Hadamard matrices.

3.3   Three-Dimensional Hadamard Matrices of Order $4k$

Many 3-dimensional Hadamard matrices of order $4k$ have been constructed in the first section. This section concentrates on the construction of 2-dimensional PBAs. Thus 3-dimensional Hadamard matrices are found by Theorem 3.2.2. The contents of this section are mainly due to [23].

3.3.1   Recursive Constructions of Perfect Binary Arrays

We start this Subsection with the following necessary condition on the orders of perfect binary arrays.

Theorem 3.3.1 *Let $A = [A(i,j)]$, $0 \leq i,j \leq m - 1$, be a PBA(m,m). Then the m must be even.*

Proof. By the definition of PBA, we have

$$\sum_{i=0}^{m-1} \sum_{j=0}^{m-1} A(i,j)A(i+1, j+1) = 0.$$

Because each term $A(i,j)A(i+1, j+1)$ in the left hand is ± 1, so m^2 should be even, otherwise the summation in the left hand can't be zero. Hence the order m is even.   **Q.E.D.**

From Theorem 3.3.1, we know that the order of PBA(m,m) should satisfy $m = 2, 4, 6, 8, 10, 12, \ldots$. We have shown a PBA$(6,6)$ in the last subsection. Here are a few other known PBAs of small orders:

1. PBA$(2,2)$:

$$A = \begin{bmatrix} + & + \\ + & - \end{bmatrix};$$

2. PBA$(4,4)$:

$$A = \begin{bmatrix} + & + & + & - \\ + & + & + & - \\ + & + & + & - \\ - & - & - & + \end{bmatrix};$$

3. PBA$(8,8)$[28]:

$$A = \begin{bmatrix}
- & + & + & + & - & + & + & + \\
+ & - & + & + & - & + & - & - \\
+ & + & + & - & + & + & + & - \\
+ & + & - & - & - & - & + & + \\
- & - & + & - & - & - & + & - \\
+ & + & + & - & - & - & - & + \\
+ & - & + & + & + & - & + & + \\
+ & - & - & + & - & + & + & -
\end{bmatrix};$$

4. PBA$(12,12)$[14]:

$$A = \begin{bmatrix}
+ & - & + & - & + & - & - & + & - & - & - & - \\
+ & - & - & - & + & - & - & - & - & - & + & + \\
+ & + & - & + & + & + & + & - & - & + & - & - \\
- & + & - & + & - & + & + & - & + & + & + & + \\
+ & - & - & - & + & - & - & - & - & - & + & + \\
+ & + & - & + & + & + & + & - & - & + & - & - \\
+ & - & + & - & + & - & - & + & - & - & - & - \\
- & + & + & + & - & + & + & + & + & + & - & - \\
+ & + & - & + & + & + & + & - & - & + & - & - \\
+ & - & + & - & + & - & - & + & - & - & - & - \\
+ & - & - & - & + & - & - & - & - & - & + & + \\
- & - & + & - & - & - & - & + & + & - & + & +
\end{bmatrix}.$$

In addition, some other PBAs of small size, e.g., PBA(16, 16), PBA(24, 24), PBA(32, 32), and PBA(64, 64), have been constructed [29]. These small PBAs are useful in the general recursive constructions.

Definition 3.3.1 ([23], [9]) *Let* $A = [A(i,j)]$, $B = [B(i,j)]$, $0 \leq i \leq s-1$, $0 \leq j \leq t-1$, *be two* (± 1)-*valued arrays. The column interleaving* $ic(A, B)$ *of* A *with* B *is the* $s \times (2t)$ (± 1)-*valued array* $C = [C(i,j)]$ *given by*

$$C(i, 2j) = A(i,j), \quad \text{and} \quad C(i, 2j+1) = B(i,j),$$
$$0 \leq i \leq s-1 \text{ and } 0 \leq j \leq t-1.$$

The row interleaving $ir(A, B)$ *of* A *with* B *is the* $(2s) \times t$ (± 1)-*valued array* $D = [d(i,j)]$ *given by*

$$d(2i, j) = A(i,j), \quad \text{and} \quad d(2i+1, j) = B(i,j), \quad 0 \leq i \leq s-1 \text{ and } 0 \leq j \leq t-1.$$

The 2-dimensional cyclic cross-correlation between two matrices, say A and B, of the same size, say $n \times m$, is another matrix $R_{AB} = [R_{AB}(u,v)]$, $0 \leq u \leq n-1$, $0 \leq v \leq m-1$, defined by ([23])

$$R_{AB}(u, v) = \sum_{i=0}^{n-1} \sum_{j=0}^{m-1} A(i,j)B(i+u, j+v),$$

where $i + u \equiv (i+u) \bmod n$ and $j + v \equiv (j+v) \bmod m$. If $A = B$, then $R_{AA}(u, v)$ is the auto-correlation and abbreviated as $R_A(u, v)$.

By Definition 3.3.1 and the definition of 2-dimensional cyclic auto- and cross-correlations we have the following identities:

Lemma 3.3.1 ([23], [9]) *Let* $A = [A(i,j)]$, *and* $B = [B(i,j)]$, $0 \leq i \leq s-1$, $0 \leq j \leq t-1$, *be two* (± 1)-*valued arrays. Let* $C = ic(A, B)$ *and* $D = ir(A, B)$. *Then*

$$R_C(u, 2v) = R_A(u, v) + R_B(u, v)$$
$$R_C(u, 2v+1) = R_{AB}(u, v) + R_{AB}(s-u, t-v-1)$$
$$R_D(2u, v) = R_A(u, v) + R_B(u, v)$$
$$R_D(2u+1, v) = R_{AB}(u, v) + R_{AB}(s-u-1, t-v)$$

for all $0 \leq u \leq s-1$ *and* $0 \leq v \leq t-1$.

Theorem 3.3.2 ([23], [9]) *Let $A = [A(i,j)]$, $B = [B(i,j)]$, $0 \le i \le s-1$, $0 \le j \le t-1$, be two (± 1)-valued arrays. Any two of the following imply the third:*

1. $C = ic(A,B)$ *is a PBA$(s, 2t)$;*

2. $D = ir(A,B)$ *is a PBA$(2s, t)$;*

3. $R_{AB}(u, v) = R_{AB}(u+1, v-1)$ *for all $0 \le u \le s-1$, $0 \le v \le t-1$.*

Proof. If the first two equations are satisfied, then

$$R_C(u, 2v+1) = R_D(2u+1, v)(= 0)$$

for all $0 \le u \le s-1$, $0 \le v \le t-1$. Hence by Lemma 3.3.1,

$$R_{AB}(s-u, t-v-1) = R_{AB}(s-u-1, t-v)$$

for all $0 \le u \le s-1$, $0 \le v \le t-1$. Replacing $s-u-1$ by u and $t-v$ by v, we obtain the third equation.

Conversely if the third equation holds, then by Lemma 3.3.1

$$R_C(u, 2v) = R_D(2u, v) \text{ and } R_C(u, 2v+1) = R_D(2u+1, v)$$

for all $0 \le u \le s-1$, $0 \le v \le t-1$. Hence the first equation holds if and only if the second equation holds.   **Q.E.D.**

Definition 3.3.2 ([23], [25]) *Let $A = [A(i,j)]$, $B = [B(i,j)]$, $0 \le i \le s-1$, $0 \le j \le t-1$, be two (± 1)-valued arrays. A and B are called complementary if*

$$R_A(u, v) + R_B(u, v) = 0 \text{ for all } (u, v) \ne (0, 0)$$

and uncorrelated if
$$R_{AB}(u, v) = 0 \text{ for all } u, v.$$

Now, we are ready to construct PBAs C and D by interleaving appropriate arrays A and B.

Theorem 3.3.3 ([23], [9]) *Let $A = [A(i,j)]$, $B = [B(i,j)]$, $0 \le i \le s-1$, $0 \le j \le t-1$, be two (± 1)-valued arrays. Then $C = ic(A,B)$ is a*

PBA$(s, 2t)$ *(resp., $D = ir(A, B)$ is a PBA$(2s, t)$) if and only if A and B are complementary arrays such that*

$$R_{AB}(u, v) + R_{AB}(s - u, t - v - 1) = 0 \text{ for all } u, v$$

$$(\text{resp., } R_{AB}(u, v) + R_{AB}(s - u - 1, t - v) = 0 \text{ for all } u, v).$$

Proof. This theorem follows immediately from Definition 3.3.2 and Lemma 3.3.1.   **Q.E.D.**

Corollary 3.3.1 ([23], [9]) *Let $A = [A(i, j)]$, $B = [B(i, j)]$, $0 \le i \le s - 1$, $0 \le j \le t - 1$, be two (± 1)-valued arrays which are complementary and uncorrelated. Then $C = ic(A, B)$ is a PBA$(s, 2t)$ and $D = ir(A, B)$ is a PBA$(2s, t)$.*

Let $A = [A(i, j)]$, $B = [B(i, j)]$, $0 \le i \le s - 1$, $0 \le j \le t - 1$, be two (± 1)-valued arrays. Define a $(2s) \times t$ binary array $E = [E(i, j)] = \begin{bmatrix} A \\ B \end{bmatrix}$ by:

$E(i, j) = A(i, j)$ and $E(i+s, j) = B(i, j)$ for all $0 \le i \le s-1$, $0 \le j \le t-1$,

and a $s \times (2t)$ binary array $F = [F(i, j)] = [AB]$ by:

$F(i, j) = A(i, j)$ and $F(i, j+t) = B(i, j)$ for all $0 \le i \le s-1$, $0 \le j \le t-1$.

The following lemma states how to construct uncorrelated binary arrays.

Lemma 3.3.2 ([23], [9], [15]) *Let $A = [A(i, j)]$, $B = [B(i, j)]$, $0 \le i \le s - 1$, $0 \le j \le t - 1$, be two (± 1)-valued arrays. Then:*

1. $A' := \begin{bmatrix} A \\ A \end{bmatrix}$ *and* $B' := \begin{bmatrix} B \\ -B \end{bmatrix}$ *are uncorrelated;*

2. $A'' := [A \ A]$ *and* $B'' := [B \ -B]$ *are uncorrelated.*

This lemma can be proved by directly verification.

Definition 3.3.3 ([25], [23], [15]) *Let $B = [B(i, j)]$, $0 \le i \le s - 1$, $0 \le j \le t - 1$, be a (± 1)-valued array. B is called row-wise quasiperfect if* $B' := \begin{bmatrix} B \\ -B \end{bmatrix}$ *satisfies $R_{B'}(u, v) = 0$ for all $(u, v) \ne (0, 0)$ or $(s, 0)$.*

B *is called column-wise quasi-perfect if* $B'' := [B \ -B]$ *satisfies* $R_{B''}(u, v)$
$= 0$ *for all* $(u, v) \neq (0, 0)$ *or* $(0, t)$.

We write RQPBA(s, t) *(resp.,* CQPBA$(s, t))$ *for an* $s \times t$ *array which is either row-wise or column-wise quasi-perfect.*

Note that B is row-wise quasi-perfect if and only if B^T is column-wise quasi-perfect. Note also that if $A = [A(i, j)]$, $B = [B(i, j)]$ and $C = [C(i, j)]$, where $B(i, j) = (-1)^i A(i, j)$ and $C(i, j) = (-1)^j A(i, j)$, then

1. for s odd, A is a PBA(s, t) if and only if B is a RQPBA(s, t);

2. for t odd, A is a PBA(s, t) if and only if C is a CQPBA(s, t).

Lemma 3.3.3 ([23], [25]) *Let A be a PBA(s, t) and B a quasi-perfect binary array. If B is row-wise quasi-perfect, then $A' := \begin{bmatrix} A \\ A \end{bmatrix}$ is complementary to $B' := \begin{bmatrix} B \\ -B \end{bmatrix}$. If B is column-wise quasi-perfect, then $A'' := [A \ A]$ is complementary to $B'' := [B \ -B]$.*

Proof. By Definition 3.3.3, we have

$$R_{B'}(u, v) = \begin{cases} 0 & \text{for all } (u, v) \neq (0, 0), (s, 0), \\ -2st & \text{for } (u, v) = (s, 0), \end{cases}$$

and

$$R_{A'}(u, v) = \begin{cases} 0 & \text{for all } (u, v) \neq (0, 0), (s, 0), \\ 2st & \text{for } (u, v) = (s, 0). \end{cases}$$

Hence

$$R_{A'}(u, v) + R_{B'}(u, v) = 0 \text{ for all } (u, v) \neq (0, 0).$$

Similarly,

$$R_{A''}(u, v) + R_{B''}(u, v) = 0 \text{ for all } (u, v) \neq (0, 0).$$

Q.E.D.

Theorem 3.3.4 ([23], [25]) *If there exist a PBA(s, t), A, and a quasi-perfect binary array, B, of size $s \times t$, then there exists a PBA$(2s, 2t)$. Moreover, if the quasi-perfect binary array B is row-wise quasi-perfect, there also exists a PBA$(4s, t)$, and if B is column-wise quasi-perfect, there also exists a PBA$(s, 4t)$.*

Proof. If B is row-wise quasi-perfect, then, by Lemmas 3.3.2 and 3.3.3, $A' := \begin{bmatrix} A \\ A \end{bmatrix}$ and $B' := \begin{bmatrix} B \\ -B \end{bmatrix}$ are complementary uncorrelated $(2s) \times t$ (± 1)-valued arrays. Hence $ic(A', B')$ is a PBA$(2s, 2t)$ and $ir(A', B')$ is a PBA$(4s, t)$, by Corollary 3.3.1.

Similarly, if B is column-wise quasi-perfect and $A'' := [A \ A]$ and $B'' := [B \ -B]$, then $ir(A'', B'')$ is a PBA$(2s, 2t)$ and $ic(A'', B'')$ is a PBA$(s, 4t)$. **Q.E.D.**

It should be remarked that with the definitions used in the above proof, if a PBA takes one of the forms constructed, namely, $ic(A', B')$ or $ir(A', B')$ (resp., $ic(A'', B'')$ or $ir(A'', B'')$) for any PBAs, A and B, of size $s \times t$, then A is perfect and B is row-wise (resp. column-wise) quasi-perfect. This follows easily from Lemma 3.3.1 and the following identities: for all $0 \le u \le s - 1$, $0 \le v \le t - 1$, $(u, v) \ne (0, 0)$,

$$R_{A'}(u + s, v) = R_{A'}(u, v) \quad \text{and}$$
$$R_{B'}(u + s, v) = -R_{B'}(u, v)$$
$$(reps. \quad R_{A''}(u, v + t) = R_{A''}(u, v) \quad \text{and}$$
$$R_{B''}(u, v + t) = -R_{B''}(u, v)).$$

3.3.2 Quasi-Perfect Binary Arrays

In order to make use of Theorem 3.3.4 to construct PBAs, we need as many quasi-perfect binary arrays as possible. This subsection aims at constructing quasi-perfect and doubly quasi-perfect binary arrays [23], [25].

Definition 3.3.4 ([23], [25]) *Let $A = [A(i, j)]$ and $B = [B(i, j)]$, $0 \le i \le s - 1$, $0 \le j \le t - 1$, be (± 1)-valued arrays. Let $A' := \begin{bmatrix} A \\ -A \end{bmatrix}$, $B' := \begin{bmatrix} B \\ -B \end{bmatrix}$, and $A'' := [A \ -A]$, $B'' := [B \ -B]$. A and B are called row-wise quasi-complementary if*

$$R_{A'}(u, v) + R_{B'}(u, v) = 0 \text{ for all } (u, v) \ne (0, 0), (s, 0),$$

and column-wise quasi-complementary if

$$R_{A''}(u, v) + R_{B''}(u, v) = 0 \text{ for all } (u, v) \ne (0, 0), (0, t).$$

A and B are called *row-wise quasi-uncorrelated* if A' and B' are uncorrelated, and *column-wise quasi-uncorrelated* if A'' and B'' are uncorrelated.

Theorem 3.3.5 ([23], [25]) *Let* $A = [A(i,j)]$ *and* $B = [B(i,j)]$, $0 \leq i \leq s-1$, $0 \leq j \leq t-1$, *be* (± 1)*-valued arrays.*

1. *Let* $A' := \begin{bmatrix} A \\ -A \end{bmatrix}$, $B' := \begin{bmatrix} B \\ -B \end{bmatrix}$. *Then* $C = ic(A,B)$ *(resp.,* $D = ir(A,B)$*) is a* RQPBA$(s, 2t)$ *(resp.,* RQPBA$(2s, t)$*) if and only if* A *and* B *are row-wise quasi-complementary arrays such that*

$$R_{A'B'}(u,v) + R_{A'B'}(2s - u, t - v - 1) = 0 \text{ for all } u, v$$

(resp., $R_{A'B'}(u,v) + R_{A'B'}(2s - u - 1, t - v) = 0$ *for all* u, v*).*

2. *Let* $A'' := [A \ -A]$, $B'' := [B \ -B]$. *Then* $C = ic(A,B)$ *(resp.,* $D = ir(A,B)$*) is a* CQPBA$(s, 2t)$ *(resp.,* CQPBA$(2s, t)$*) if and only if* A *and* B *are column-wise quasi-complementary arrays such that*

$$R_{A''B''}(u,v) + R_{A''B''}(s - u, 2t - v - 1) = 0 \text{ for all } u, v$$

(resp., $R_{A''B''}(u,v) + R_{A''B''}(s - u - 1, 2t - v) = 0$ *for all* u, v*).*

Proof. The first statement follows by applying Lemma 3.3.1 to $C' := \begin{bmatrix} C \\ -C \end{bmatrix} = ic(A', B')$ (resp., $D' := \begin{bmatrix} D \\ -D \end{bmatrix}$).

The second statement follows by applying Lemma 3.3.1 to $C'' := [C \ -C] = ic(A'', B'')$ (resp., $D'' := [D \ -D]$). **Q.E.D.**

Corollary 3.3.2 ([23], [25], [15]) *Let* A *and* B *be row-wise (resp., column-wise) quasi-complementary, row-wise (resp., column-wise) quasi-uncorrelated* $s \times t$ (± 1)*-valued arrays. Then* $C = ic(A,B)$ *is a row-wise (resp., column-wise) quasi-perfect binary array of size* $s \times (2t)$ *and* $D = ir(A,B)$ *is a row-wise (resp., column-wise) quasi-perfect binary array of size* $(2s) \times t$.

The following lemma shows how to construct quasi-uncorrelated binary arrays.

Lemma 3.3.4 ([23], [25], [15]) *Let* $A = [A(i,j)]$ *and* $B = [B(i,j)]$, $0 \leq i \leq s-1$, $0 \leq j \leq t-1$, *be* (± 1)*-valued arrays. Then*

1. $A' := \begin{bmatrix} A \\ A \end{bmatrix}$ and $B' := \begin{bmatrix} B \\ -B \end{bmatrix}$ are column-wise quasi-uncorrelated;

   and

2. $A'' := [A \ A]$ and $B'' := [B \ -B]$ are row-wise quasi-uncorrelated.

Proof. The first statement follows by applying the first statement of Lemma 3.3.2 to $[A \ -A]$ and $[B \ -B]$.

The second statement follows by applying the second statement of

Lemma 3.3.2 to $\begin{bmatrix} A \\ -A \end{bmatrix}$ and $\begin{bmatrix} B \\ -B \end{bmatrix}$.   **Q.E.D.**

Definition 3.3.5 ([23], [25]) *Let* $C = [C(i,j)]$, $0 \le i \le s - 1$, $0 \le j \le t - 1$, *be a* (± 1)-*valued array.* C *is called doubly quasi-perfect, written by* DQPBA(s,t), *if and only if the cyclic auto-correlation of the matrix*

$$C' = \begin{bmatrix} C & -C \\ -C & C \end{bmatrix}$$

satisfies $R_{C'}(u,v) = 0$ *for all* $(u,v) \ne (0,0), (s,0), (0,t), (s,t)$.

Note that C is doubly quasi-perfect if and only if C^T is doubly quasi-perfect. Note also that if $A = [A(i,j)]$, $B = [B(i,j)]$ and $C = [C(i,j)]$ where $B(i,j) = (-1)^i C(i,j)$ and $A(i,j) = (-1)^j C(i,j)$, then

1. *For s odd,* C *is a* DQPBA(s,t) *if and only if* B *is a* CQPBA(s,t);

2. *For t odd,* C *is a* DQPBA(s,t) *if and only if* A *is a* RQPBA(s,t).

Lemma 3.3.5 ([23], [25]) *Let* B *be a quasi-perfect binary array of size* $s \times t$ *and* C *a* DQPBA(s,t).

1. *If* B *is row-wise quasi-perfect, then* $[B \ B]$ *and* $[C \ -C]$ *are row-wise quasi-complementary; and,*

2. *If* B *is column-wise quasi-perfect, then* $\begin{bmatrix} B \\ B \end{bmatrix}$ *and* $\begin{bmatrix} C \\ -C \end{bmatrix}$ *are column-wise quasi-complementary.*

Proof. The proof is similar to that of Lemma 3.3.3.   **Q.E.D.**

Theorem 3.3.6 ([23], [25]) *If there exist a row-wise (resp., column-wise) quasi-perfect binary array, say B, of size $s \times t$, and a DQPBA(s,t), say C, then there exist a row-wise (resp., column-wise) quasi-perfect binary array of size $(2s) \times (2t)$ and a RQPBA$(s, 4t)$ (resp., CQPBA$(4s, t)$).*

Proof. The proof is similar to that of Theorem 3.3.4. In fact, if B is row-wise quasi-perfect, then $ir\left([B\ B], [C\ -C]\right)$ is a RQPBA$(2s, 2t)$ and $ic\left([B\ B], [C\ -C]\right)$ is a RQPBA$(s, 4t)$.

If B is column-wise quasi-perfect, then

$$ic\left(\begin{bmatrix} B \\ B \end{bmatrix}, \begin{bmatrix} C \\ -C \end{bmatrix}\right) \text{ is a CQPBA}(2s, 2t)$$

and

$$ir\left(\begin{bmatrix} B \\ B \end{bmatrix}, \begin{bmatrix} C \\ -C \end{bmatrix}\right) \text{ is a CQPBA}(4s, t).$$

Q.E.D.

We remark that if a row-wise (resp., column-wise) quasi-perfect binary array takes one of the forms constructed in the above proof, namely $ir\left([B\ B], [C\ -C]\right)$ or $ic\left([B\ B], [C\ -C]\right)$ (resp., $ic\left(\begin{bmatrix} B \\ B \end{bmatrix}, \begin{bmatrix} C \\ -C \end{bmatrix}\right)$ or $ir\left(\begin{bmatrix} B \\ B \end{bmatrix}, \begin{bmatrix} C \\ -C \end{bmatrix}\right)$) for any $s \times t$ binary arrays B and C. Then B is row-wise (resp., column-wise) quasi-perfect and C is doubly quasi-perfect.

Definition 3.3.6 ([23], [25]) *Let $A = [A(i,j)]$ and $B = [B(i,j)]$, $0 \le i \le s-1$, $0 \le j \le t-1$, be (± 1)-valued arrays and let c be an integer. B is called the row-wise c-shear of A if $ct \equiv 0 \ (\mathrm{mod}\, s)$ and*

$$B(i,j) = A(i - cj, j), \text{ for all } i, j,$$

or B is called the column-wise c-shear of A if $cs \equiv 0 (\mathrm{mod}\, t)$ and

$$B(i,j) = A(i, j - ci), \text{ for all } i, j.$$

(We regard the sums $i - cj$ and $j - ci$ to be reduced modulo s and t, respectively.)

The condition on c is necessary for the c-shear to be well defined. Note that if B is the row-wise (resp., column-wise) c-shear of A, then A is the row-wise (resp., column-wise) $(-c)$-shear of B. With this definition in mind, it is easy to verify the following lemma.

Lemma 3.3.6 ([23], [25]) *Let* $A = [A(i,j)]$, $0 \le i \le s-1$, $0 \le j \le t-1$, *be a* (± 1)-*valued array.*

1. *If* A' *is the row-wise c-shear of* A, *then*

$$R_{A'}(u,v) = R_A(u - cv, v) \text{ for all } u, v;$$

2. *If* A'' *is the column-wise c-shear of* A, *then*

$$R_{A''}(u,v) = R_A(u, v - cu) \text{ for all } u, v;$$

Corollary 3.3.3 ([23], [25]) *Let* A *be an* $(rs) \times (rt)$ *binary array for some positive integer* r. *Let* B *be the c-shear of* A *(row-wise or column-wise). Then*

$$R_A(u,v) = 0 \text{ for } u \not\equiv 0 \text{ (mod} s) \text{ or } v \not\equiv 0 \text{ (mod} t)$$

if and only if

$$R_B(u,v) = 0 \text{ for } u \not\equiv 0 \text{ (mod} s) \text{ or } v \not\equiv 0 \text{ (mod} t).$$

Proof. This corollary follows from Lemma 3.3.6, together with the necessary condition on c for the c-shear to be well defined.   **Q.E.D.**

Definition 3.3.7 ([23], [9]) *Let* B *be a* $(2s) \times (2t)$ *binary array. Let* $B' = [B'(i,j)]$ *be the row-wise (resp., column-wise) c-shear of* B *and define an* $s \times t$ *binary array* $A' = [A'(i,j)]$ *by* $A'(i,j) = B'(i,j)$, *for all* $0 \le i \le s-1$, $0 \le j \le t-1$. A' *is called the row-wise (resp., column-wise) c-transform of* B.

Theorem 3.3.7 ([23], [25]) *Let* A *be an* $s \times t$ *binary array.*

1. *Let* B *be the c-shear of* A *(row-wise or column-wise). Then* A *is perfect if and only if* B *is perfect;*

2. *Let c satisfy $ct \equiv s(\text{mod}2s)$ (resp., $cs \equiv t(\text{mod}2t)$). Then A is row-wise (resp., column-wise) quasi-perfect if and only if the row-wise (resp., column-wise) c-transform of $\begin{bmatrix} A & A \\ -A & -A \end{bmatrix}$ (resp., $\begin{bmatrix} A & -A \\ A & -A \end{bmatrix}$) is doubly quasi-perfect;*

3. *Let c satisfy $ct \equiv 0(\text{mod}2s)$ (resp., $cs \equiv 0(\text{mod}2t)$). Then A is row-wise (resp., column-wise) quasi-perfect if and only if the row-wise (resp., column-wise) c-transform of $\begin{bmatrix} A & A \\ -A & -A \end{bmatrix}$ (resp., $\begin{bmatrix} A & -A \\ A & -A \end{bmatrix}$) is row-wise (resp., column-wise) quasi-perfect. Furthermore A is doubly quasi-perfect if and only if the row-wise (resp. column-wise) c-transform of $\begin{bmatrix} A & -A \\ -A & A \end{bmatrix}$ is doubly quasi-perfect.*

Proof. The first statement follows immediately from Corollary 3.3.3 with $r = 1$. To prove the second statement, suppose $ct \equiv s(\text{mod}2s)$. Let $B = \begin{bmatrix} A & A \\ -A & -A \end{bmatrix}$ and let B' be the row-wise c-shear of B. By Definition 3.3.6, we know that

$$B'(i,j) = B(i - cj, j),$$
$$B'(i,j+t) = B(i - cj - ct, j + t),$$
$$B'(i+s,j) = B(i + s - cj, j),$$
$$B'(i+s,j+t) = B(i + s - cj - ct, j + t)$$

for all $0 \leq i \leq s - 1$, $0 \leq j \leq t - 1$. From the given form of B,

$$B(i,j) = B(i,j+t) = -B(i+s,j) = -B(i+s,j+t)$$

for all $0 \leq i \leq s - 1$, $0 \leq j \leq t - 1$. Hence, using $ct \equiv s(\text{mod}2s)$,

$$B'(i,j) = -B'(i,j+t) = -B'(i+s,j) = B'(i+s,j+t)$$

for all $0 \leq i \leq s - 1$, $0 \leq j \leq t - 1$. Therefore $B' = \begin{bmatrix} A' & -A' \\ -A' & A' \end{bmatrix}$, where A' is the row-wise c-transform of B. By Corollary 3.3.3 with $r = 2$, A is row-wise quasi-perfect if and only if A' is doubly quasi-perfect.

Similarly, we may show, given $cs \equiv t \pmod{2t}$, that A is column-wise quasi-perfect if and only if the column-wise c-transform of $\begin{bmatrix} A & -A \\ A & -A \end{bmatrix}$ is doubly quasi-perfect.

The third statement follows from simple modifications of the above arguments used in the proof of the second statement.   **Q.E.D.**

Thus we have now established the equivalence between quasi-perfect and doubly quasi-perfect binary arrays.

Corollary 3.3.4 ([23], [9]) *If $t/\gcd(s,t)$ is odd, then there exists a DQP BA(s,t) if and only if there exists a RQPBA(s,t). If $s/\gcd(s,t)$ is odd, then there exists a DQPBA(s,t) if and only if there exists a CQPBA(s,t).*

Proof. We note that $ct \equiv s \pmod{2s}$ (resp., $cs \equiv t \pmod{2t}$) if and only if $t/\gcd(s,t)$ (resp., $s/\gcd(s,t)$) is odd and c is an odd multiple of $s/\gcd(s,t)$ (resp., $t/\gcd(s,t)$). The corollary follows from the second statement of Theorem 3.3.7.   **Q.E.D.**

Suppose that $t/\gcd(s,t)$ is odd and A is a RQPBA(s,t). The above proof gives a procedure for obtaining a DQPBA(s,t) A'. Put

$$B = \begin{bmatrix} A & A \\ -A & -A \end{bmatrix}$$

and $c = s/\gcd(s,t)$. Form B', the row-wise c-shear of B, by cycling column j of B by cj places for $j = 0, \ldots, 2t - 1$. Then the first s rows and t columns of B' are A', the row-wise c-transform of B.

3.3.3   Three-Dimensional Hadamard Matrices Based on PBA$(2^m, 2^m)$ and PBA$(3.2^m, 3.2^m)$

This subsection recursively applies the construction theorems to yield infinite families of 2-dimensional PBAs, and thus 3-dimensional Hadamard matrices are produced by Theorem 3.2.2.

Theorem 3.3.8 ([23], [25]) *If there exist a PBA(s,t) and a DQPBA(s,t), then there exist a PBA$(2s,2t)$ and a DQPBA$(2s,2t)$. If $t/\gcd(s,t)$ is odd there also exists a PBA$(4s,t)$ and a RQPBA$(s,4t)$. If $s/\gcd(s,t)$ is odd there also exists a PBA$(s,4t)$ and a CQPBA$(4s,t)$.*

Proof. This theorem is the restatement of Theorems 3.3.4 and 3.3.6, by using the equivalence between quasi-perfect binary arrays and doubly quasi-perfect binary arrays described in Corollary 3.3.4. **Q.E.D.**

Corollary 3.3.5 ([23], [9]) *If there exist a* PBA(s,t) *and a* DQPBA(s,t), *then there exist* PBA$(2^y s, 2^y t)$ *and* DQPBA$(2^y s, 2^y t)$ *for each* $y \geq 0$. *If* $t/\gcd(s,t)$ *is odd there also exist a* PBA$(2^{y+2}s, 2^y t)$ *and a* RQPBA$(2^y s,$ $2^{y+2}t)$ *for each* $y \geq 0$. *If* $s/\gcd(s,t)$ *is odd there also exist a* PBA$(2^y s,$ $2^{y+2}t)$ *and a* CQPBA$(2^{y+2}s, 2^y t)$ *for each* $y \geq 0$.

Proof. It can be proved by repeating Theorem 3.3.8. **Q.E.D.**

Corollary 3.3.6 ([23], [25]) *There exist the following infinite families of 2-dimensional* PBAs:

F1.1: PBA$(2^y, 2^y)$, $y \geq 1$;

F1.2: PBA$(2^{y+1}, 2^{y-1})$, $y \geq 1$;

F1.3: PBA$(2^y \cdot 3, 2^y \cdot 3)$, $y \geq 1$;

F1.4: PBA$(2^{y+1} \cdot 3, 2^{y-1} \cdot 3)$, $y \geq 1$.

There exist the following infinite families of doubly quasi-perfect and row-wise quasi-perfect binary arrays:

F2.1: DQPBA$(2^y, 2^y)$, $y \geq 1$;

F2.2: DQPBA$(2^y \cdot 3, 2^y \cdot 3)$, $y \geq 1$;

F2.3: RQPBA$(2^{y-1}, 2^{y+1})$, $y \geq 1$;

F2.4: RQBA$(2^{y-1} \cdot 3, 2^{y+1} \cdot 3)$, $y \geq 1$.

Proof. The proof of this corollary is finished by using Corollary 3.3.5 to the following:

$$\text{PBA}(2,2): \begin{bmatrix} + & + \\ + & - \end{bmatrix};$$

$$\text{DQPBA}(2,2): \begin{bmatrix} + & + \\ + & + \end{bmatrix};$$

PBA$(6,6)$ stated in the last section;

$$
\text{DQPBA}(6,6): \begin{bmatrix}
- & + & - & - & + & + \\
+ & - & + & - & + & + \\
- & + & - & - & - & + \\
- & - & - & + & - & + \\
+ & + & - & - & - & + \\
+ & + & + & + & + & +
\end{bmatrix};
$$

$$
\text{PBA}(3,12): \begin{bmatrix}
+ & + & - & + & + & + & + & - & + & - & - & - \\
- & + & - & + & - & - & - & - & - & + & + & - \\
- & - & + & + & - & + & - & - & - & + & - & -
\end{bmatrix}.
$$

Thus by the second property following Definition 3.3.3, the existence of PBA$(3,12)$ implies the existence of a RQPBA$(12,3)$.   **Q.E.D.**

Applying Theorem 3.2.2 to the families F1.1 and F1.3 in Corollary 3.3.6, we have finally constructed 3-dimensional Hadamard matrices of orders 2^m and $2^m.3$. It should be noted that 3-dimensional Hadamard matrices of order $2^{m+n}.3^m$ can also be simply constructed by the direct multiplication of 3-dimensional Hadamard matrices of orders 2 and 6 which have known to exist.

Theorem 3.3.9 ([23], [25]) *If there exists a* DQPBA$(2t,t)$, *then there exist a* DQPBA$(4t,2t)$, *a* RQPBA$(2t,4t)$ *and a* RQPBA$(2t,16t)$.

Proof. Given a DQPBA$(2t,t)$, by Corollary 3.3.4 there exists a RQPBA$(2t, t)$. Then by Theorem 3.3.6 there exists a RQPBA$(4t, 2t)$ and a PQPBA$(2t, 4t)$. Therefore from Corollary 3.3.4 there exists a DQPBA$(4t, 2t)$. Transposing gives a DQPBA$(2t,4t)$, which we combine with the RQPBA$(2t,4t)$ to give a RQPBA$(2t,16t)$ by Theorem 3.3.6.   **Q.E.D.**

Corollary 3.3.7 ([23], [25]) *If there exists a* DQPBA$(2t,t)$, *then for each* $y \geq 0$ *there exists a* DQPBA$(2^{y+1}t, 2^y t)$, *a* RQPBA$(2^{y+1}t, 2^{y+2}t)$, *and a* RQPBA$(2^{y+1}t, 2^{y+4}t)$.

Corollary 3.3.8 ([23], [9]) *There exists the following infinite families of doubly quasi-perfect and row-wise quasi-perfect binary arrays:*

1. DQPBA$(2^y, 2^{y-1})$, $y \geq 1$;

2. RQPBA$(2^y, 2^{y+1})$, $y \geq 1$;

3. RQPBA$(2^y, 2^{y+3})$, $y \geq 1$.

Proof. $\begin{bmatrix} + \\ + \end{bmatrix}$ is a DQPBA$(2, 1)$.   **Q.E.D.**

When we recursively applying Theorems 3.3.4 and 3.3.6 and Corollary 3.3.4 to the trivial array $[+]$, we can obtain a family of perfect, row-wise quasi-perfect and doubly quasi-perfect binary arrays of size $2^m \times 2^m$, $m \geq 1$. Precisely, we have the following:

Theorem 3.3.10 ([23], [25]) *Let* $t = 2^y$, $y \geq 0$. *There exists arrays* $A = [A(i,j)]$, $B = [B(i,j)]$, *and* $C = [C(i,j)]$, *which are respectively a* PBA(t, t), *a* RQPBA(t, t), *and a* DQPBA(t, t), *for which the following properties are satisfied for all* $0 \leq i, j \leq t-1$:

1. $A(i,j) = A(t-i, t-j)$;

2. $B(0,j) = B(0, t-j-1)$;

3. $B(i,j) = -B(t-i, t-j-1)$, $i \neq 0$;

4. $B(i,j) = B(t-i, j)$, $i \neq 0$;

5. $C(i,j) = C(i, t-j-1)$;

6. $C(i,j) = C(t-i-1, j)$.

Proof. It can be proved by using induction on y. The case $y = 0$ is trivial by choose $A = B = C = [+]$. Assume that arrays A, B, and C with the desired properties exist for some $y \geq 0$ and let $t = 2^y$. Assume also that $D' = [D'(i,j)] = \begin{bmatrix} C & -C \\ -C & C \end{bmatrix}$ is the row-wise 1-shear of $D = [D(i,j)] = \begin{bmatrix} B & B \\ -B & -B \end{bmatrix}$. From the proof of Theorems 3.3.4 and 3.3.6, $A' = [A'(i,j)] = ic\left(\begin{bmatrix} A \\ A \end{bmatrix}, \begin{bmatrix} B \\ -B \end{bmatrix} \right)$ is a PBA$(2t, 2t)$ and $B' = [B'(i,j)] = ir\left([B\ B], [C\ -C] \right)$ is a RQPBA$(2t, 2t)$. From the proof of Corollary 3.3.4, $C' = [C'(i,j)]$ is a DQPBA$(2t, 2t)$, where $E' = [E'(i,j)] =$

$\begin{bmatrix} C' & -C' \\ -C' & C' \end{bmatrix}$ is the row-wise 1-shear of $E = [E(i,j)] = \begin{bmatrix} B' & B' \\ -B' & -B' \end{bmatrix}$. By
the construction of A', for all $0 \le i, j \le t - 1$,

$$A'(i + t, 2j) = A'(i, 2j) = A(i, j)$$
$$A'(i + t, 2j + 1) = -A'(i, 2j + 1) = -B(i, j).$$

To establish the first property for A' we must show that for all $0 \le i, j \le t - 1$,

$$A'(i, 2j) = A'(2t - i, 2t - 2j)$$
$$A'(i, 2j + 1) = A'(2t - i, 2t - 2j - 1),$$

which is equivalent to

$$A(i, j) = A(t - i, t - j)$$
$$B(0, j) = B(0, t - j - 1)$$
$$B(i, j) = -B(t - i, t - j - 1), i \ne 0.$$

These relations are given by the first three properties. Similarly, the second, third and fourth properties for B' are given respectively by the second, by the third, fifth and sixth, and by the fourth and sixth properties.

By the definition of E', for all $0 \le i, j \le 2t - 1$,

$$E'(i + 2t, j + 2t) = -E'(i + 2t, j) = -E'(i, j + 2t) = E'(i, j).$$

To establish the fifth property for C' we must therefore show that for all $0 \le i \le 4t - 1$, $0 \le j \le 2t - 1$,

$$E'(i, j) = -E'(i, 4t - j - 1).$$

Now by Definition 3.3.6, $E'(i, j) = E(i - j, j)$ and so this is equivalent to

$$E(i - j, j) = -E(i + j + 1, 4t - j - 1).$$

Replacing $i - j$ by $2j$ and then by $2i + 1$, we require that for all $0 \le i, j \le 2t - 1$,

$$E(2i, j) = -E(2i + 2j + 1, 4t - j - 1)$$
$$E(2i + 1, j) = -E(2i + 2j + 2, 4t - j - 1).$$

But by the construction of B', $E = ir\,([DD],[D'D'])$ and so this is equivalent to

$$D(i,j) = -D'(i+j, 2t-j-1)$$
$$D'(i,j) = -D(i+j+1, 2t-j-1).$$

Both of these hold provided

$$D'(i,j) = -D'(i, 2t-j-1)$$

for all $0 \le i,j \le 2t-1$, since $D(i,j) = D'(i+j,j)$, by Definition 3.3.6. By the definition of D' this relation is given by the fifth property. A similar argument gives the sixth property for C'.   **Q.E.D.**

3.4   Three-Dimensional Walsh Matrices

3-dimensional Hadamard matrices of order 2^n have been constructed in the last section, whilst this section will continue to investigate a special class of 3-dimensional Hadamard matrices of order 2^n which are called 3-dimensional Walsh matrices. Their definitions, constructions, and analytic representations will be presented.

Recall that every integer i, $0 \le i \le 2^n - 1$, can be uniquely expended as $i = \sum_{k=0}^{n-1} i_k 2^k$, which is uniquely determined by an n-dimensional binary (0 or 1) vector $(i_0, i_1, \ldots, i_{n-1})$. In this section, we use the same symbol i to represent both the integer i and its corresponding vector $(i_0, i_1, \ldots, i_{n-1})$.

3.4.1   Generalized 2-Dimensional Walsh Matrices

Definition 3.4.1 *A (± 1)-valued matrix $H = [h(i,j)]$, $0 \le i,j \le 2^n - 1$, of order 2^n is called a generalized 2-dimensional Walsh matrix if and only if there exist two permutation matrices, say A and B, such that AHB is a regular Walsh matrix.*

Clearly the generalized Walsh matrices are also orthogonal, i.e., $2^n H^{-1} = H$. In addition, their rows are closed under the operation of bitwise multiplication, i.e., for any two rows $(h(i,0), h(i,1), \ldots, h(i,n-1))$ and $(h(j,0), h(j,1), \ldots, h(j,n-1))$ there exists exactly one row $(h(k,0), h(k,1), \ldots, h(k,n-1))$ such that for each $0 \le s \le n-1$, $h(k,s) =$

$h(i, s)h(j, s)$. The columns of a generalized Walsh matrix are also closed under the operation of bit-wise multiplication. The regular Walsh matrices are special cases of the generalized matrices corresponding to $A = B = I_n$, the unit matrix.

Lemma 3.4.1 *Let*

$$a = (\overbrace{1\ldots1}^{2k}\overbrace{-1\ldots-1}^{2k}),$$

and

$$b = (b_0, b_1, \ldots, b_{4k-1}), \qquad b_i = \pm1, \qquad \sum_{i=0}^{4k-1} b_i = 0$$

are two vectors of length $4k$. If the vector $c = (c_0, \ldots, c_{4k-1})$ formed by the bit-wise multiplication of the vectors a and b satisfies $\sum_{i=0}^{4k-1} c_i = 0$, then the vector b satisfies

$$\sum_{i=0}^{2k-1} b_i = \sum_{i=0}^{2k-1} b_{2k+i} = 0.$$

Proof. The equation $\sum_{i=0}^{4k-1} c_i = 0$ implies

$$\sum_{i=0}^{2k-1} b_i - \sum_{i=0}^{2k-1} b_{2k+i} = 0.$$

On the other hand, we have assumed that

$$\sum_{i=0}^{2k-1} b_i + \sum_{i=0}^{2k-1} b_{2k+i} = \sum_{i=0}^{4k-1} b_i = 0.$$

The lemma follows.   **Q.E.D.**

With the help of Lemma 3.4.1 we can prove the following very important and basic theorem about the generalized Walsh matrix.

Theorem 3.4.1 *A (±1)-valued matrix $H = [h(i, j)]$, $0 \leq i, j \leq 2^n - 1$, of order 2^n is a generalized 2-dimensional Walsh matrix if and only if the following conditions are satisfied:*

1. *The rows of H are closed under the operation of bit-wise multiplication;*

2. *Except the all one row $(1, 1, \ldots, 1)$, in each other row there are 2^{n-1} '1's and 2^{n-1} '-1's.*

Proof. $\Longrightarrow$: Permutations among rows and columns keep the properties of: (1) closed bit-wise multiplication and (2) orthogonality of the original Walsh matrix. The balance between '1's and '-1's in each non-all-one row is owed to the orthogonality between that row and the all one row.

$\Longleftarrow$: Because of the property of closed bit-wise multiplication, we know the existence of the all one row. In fact, the bit-wise multiplication of each row by itself produces the all one row.

Without loss of the generality, we assume that

$$H = \begin{pmatrix} 1 & 1 & \cdots & 1 \\ h(1,0) & h(1,1) & \cdots & h(1,2^n-1) \\ h(2,0) & h(2,1) & \cdots & h(2,2^n-1) \\ \vdots & \vdots & \vdots & \vdots \\ h(2^n-1,0) & h(2^n-1,1) & \cdots & h(2^n-1,2^n-1) \end{pmatrix}$$

$$:= \begin{pmatrix} h(0) \\ h(1) \\ \vdots \\ h(2^n-1) \end{pmatrix},$$

where $\sum_{j=0}^{2^n-1} h(i,j) = 0$ for each i, $1 \le i \le 2^n - 1$.

This matrix H can be changed, by column-permutations, to the following form (for simplicity we still use the symbol H for the renewed matrix.)

$$H = \begin{pmatrix} 1 & \cdots & 1 & 1 & \cdots & 1 \\ 1 & \cdots & 1 & -1 & \cdots & -1 \\ h(2,0) & & \cdots & \cdots & & h(2,2^n-1) \\ h(3,0) & & \cdots & \cdots & & h(3,2^n-1) \\ \vdots & & \vdots & \vdots & \vdots & \vdots & \vdots \\ h(2^n-1,0) & & \cdots & \cdots & & h(2^n-1,2^n-1) \end{pmatrix}$$

$$:= \begin{pmatrix} h(0) \\ h(1) \\ \vdots \\ h(2^n-1) \end{pmatrix}.$$

Because of Lemma 3.4.1 and because the rows are closed under bit-wise multiplication, we know that for each j, $j \geq 2$, both the first half and the second half of $h(j)$ are (± 1)-balanced. Hence the matrix H can be transformed, still by column permutations, to the following form:

$$
H = \begin{pmatrix}
1 & \cdots & 1 & 1 & \cdots & 1 & 1 & \cdots & 1 & 1 & \cdots & 1 \\
1 & \cdots & 1 & 1 & \cdots & 1 & -1 & \cdots & -1 & -1 & \cdots & -1 \\
1 & \cdots & 1 & -1 & \cdots & -1 & 1 & \cdots & 1 & -1 & \cdots & -1 \\
h(3,0) & & & & \cdots & & & & & & & h(3, 2^n - 1) \\
h(4,0) & & & & \cdots & & & & & & & h(4, 2^n - 1) \\
\vdots & & \vdots & \vdots & \vdots & \vdots & \vdots & \vdots & \vdots & \vdots & & \vdots \\
h(2^n - 1, 0) & & & & \cdots & & & & & & & h(2^n - 1, 2^n - 1)
\end{pmatrix}
$$

$$
:= \begin{pmatrix}
h(0) \\
h(1) \\
\vdots \\
h(2^n - 1)
\end{pmatrix}.
$$

Similarly, because of Lemma 3.4.1 and because the rows are closed under bit-wise multiplication, we know that for each j, $j \geq 3$, the first quarter, the second quarter, the third quarter, and the fourth quarter of $h(j)$ are all (± 1)-balanced. Hence the matrix H can be transformed, by column-permutations, to the following form:

$$
H = \begin{pmatrix}
1 & \cdots & 1 & 1 & \cdots & 1 & 1 & \cdots & 1 & 1 & \cdots & 1 \\
1 & \cdots & 1 & 1 & \cdots & 1 & -1 & \cdots & -1 & -1 & \cdots & -1 \\
1 & \cdots & 1 & -1 & \cdots & -1 & -1 & \cdots & -1 & 1 & \cdots & 1 \\
1 & \cdots & 1 & -1 & \cdots & -1 & 1 & \cdots & 1 & -1 & \cdots & -1 \\
h(4,0) & & & & \cdots & & & & & & & h(4, 2^n - 1) \\
h(5,0) & & & & \cdots & & & & & & & h(5, 2^n - 1) \\
\vdots & & \vdots & \vdots & \vdots & \vdots & \vdots & \vdots & \vdots & \vdots & & \vdots \\
h(2^n - 1, 0) & & & & \cdots & & & & & & & h(2^n - 1, 2^n - 1)
\end{pmatrix}
$$

$$
:= \begin{pmatrix}
h(0) \\
h(1) \\
\vdots \\
h(2^n - 1)
\end{pmatrix}.
$$

Thus for $0 \leq i \leq 3$, $h(i) = wal_{i-1}(x)$, the i-th Walsh function. Repeating the above process, we can finally transform the matrix H to the Walsh matrix.   **Q.E.D.**

Theorem 3.4.2 *The following two conditions are equivalent to each other:*

C1 : $H = [h(i,j)]$, $0 \le i, j \le 2^n - 1$, *is a generalized 2-dimensional Walsh matrix satisfying*

$$h(i_1, j).h(i_2, j) = h(i_1 \oplus i_2, j) \text{ and } h(i, j_1).h(i, j_2) = h(i, j_1 \oplus j_2),$$

where $u \oplus v$ stands for the dyadic summation between the two vectors $u = (u_0, \ldots, u_{n-1})$ and $v = (v_0, \ldots, v_{n-1})$, i.e., $u \oplus v = ((u_0 + v_0))\mathrm{mod}2, \ldots, (u_{n-1} + v_{n-1})\mathrm{mod}2)$.

C2 : *There exists, in $GF(2)$, a non-singular matrix R of size $n \times n$ such that*

$$h(i,j) = (-1)^{iRj'}.$$

Where the symbols i and j in $(-1)^{iRj'}$ stands for the expended vectors of the integers i and j, and j' the transpose of the vector j.

Proof. $C2 \Rightarrow C1$: The identities

$$h(i_1, j).h(i_2, j) = h(i_1 \oplus i_2, j) \text{ and } h(i, j_1).h(i, j_2) = h(i, j_1 \oplus j_2).$$

are straightforward of the assumption $h(i,j) = (-1)^{iRj'}$.

From Theorem 3.4.1 and the equation:

$$\sum_{j=0}^{2^n-1} h(i,j) = \sum_{j=0}^{2^n-1} (-1)^{iRj'} = 0, \text{ for } i \ne 0$$

we know that H is indeed a generalized Walsh matrix.

$C1 \Rightarrow C2$: At first $h(i,j)$ can be formulated by

$$h(i,j) = (-1)^{B(i,j)}, \ 0 \le i, j \le 2^n - 1,$$

where $B(i,j)$ is a $(0,1)$-valued function of i and j.

Since

$$h(i_1, j).h(i_2, j) = h(i_1 \oplus i_2, j) \text{ and } h(i, j_1).h(i, j_2) = h(i, j_1 \oplus j_2),$$

we have

$$B(u,v) \oplus B(w,v) = B(u \oplus w, v) \text{ and } B(u,v) \oplus B(u,w) = B(u, v \oplus w).$$

Hence $B(0,v) = B(v,0) = 0$, and for each $c = 0$ or 1, holds

$$B(cu, v) = B(u, cv) = cB(u,v).$$

Up to now, we have proved that $B(u, v)$ is a bi-linear function. Thus there exists a matrix of size $n \times n$ such that $B(u, v) = uRv'$. It is now sufficient to prove that the matrix R is non-singular.

From the orthogonality we know that

$$\sum_{v=0}^{2^n-1} h(u, v)h(w, v) = \sum_{v=0}^{2^n-1} h(u \oplus w, v) = 2^n \delta_{w,u}$$

or, equivalently,

$$\sum_{v=0}^{2^n-1} (-1)^{uRv'} = \begin{cases} 2^n & \text{if } u = 0, \\ 0 & \text{otherwise.} \end{cases}$$

In other words, we have $uR \neq (0, \ldots 0)$ for each non-zero u. Thus the matrix R is indeed non-singular.   **Q.E.D.**

The previous Theorem 3.4.2 states that the Walsh functions in universal orders are special cases of the generalized Walsh matrices.

Theorem 3.4.3 *The direct multiplication of two 2-dimensional generalized Walsh matrices is also a 2-dimensional generalized Walsh matrix.*

Proof. Let A and B be two 2-dimensional generalized Walsh matrices. If $B = W_2 = \begin{pmatrix} 1 & 1 \\ 1 & -1 \end{pmatrix}$ then $B \otimes A = \begin{pmatrix} A & A \\ A & -A \end{pmatrix}$, which is clearly a 2-dimensional generalized Walsh matrix.

If $B = W_{2^n} = W_2 \otimes W_2 \otimes \ldots \otimes W_2$, by induction on n, it can be proved that $B \otimes A$ is also a 2-dimensional generalized Walsh matrix.

In general, let $B = EW_{2^n}D$, where E and D are two permutation matrices of size $2^n \times 2^n$. Then

$$B \otimes A = (EW_{2^n}D) \otimes A = (E \otimes I_{2^n})(W_{2^n} \otimes A)(D \otimes I_{2^n}).$$

Hence their direct multiplication is indeed a 2-dimensional generalized Walsh matrix.   **Q.E.D.**

3.4.2   3-Dimensional Walsh Matrices

The 3-dimensional Walsh matrices studied in this subsection are special cases of 3-dimensional Hadamard matrices of size $2^n \times 2^n$ which are defined by the following definition.

Definition 3.4.2 *A 3-dimensional Hadamard matrix of size $2^n \times 2^n$ is called a 3-dimensional Walsh matrix if all 2-dimensional layers in at least one axis-normal orientation are 2-dimensional generalized Walsh matrices. A 3-dimensional Walsh matrix is said to be proper in the x- (resp., y- and z-) direction if all layers in the x- (resp., y- and z-) direction are 2-dimensional generalized Walsh matrices. A 3-dimensional Walsh matrix is said to be absolutely proper if all of the possible 2-dimensional layers are generalized Walsh matrices, otherwise it is said to be improper.*

All of the absolutely proper 3-dimensional Walsh matrices of size 2×2 are described by

$$A_1(i,j,0) = \begin{bmatrix} 1 & -1 \\ 1 & 1 \end{bmatrix} \text{ and } A_1(i,j,1) = \begin{bmatrix} 1 & 1 \\ -1 & 1 \end{bmatrix}$$

$$A_2(i,j,0) = \begin{bmatrix} 1 & 1 \\ 1 & -1 \end{bmatrix} \text{ and } A_2(i,j,1) = \begin{bmatrix} -1 & 1 \\ 1 & 1 \end{bmatrix}$$

$$A_3(i,j,0) = \begin{bmatrix} 1 & 1 \\ -1 & 1 \end{bmatrix} \text{ and } A_3(i,j,1) = \begin{bmatrix} 1 & -1 \\ 1 & 1 \end{bmatrix}$$

and

$$A_4(i,j,0) = \begin{bmatrix} -1 & 1 \\ 1 & 1 \end{bmatrix} \text{ and } A_4(i,j,1) = \begin{bmatrix} 1 & 1 \\ 1 & -1 \end{bmatrix}.$$

All of the improper 3-dimensional Walsh matrices of size 2×2 that are proper in two directions are described by

$$B_1(i,j,0) = \begin{bmatrix} 1 & 1 \\ -1 & 1 \end{bmatrix} \text{ and } B_1(i,j,1) = \begin{bmatrix} 1 & 1 \\ 1 & -1 \end{bmatrix}$$

$$B_2(i,j,0) = \begin{bmatrix} 1 & 1 \\ 1 & -1 \end{bmatrix} \text{ and } B_2(i,j,1) = \begin{bmatrix} 1 & 1 \\ -1 & 1 \end{bmatrix}$$

$$B_3(i,j,0) = \begin{bmatrix} 1 & 1 \\ -1 & 1 \end{bmatrix} \text{ and } B_3(i,j,1) = \begin{bmatrix} -1 & 1 \\ 1 & 1 \end{bmatrix}$$

and

$$B_4(i,j,0) = \begin{bmatrix} -1 & 1 \\ 1 & 1 \end{bmatrix} \text{ and } B_4(i,j,1) = \begin{bmatrix} 1 & -1 \\ 1 & 1 \end{bmatrix}.$$

Theorem 3.4.4 *The direct multiplication of two 3-dimensional Walsh matrices is also a 3-dimensional Walsh matrix. Furthermore, the product matrix is proper in those directions in which both the parent matrices are proper.*

Proof. Let A and B be the two parent 3-dimensional Walsh matrices. Every layer in the x- (resp., y- and z-) direction of $A \otimes B$ is the direct multiplication of two layers of their parent matrices also in the x- (resp., y- and z-)direction. The theorem follows from Definition 3.4.2 and Theorem 3.4.3.   **Q.E.D.**

For example, the direct multiplication $C_1 = A_1 \otimes B_1$ is absolutely proper which is described

$$C_1(i,j,0) = \begin{bmatrix} 1 & -1 & 1 & -1 \\ 1 & 1 & 1 & 1 \\ 1 & -1 & -1 & 1 \\ 1 & 1 & -1 & -1 \end{bmatrix} ; \quad C_1(i,j,1) = \begin{bmatrix} 1 & 1 & 1 & 1 \\ -1 & 1 & -1 & 1 \\ 1 & 1 & -1 & -1 \\ -1 & 1 & 1 & -1 \end{bmatrix}$$

and

$$C_1(i,j,2) = \begin{bmatrix} -1 & 1 & 1 & -1 \\ -1 & -1 & 1 & 1 \\ 1 & -1 & 1 & -1 \\ 1 & 1 & 1 & 1 \end{bmatrix} ; \quad C_1(i,j,3) = \begin{bmatrix} -1 & -1 & 1 & 1 \\ 1 & -1 & -1 & 1 \\ 1 & 1 & 1 & 1 \\ -1 & 1 & -1 & 1 \end{bmatrix} .$$

The direct multiplication $C_2 = A_1 \otimes B_2$ is proper in two directions which is described

$$C_2(i,j,0) = \begin{bmatrix} 1 & 1 & 1 & 1 \\ -1 & 1 & -1 & 1 \\ 1 & 1 & -1 & -1 \\ -1 & 1 & 1 & -1 \end{bmatrix} ; \quad C_2(i,j,1) = \begin{bmatrix} 1 & 1 & 1 & 1 \\ 1 & -1 & 1 & -1 \\ 1 & 1 & -1 & -1 \\ 1 & -1 & -1 & 1 \end{bmatrix}$$

and

$$C_2(i,j,2) = \begin{bmatrix} 1 & 1 & 1 & 1 \\ 1 & 1 & 1 & 1 \\ -1 & -1 & 1 & 1 \\ -1 & -1 & 1 & 1 \end{bmatrix} ; \quad C_2(i,j,3) = \begin{bmatrix} -1 & -1 & 1 & 1 \\ -1 & -1 & 1 & 1 \\ 1 & 1 & 1 & 1 \\ 1 & 1 & 1 & 1 \end{bmatrix} .$$

The following D is a 3-dimensional Walsh matrix of size 4×4 which is proper in z-direction but improper in both x- and y-directions.

$$D(i,j,0) = \begin{bmatrix} 1 & 1 & 1 & 1 \\ -1 & 1 & -1 & 1 \\ -1 & -1 & 1 & 1 \\ 1 & -1 & -1 & 1 \end{bmatrix} ; \quad D(i,j,1) = \begin{bmatrix} 1 & 1 & 1 & 1 \\ -1 & 1 & -1 & 1 \\ 1 & 1 & -1 & -1 \\ -1 & 1 & 1 & -1 \end{bmatrix}$$

and

$$D(i,j,2) = \begin{bmatrix} -1 & 1 & -1 & 1 \\ 1 & 1 & 1 & 1 \\ 1 & -1 & -1 & 1 \\ -1 & -1 & 1 & 1 \end{bmatrix} ; \quad D(i,j,3) = \begin{bmatrix} -1 & 1 & -1 & 1 \\ 1 & 1 & 1 & 1 \\ -1 & 1 & 1 & -1 \\ 1 & 1 & -1 & -1 \end{bmatrix} .$$

Lemma 3.4.2

$$\sum_{i=0}^{2^n-1} (-1)^{i \cdot j'} = 2^n \delta(j) = \begin{cases} 2^n & j = 0, \\ 0 & j \neq 0, \end{cases}$$

where $i \cdot j'$ refers to the dot production of the vectors i and j.

This lemma can be easily proved.

Theorem 3.4.5 *Let R_1 and R_2 be two non-singular $n \times n$ matrices in $GF(2)$. Then the matrix $H = [h(i,j,k)]$ defined by*

$$h(i,j,k) = (-1)^{iR_1 j' + kR_2 j'}, \quad 0 \leq i,j,k \leq 2^n - 1,$$

is a 3-dimensional Walsh matrix which is proper in x- and z-directions.

Proof. At first we prove that H is a 3-dimensional Hadamard matrix. In fact,

$$\sum_{i=0}^{2^n-1} \sum_{j=0}^{2^n-1} h(i,j,a)h(i,j,b) = \sum_{i=0}^{2^n-1} \sum_{j=0}^{2^n-1} (-1)^{(a \oplus b)R_2 j'}$$

$$= 2^n \sum_{j=0}^{2^n-1} (-1)^{(a \oplus b)R_2 j'}$$

$$= 4^n \delta(a,b).$$

The last equation is due to Lemma 3.4.2 and the non-singular property of R_2.

Similarly, it can be proved that

$$\sum_{j=0}^{2^n-1}\sum_{k=0}^{2^n-1} h(a,j,k)h(b,j,k) = 4^n\delta(a,b).$$

Furthermore,

$$\sum_{i=0}^{2^n-1}\sum_{k=0}^{2^n-1} h(i,a,k)h(i,b,k) = \sum_{i=0}^{2^n-1}\sum_{k=0}^{2^n-1}(-1)^{iR_1(a\oplus b)'kR_2(a\oplus b)'}$$

$$= [\sum_{i=0}^{2^n-1}(-1)^{iR_1(a\oplus b)'}][\sum_{k=0}^{2^n-1}(-1)^{kR_2(a\oplus b)'}]$$

$$= 2^n\delta(a,b)2^n\delta(a,b)$$

$$= 4^n\delta(a,b).$$

Therefore we have proved that H is a 3-dimensional Hadamard matrix.

Let $A = [a(j,k)] = [h(i_0,j,k)]$, $0 \le i_0 \le 2^n - 1$, be the i_0-th layer in the x-direction of the matrix H. Then

$$a(j_1,k)a(j_2,k) = a(j_1 \oplus j_2, k).$$

In other word, the first condition in Theorem 3.4.1 is satisfied.

Since

$$\sum_{k=0}^{2^n-1} a(j,k) = (-1)^{i_0 R_1 j'}\sum_{k=0}^{2^n-1}(-1)^{kR_2 j'} = 2^n\delta(j),$$

the second condition in Theorem 3.4.1 is also satisfied. Thus the layer $[a(j,k)]$ is a generalized 2-dimensional Walsh matrix.

By the same way, it can be proved that each layer in z-direction is also a generalized 2-dimensional Walsh matrix.

Since, there is an all one layer in y-direction, H is improper in y-direction. In a word, the matrix H is a 3-dimensional Walsh matrix that is proper in x- and z-direction. The theorem follows.   **Q.E.D.**

Theorem 3.4.6 $H = [h(i,j,k)]$ *is a matrix defined by*

$$h(i,j,k) = (-1)^{iR_1 j'+kR_2 j'}, \quad 0 \le i,j,k \le 2^n - 1,$$

where R_1 and R_2 are two non-singular $n \times n$ matrices in $GF(2)$, if and only if H is a 3-dimensional Hadamard matrix satisfying

$$h(i,j_1,k)h(i,j_2,k) = h(i,j_1 \oplus j_2,k) \text{ and } h(i_1,j,k_1)h(i_2,j,k_2)$$
$$= h(i_1 \oplus i_2, j, k_1 \oplus k_2)$$

$$for\ all\ 0 \le i, i_1, i_2, j, j_1, j_2, k, k_1, k_2 \le 2^n - 1.$$

Proof. $\Longrightarrow$: By direct verification.

$\Longleftarrow$: Let

$$h(i, j, k) = (-1)^{B(i,j,k)},\ 0 \le i, j, k \le 2^n - 1,$$

where $B(i, j, k)$ is a $(0, 1)$-valued function of i, j, and k.

$h(i, j, k) = h(0, j, k)h(i, j, 0)$ implies $B(i, j, k) = B(0, j, k) \oplus B(i, j, 0)$;

$$h(0, j_1, k)h(0, j_2, k) = h(0, j_1 \oplus j_2, k)$$
$$\text{implies}\, B(0, j_1 \oplus j_2, k) = B(0, j_1, k) \oplus B(0, j_2, k);$$

and

$$h(0, j, k_1)h(0, j, k_2) = h(0, j, k_1 \oplus k_2)$$
$$\text{implies}\ B(0, j, k_1 \oplus k_2) = B(0, j, k_1) \oplus B(0, j, k_2).$$

Thus $h(0, 0, k) = 1$ and $h(0, j, 0) = 1$ imply

$$B(0, cj, k) = cB(0, j, k) \text{ and } B(0, j, ck) = cB(0, j, k)$$

for each $c \in GF(2)$. Therefore $B(0, j, k)$ is bilinear in both j and k.

Similarly, it can be proved that $B(i, j, 0)$ is bilinear in both i and j.

By the same way as that used in the proof of Theorem 3.4.2, it can be proved that there exist two $n \times n$ matrices, say R_1 and R_2, in $GF(2)$, such that

$$B(0, j, k) = kR_1 j' \text{ and } B(i, j, 0) = iR_2 j'$$

Thus we have proved that

$$h(i, j, k) = (-1)^{iR_1 j' + kR_2 j'},\quad 0 \le i, j, k \le 2^n - 1.$$

Finally, it is sufficient to prove that both R_1 and R_2 are non-singular. Because H is a 3-dimensional Hadamard matrix,

$$2^n \sum_{j=0}^{2^n-1} h(0, j, a) = \sum_{i=0}^{2^n-1} \sum_{j=0}^{2^n-1} h(i, j, 0)h(i, j, a) = 4^n \delta(a),$$

thus $\sum_{j=0}^{2^n-1} h(0, j, a) = 2^n \delta(a)$, or equivalently,

$$\sum_{j=0}^{2^n-1} (-1)^{aR_2 j'} = 2^n \delta(a).$$

Hence R_2 is non-singular. Similarly, it can be proved that R_1 is also non-singular.   **Q.E.D.**

Theorem 3.4.7 *Let $D = [D(i, j)]$, $0 \le i, j \le 2^n - 1$, be a 2-dimensional generalized Walsh matrix satisfying*

$$D(i_1, j)D(i_2, j) = D(i_1 \oplus i_2, j) \text{ and } D(i, j_1)D(i, j_2) = D(i, j_1 \oplus j_2),$$

for all $0 \le i, j, i_1, i_2, j_1, j_2 \le 2^n - 1$. And let r be an integer such that $\gcd(r, 2^n) = 1$. Then the following matrix $H = [h(i, j, k)]$, $0 \le i, j, k \le 2^n - 1$, is a 3-dimensional Walsh matrix that is proper in both y- and z-directions, where

$$H(i, j, k) = D(i, (j + kr) \bmod 2^n), \quad 0 \le i, j, k \le 2^n - 1.$$

Proof. All layers of H in y- and z-directions are clearly the generalized Walsh matrices, because they are the shift forms of D. Since there is an all one layer in the x-direction of H, it is sufficient to prove that H is a 3-dimensional Hadamard matrix.

Let $A = [a(i, j)]$ and $B = [b(i, j)]$ be two layers in z-direction. Then for each $0 \le i \le 2^n - 1$, the bit-wise multiplication between the rows $(a(i, 0), a(i, 1), \ldots, a(i, 2^n - 1))$ and $(b(i, 0), b(i, 1), \ldots, b(i, 2^n - 1))$ is the vector

$$(a(i, 0)b(i, 0), a(i, 1)b(i, 1), \ldots, a(i, 2^n - 1)b(i, 2^n - 1)),$$

which is a non-zero row of D, because of the closed bit-wise multiplication property of D. Therefore

$$\sum_{j=0}^{2^n-1} a(i, j)b(i, j) = 0, \quad 0 \le i \le 2^n - 1,$$

or equivalently,

$$\sum_{i=0}^{2^n-1} \sum_{j=0}^{2^n-1} h(i, j, a)h(i, j, b) = 4^n \delta(a, b).$$

Similarly, it can be proved that

$$\sum_{i=0}^{2^n-1} \sum_{k=0}^{2^n-1} h(i, a, k)h(i, b, k) = 4^n \delta(a, b).$$

Finally, the equation

$$\sum_{j=0}^{2^n-1} \sum_{k=0}^{2^n-1} h(a, j, k)h(b, j, k) = 4^n \delta(a, b)$$

is the result of the facts (1): the rows of D are closed under bit-wise multiplication; and (2): the sum of each non-all-ones row is zero. Therefore the matrix H is a 3-dimensional Hadamard matrix.   **Q.E.D.**

For example, if $r = 1$ and

$$D = \begin{bmatrix} 1 & 1 & 1 & 1 \\ 1 & -1 & 1 & -1 \\ 1 & 1 & -1 & -1 \\ 1 & -1 & -1 & 1 \end{bmatrix},$$

then the matrix $H = [h(i,j,k)]$, $0 \le i,j,k \le 2^n - 1$, produced by Theorem 3.4.7 is described by

$$H(i,j,0) = \begin{bmatrix} 1 & 1 & 1 & 1 \\ 1 & -1 & 1 & -1 \\ 1 & 1 & -1 & -1 \\ 1 & -1 & -1 & 1 \end{bmatrix}, \quad H(i,j,1) = \begin{bmatrix} 1 & 1 & 1 & 1 \\ -1 & 1 & -1 & 1 \\ -1 & 1 & 1 & -1 \\ 1 & 1 & -1 & -1 \end{bmatrix},$$

and

$$H(i,j,2) = \begin{bmatrix} 1 & 1 & 1 & 1 \\ 1 & -1 & 1 & -1 \\ -1 & -1 & 1 & 1 \\ -1 & 1 & 1 & -1 \end{bmatrix}, \quad H(i,j,3) = \begin{bmatrix} 1 & 1 & 1 & 1 \\ -1 & 1 & -1 & 1 \\ 1 & -1 & -1 & 1 \\ -1 & -1 & 1 & 1 \end{bmatrix}.$$

This matrix is also the special case of Theorem 3.4.5 for $R_1 = R_2$.

3.4.3   3-Dimensional Pan–Walsh Matrices

In the definition of 3-dimensional Walsh matrix, it is required that each layer in some direction is 2-dimensional generalized Walsh matrix. This subsection will study a generalization of the 3-dimensional Walsh matrix by allowing the layers to be a 2-dimensional generalized Walsh matrix H or its minus form $-H$.

Lemma 3.4.3 *Let A and B be 2-dimensional generalized Walsh matrices. Then $(-A) \otimes B$ and $A \otimes (-B)$ are minus 2-dimensional generalized Walsh matrices; and $(-A) \otimes (-B)$ is a 2-dimensional generalized Walsh matrix.*

The proof of this lemma is straightforward and is omitted here.

Definition 3.4.3 *A 3-dimensional Hadamard matrix of size $2^n \times 2^n$ is called a 3-dimensional Pan–Walsh matrix if all 2-dimensional layers in at least one axis-normal orientation are 2-dimensional generalized Walsh matrices or their minus forms. A 3-dimensional Pan–Walsh matrix is said to be proper in the x- (resp., y- and z-) direction if all layers in the x- (resp., y- and z-) direction are 2-dimensional generalized Walsh matrices or their minus forms. A 3-dimensional Pan–Walsh matrix is said to be absolutely proper if all of the possible 2-dimensional layers are generalized Walsh matrices or their minus forms, otherwise it is said to be improper.*

In the same way as that used in the proof of Theorem 3.4.4, the following theorem is proved to be true.

Theorem 3.4.8 *The direct multiplication of two 3-dimensional Pan–Walsh matrices is also a 3-dimensional Pan–Walsh matrix. Furthermore, the product matrix is proper in those directions in which both the parent matrices are proper.*

There are 16 3-dimensional Pan–Walsh matrices of size $2 \times 2 \times 2$ that are not the regular 3-dimensional Walsh matrices. They are

$$A_0(i,j,0) = \begin{bmatrix} 1 & -1 \\ 1 & 1 \end{bmatrix}; \text{ and } A_0(i,j,1) = \begin{bmatrix} -1 & -1 \\ 1 & -1 \end{bmatrix}.$$

$$A_1(i,j,0) = \begin{bmatrix} -1 & 1 \\ 1 & 1 \end{bmatrix}; \text{ and } A_1(i,j,1) = \begin{bmatrix} -1 & -1 \\ -1 & 1 \end{bmatrix}.$$

$$A_2(i,j,0) = \begin{bmatrix} 1 & 1 \\ -1 & 1 \end{bmatrix}; \text{ and } A_2(i,j,1) = \begin{bmatrix} -1 & 1 \\ -1 & -1 \end{bmatrix}.$$

$$A_3(i,j,0) = \begin{bmatrix} 1 & 1 \\ 1 & -1 \end{bmatrix}; \text{ and } A_3(i,j,1) = \begin{bmatrix} 1 & -1 \\ -1 & -1 \end{bmatrix}.$$

$$A_4(i,j,0) = \begin{bmatrix} 1 & -1 \\ -1 & -1 \end{bmatrix}; \text{ and } A_4(i,j,1) = \begin{bmatrix} 1 & 1 \\ 1 & -1 \end{bmatrix}.$$

$$A_5(i,j,0) = \begin{bmatrix} -1 & -1 \\ 1 & -1 \end{bmatrix}; \text{ and } A_5(i,j,1) = \begin{bmatrix} 1 & -1 \\ 1 & 1 \end{bmatrix}.$$

$$A_6(i,j,0) = \begin{bmatrix} -1 & -1 \\ -1 & 1 \end{bmatrix} ; \text{ and } A_6(i,j,1) = \begin{bmatrix} -1 & 1 \\ 1 & 1 \end{bmatrix} .$$

$$A_7(i,j,0) = \begin{bmatrix} -1 & 1 \\ -1 & -1 \end{bmatrix} ; \text{ and } A_7(i,j,1) = \begin{bmatrix} 1 & 1 \\ -1 & 1 \end{bmatrix} .$$

$$A_8(i,j,0) = \begin{bmatrix} 1 & 1 \\ -1 & 1 \end{bmatrix} ; \text{ and } A_8(i,j,1) = \begin{bmatrix} -1 & -1 \\ -1 & 1 \end{bmatrix} .$$

$$A_9(i,j,0) = \begin{bmatrix} 1 & 1 \\ 1 & -1 \end{bmatrix} ; \text{ and } A_9(i,j,1) = \begin{bmatrix} -1 & 1 \\ -1 & -1 \end{bmatrix} .$$

$$A_{10}(i,j,0) = \begin{bmatrix} 1 & -1 \\ 1 & 1 \end{bmatrix} ; \text{ and } A_{10}(i,j,1) = \begin{bmatrix} 1 & -1 \\ -1 & -1 \end{bmatrix} .$$

$$A_{11}(i,j,0) = \begin{bmatrix} -1 & 1 \\ 1 & 1 \end{bmatrix} ; \text{ and } A_{11}(i,j,1) = \begin{bmatrix} -1 & -1 \\ 1 & -1 \end{bmatrix} .$$

$$A_{12}(i,j,0) = \begin{bmatrix} -1 & 1 \\ -1 & -1 \end{bmatrix} ; \text{ and } A_{12}(i,j,1) = \begin{bmatrix} -1 & 1 \\ 1 & 1 \end{bmatrix} .$$

$$A_{13}(i,j,0) = \begin{bmatrix} -1 & -1 \\ 1 & -1 \end{bmatrix} ; \text{ and } A_{13}(i,j,1) = \begin{bmatrix} 1 & 1 \\ 1 & -1 \end{bmatrix} .$$

$$A_{14}(i,j,0) = \begin{bmatrix} -1 & 1 \\ -1 & 1 \end{bmatrix} ; \text{ and } A_{14}(i,j,1) = \begin{bmatrix} 1 & 1 \\ -1 & -1 \end{bmatrix} .$$

and

$$A_{15}(i,j,0) = \begin{bmatrix} -1 & -1 \\ 1 & 1 \end{bmatrix} ; \text{ and } A_{15}(i,j,1) = \begin{bmatrix} -1 & 1 \\ -1 & 1 \end{bmatrix} .$$

The direct multiplication of the above A_0 and A_1 is the following $B = [B(i,j,k)] = A_0 \otimes A_1$, which is described by

$$B(i,j,0) = \begin{bmatrix} 1 & 1 & 1 & 1 \\ 1 & -1 & 1 & -1 \\ -1 & -1 & 1 & 1 \\ -1 & 1 & 1 & -1 \end{bmatrix}, \quad B(i,j,1) = \begin{bmatrix} 1 & -1 & 1 & -1 \\ -1 & -1 & -1 & -1 \\ -1 & 1 & 1 & -1 \\ 1 & 1 & -1 & -1 \end{bmatrix},$$

and

$$B(i,j,2) = \begin{bmatrix} -1 & -1 & 1 & 1 \\ -1 & 1 & 1 & -1 \\ -1 & -1 & -1 & -1 \\ -1 & 1 & -1 & 1 \end{bmatrix}, \quad B(i,j,3) = \begin{bmatrix} -1 & 1 & 1 & -1 \\ 1 & 1 & -1 & -1 \\ -1 & 1 & -1 & 1 \\ 1 & 1 & 1 & 1 \end{bmatrix}.$$

It is not difficult to see that a minus 3-dimensional Pan–Walsh matrix is also a 3-dimensional Pan–Walsh matrix; and the matrices produced by minus some layers in z-direction of the matrices in Theorem 3.4.7 are also 3-dimensional Pan–Walsh matrices.

3.4.4   Analytic Representations

Sometimes the analytic formulas are much more convenient than others (e.g., the representation based on layers) for the study and application of 3-dimensional Walsh matrices. Thus this subsection concentrates on the list of analytic formulas.

Theorem 3.4.9 *Let M be an $n \times n$ non-singular matrix in $GF(2)$, r an integer satisfying $\gcd(r, 2^n) = 1$, and*

$$R = [R(i,j)] = [(-1)^{iMj'}], \ 0 \le i,j \le 2^n - 1.$$

Then the following matrix $H = [h(i,j,k)]$, $0 \le i,j,k \le 2^n - 1$, is a 3-dimensional Walsh matrix that is proper in both y- and z-directions, where

$$H(i,j,k) = R(i, (j + kr) \bmod 2^n), \ 0 \le i,j,k \le 2^n - 1.$$

This theorem is a direct corollary of Theorem 3.4.7. In general, we have the following theorem.

Theorem 3.4.10 *Let $f(x,y)$ be a mapping from the set $A = \{(x,y) : 0 \le x,y \le 2^n - 1\}$ to the set $B = \{z : 0 \le z \le 2^n - 1\}$ such that for each prefixed $x_0 \in B$ (resp. $y_0 \in B$), the $f(x_0, y)$ (resp., $f(x, y_0)$) is a one-to-one mapping for B to B. Let $g(x)$ be another one-to-one mapping for B to B, and R an $n \times n$ non-singular matrix in $GF(2)$. Then the matrix $H = [h(i,j,k)]$ defined by*

$$h(i,j,k) = (-1)^{f(i,k)Rg(j)'}, \quad 0 \le i,j,k \le 2^n - 1,$$

is a 3-dimensional Walsh matrix that is proper in two directions.

This theorem can be proved by the same way as that used in the proof of Theorem 3.4.5. In fact, Theorem 3.4.5, Theorem 3.4.7, and Theorem 3.4.9 are special cases of this theorem.

Theorem 3.4.11 *Let $H_1 = [H_1(i, j, k)]$, $0 \leq i, j, k \leq 1$, be a 3-dimensional matrix of size $2 \times 2 \times 2$ defined by*

$$H_1(i, j, k) = (-1)^{1+i+j+k+ij+ik+jk}, \quad 0 \leq i, j, k \leq 1.$$

Then the matrix

$$H_n := \overbrace{H_1 \otimes H_1 \otimes \ldots \otimes H_1}^{n-times}, \ n \geq 1,$$

is an absolutely proper 3-dimensional Walsh matrix of size $2^n \times 2^n \times 2^n$ represented by

$$H_n(i, j, k) = (-1)^{n+ii'+jj'+kk'+ij'+ik'+jk'}, \quad 0 \leq i, j, k \leq 2^n - 1. \quad (3.21)$$

Here and after we use the symbol xy' to simplify the inner product, $x.y'$, between the two vectors x and y.

Proof. At first, it is easy to verify that the matrix H_1 itself is an absolutely proper 3-dimensional Walsh matrix of size $2 \times 2 \times 2$. Thus, by Theorem 3.4.4, it is sufficient to proved the analytic formula in Equation 3.21. This will be prove by induction on n as follows.

The case of $n = 1$ is trivial.

Assume that Equation 3.21 works for $n = m$.

Now, consider the case of $n = m + 1$: Let

$$i = (i_m, i_{m-1}, \ldots, i_0), \quad j = (j_m, j_{m-1}, \ldots, j_0),$$

and

$$k = (k_m, k_{m-1}, \ldots, k_0)$$

are the expanded vectors of the integers i, j, and k, respectively.

Let

$$I := (0, i_{m-1}, \ldots, i_0), \quad J := (0, j_{m-1}, \ldots, j_0)$$

and

$$K := (0, k_{m-1}, \ldots, k_0).$$

Because of $H_{m+1} = H_m \otimes H_1$, we have

$$\begin{aligned} H_{m+1}(i, j, k) &= H_m(I, J, K)H_1(i_m, j_m, k_m) \\ &= (-1)^{m+II'+JJ'+KK'+IJ'+IK'+JK'} \\ &\quad \cdot (-1)^{1+i_m+j_m+k_m+i_m j_m+i_m k_m+j_m k_m} \\ &= (-1)^{(m+1)+ii'+jj'+kk'+ij'+ik'+jk'}. \end{aligned}$$

Thus the theorem is also true for the case of $n = m + 1$.   **Q.E.D.**

Corollary 3.4.1 *Let R be an $n \times n$ non-singular matrix in $GF(2)$, and $B = RR'$. Then the following matrix $A = [a(i, j, k)]$ is an absolutely proper 3-dimensional Walsh matrix of size $2^n \times 2^n \times 2^n$, where for each $0 \le i, j, k \le 2^n - 1$,*

$$a(i, j, k) = (-1)^{n+iBi'+jBj'+kBk'+iBj'+iBk'+jBk'}.$$

At the end of this chapter we present the analytic formulas of some basic 3-dimensional Walsh and Pan–Walsh matrices of size $2 \times 2 \times 2$ and their direct multiplications of size $2^n \times 2^n \times 2^n$.

1. The first matrix is

$$H_1 = [h(i, j, k)] = [(-1)^{1+i+j+k+ij+ik+jk}], \quad 0 \le i, j, k \le 1.$$

Its direct multiplication $H_n = H_1 \otimes H_1 \otimes \ldots \otimes H_1$ is represented by

$$H_n(i, j, k) = (-1)^{n+ii'+jj'+kk'+ij'+ik'+jk'}, \quad 0 \le i, j, k \le 2^n - 1.$$

This matrix has been stated in Theorem 3.4.11.

2. The second matrix is

$$H_1 = [h(i, j, k)] = [(-1)^{k+ij+ik+jk}], \quad 0 \le i, j, k \le 1.$$

Its direct multiplication $H_n = H_1 \otimes H_1 \otimes \ldots \otimes H_1$ is represented by

$$H_n(i, j, k) = (-1)^{kk'+ij'+ik'+jk'}, \quad 0 \le i, j, k \le 2^n - 1.$$

3. The third matrix is

$$H_1 = [h(i, j, k)] = [(-1)^{i+ij+ik+jk}], \quad 0 \le i, j, k \le 1.$$

Its direct multiplication $H_n = H_1 \otimes H_1 \otimes \ldots \otimes H_1$ is represented by

$$H_n(i, j, k) = (-1)^{ii'+ij'+ik'+jk'}, \quad 0 \le i, j, k \le 2^n - 1.$$

4. The fourth matrix is

$$H_1 = [h(i,j,k)] = [(-1)^{j+ij+ik+jk}], \quad 0 \le i,j,k \le 1.$$

Its direct multiplication $H_n = H_1 \otimes H_1 \otimes \ldots \otimes H_1$ is represented by

$$H_n(i,j,k) = (-1)^{jj'+ij'+ik'+jk'}, \quad 0 \le i,j,k \le 2^n - 1.$$

5. The fifth matrix is

$$H_1 = [h(i,j,k)] = [(-1)^{i+ij+ik}], \quad 0 \le i,j,k \le 1.$$

Its direct multiplication $H_n = H_1 \otimes H_1 \otimes \ldots \otimes H_1$ is represented by

$$H_n(i,j,k) = (-1)^{ii'+ij'+ik'}, \quad 0 \le i,j,k \le 2^n - 1.$$

6. The sixth matrix is

$$H_1 = [h(i,j,k)] = [(-1)^{ij+ik}], \quad 0 \le i,j,k \le 1.$$

Its direct multiplication $H_n = H_1 \otimes H_1 \otimes \ldots \otimes H_1$ is represented by

$$H_n(i,j,k) = (-1)^{ij'+ik'}, \quad 0 \le i,j,k \le 2^n - 1.$$

7. The seventh matrix is

$$H_1 = [h(i,j,k)] = [(-1)^{j+k+ij+ik}], \quad 0 \le i,j,k \le 1.$$

Its direct multiplication $H_n = H_1 \otimes H_1 \otimes \ldots \otimes H_1$ is represented by

$$H_n(i,j,k) = (-1)^{jj'+kk'+ij'+ik'}, \quad 0 \le i,j,k \le 2^n - 1.$$

8. The eighth matrix is

$$H_1 = [h(i,j,k)] = [(-1)^{1+i+j+k+ij+ik}], \quad 0 \le i,j,k \le 1.$$

Its direct multiplication $H_n = H_1 \otimes H_1 \otimes \ldots \otimes H_1$ is represented by

$$H_n(i,j,k) = (-1)^{n+ii'+jj'+kk'+ij'+ik'}, \quad 0 \le i,j,k \le 2^n - 1.$$

9. The ninth matrix is

$$H_1 = [h(i,j,k)] = [(-1)^{j+k+ij+ik+jk}], \quad 0 \le i,j,k \le 1.$$

Its direct multiplication $H_n = H_1 \otimes H_1 \otimes \ldots \otimes H_1$ is represented by

$$H_n(i,j,k) = (-1)^{jj'+kk'+ij'+ik'+jk'}, \quad 0 \le i,j,k \le 2^n - 1.$$

10. The tenth matrix is

$$H_1 = [h(i,j,k)] = [(-1)^{1+i+j+ij+ik+jk}], \quad 0 \le i,j,k \le 1.$$

Its direct multiplication $H_n = H_1 \otimes H_1 \otimes \ldots \otimes H_1$ is represented by

$$H_n(i,j,k) = (-1)^{n+ii'+jj'+ij'+ik'+jk'}, \quad 0 \le i,j,k \le 2^n - 1.$$

11. The eleventh matrix is

$$H_1 = [h(i,j,k)] = [(-1)^{i+k+ij+ik+jk}], \quad 0 \le i,j,k \le 1.$$

Its direct multiplication $H_n = H_1 \otimes H_1 \otimes \ldots \otimes H_1$ is represented by

$$H_n(i,j,k) = (-1)^{ii'+kk'+ij'+ik'+jk'}, \quad 0 \le i,j,k \le 2^n - 1.$$

12. The twelfth matrix is

$$H_1 = [h(i,j,k)] = [(-1)^{ij+ik+jk}], \quad 0 \le i,j,k \le 1.$$

Its direct multiplication $H_n = H_1 \otimes H_1 \otimes \ldots \otimes H_1$ is represented by

$$H_n(i,j,k) = (-1)^{ij'+ik'+jk'}, \quad 0 \le i,j,k \le 2^n - 1.$$

13. The thirteenth matrix is

$$H_1 = [h(i,j,k)] = [(-1)^{i+j+ij+ik+jk}], \quad 0 \le i,j,k \le 1.$$

Its direct multiplication $H_n = H_1 \otimes H_1 \otimes \ldots \otimes H_1$ is represented by

$$H_n(i,j,k) = (-1)^{ii'+jj'+ij'+ik'+jk'}, \quad 0 \le i,j,k \le 2^n - 1.$$

14. The fourteenth matrix is

$$H_1 = [h(i,j,k)] = [(-1)^{1+i+k+ij+ik+jk}], \quad 0 \le i,j,k \le 1.$$

Its direct multiplication $H_n = H_1 \otimes H_1 \otimes \ldots \otimes H_1$ is represented by

$$H_n(i,j,k) = (-1)^{n+ii'+kk'+ij'+ik'+jk'}, \quad 0 \le i,j,k \le 2^n - 1.$$

15. The fifteenth matrix is

$$H_1 = [h(i,j,k)] = [(-1)^{1+ij+ik+jk}], \quad 0 \le i,j,k \le 1.$$

Its direct multiplication $H_n = H_1 \otimes H_1 \otimes \ldots \otimes H_1$ is represented by

$$H_n(i,j,k) = (-1)^{n+ij'+ik'+jk'}, \quad 0 \le i,j,k \le 2^n - 1.$$

16. The sixteenth matrix is

$$H_1 = [h(i,j,k)] = [(-1)^{j+k+ij+ik+jk}], \quad 0 \le i,j,k \le 1.$$

Its direct multiplication $H_n = H_1 \otimes H_1 \otimes \ldots \otimes H_1$ is represented by

$$H_n(i,j,k) = (-1)^{jj'+kk'+ij'+ik'+jk'}, \quad 0 \le i,j,k \le 2^n - 1.$$

17. The seventeenth matrix is

$$H_1 = [h(i,j,k)] = [(-1)^{i+k+ij+ik}], \quad 0 \le i,j,k \le 1.$$

Its direct multiplication $H_n = H_1 \otimes H_1 \otimes \ldots \otimes H_1$ is represented by

$$H_n(i,j,k) = (-1)^{ii'+kk'+ij'+ik'}, \quad 0 \le i,j,k \le 2^n - 1.$$

18. The eighteenth matrix is

$$H_1 = [h(i,j,k)] = [(-1)^{k+ij+jk}], \quad 0 \le i,j,k \le 1.$$

Its direct multiplication $H_n = H_1 \otimes H_1 \otimes \ldots \otimes H_1$ is represented by

$$H_n(i,j,k) = (-1)^{kk'+ij'+jk'}, \quad 0 \le i,j,k \le 2^n - 1.$$

19. The nineteenth matrix is

$$H_1 = [h(i,j,k)] = [(-1)^{j+ij+ik}], \quad 0 \le i,j,k \le 1.$$

Its direct multiplication $H_n = H_1 \otimes H_1 \otimes \ldots \otimes H_1$ is represented by

$$H_n(i,j,k) = (-1)^{jj'+ij'+ik'}, \quad 0 \le i,j,k \le 2^n - 1.$$

20. The twentieth matrix is

$$H_1 = [h(i,j,k)] = [(-1)^{1+i+j+ij+jk}], \quad 0 \le i,j,k \le 1.$$

Its direct multiplication $H_n = H_1 \otimes H_1 \otimes \ldots \otimes H_1$ is represented by

$$H_n(i,j,k) = (-1)^{n+ii'+jj'+ij'+jk'}, \quad 0 \le i,j,k \le 2^n - 1.$$

21. The 21-st matrix is

$$H_1 = [h(i,j,k)] = [(-1)^{1+j+ij+ik}], \quad 0 \le i,j,k \le 1.$$

Its direct multiplication $H_n = H_1 \otimes H_1 \otimes \ldots \otimes H_1$ is represented by

$$H_n(i,j,k) = (-1)^{n+jj'+ij'+ik'}, \quad 0 \le i,j,k \le 2^n - 1.$$

22. The 22-nd matrix is

$$H_1 = [h(i,j,k)] = [(-1)^{1+i+k+ij+ik}], \quad 0 \le i,j,k \le 1.$$

Its direct multiplication $H_n = H_1 \otimes H_1 \otimes \ldots \otimes H_1$ is represented by

$$H_n(i,j,k) = (-1)^{n+ii'+kk'+ij'+ik'}, \quad 0 \le i,j,k \le 2^n - 1.$$

23. The 23-rd matrix is

$$H_1 = [h(i,j,k)] = [(-1)^{1+j+k+ik+jk}], \quad 0 \le i,j,k \le 1.$$

Its direct multiplication $H_n = H_1 \otimes H_1 \otimes \ldots \otimes H_1$ is represented by

$$H_n(i,j,k) = (-1)^{n+jj'+kk'+ik'+jk'}, \quad 0 \le i,j,k \le 2^n - 1.$$

24. The 24-th matrix is

$$H_1 = [h(i,j,k)] = [(-1)^{1+i+ik+jk}], \quad 0 \le i,j,k \le 1.$$

Its direct multiplication $H_n = H_1 \otimes H_1 \otimes \ldots \otimes H_1$ is represented by

$$H_n(i,j,k) = (-1)^{n+ii'+ik'+jk'}, \quad 0 \le i,j,k \le 2^n - 1.$$

For more constructions and related backgrounds the readers are recommended to the papers [13-18].

Bibliography

[1] P.J. Shlichta, *Higher-Dimensional Hadamard Matrices,* IEEE Trans. On Inform. Theory, Vol.IT-25, No.5, pp.566-572, 1979.

[2] P.J. Shlichta, *Three- And Four-Dimensional Hadamard Matrices,* Bull. Amer. Phys. Soc. Ser. 11, Vol.16, pp825-826, 1971.

[3] Y.X.Yang, *Higher-Dimensional Walsh–Hadamard Transforms,* J. Beijing Univ. of Posts and Telecomm., Vol.11,No.2, pp22-30, 1988.

[4] Y.X. Yang, *On The Perfect Binary Arrays,* J. of Electronics(China), 1990; 7(2): 175-181.

[5] J.Jedwab, *Perfect Arrays, Barker Arrays And Difference Sets,* PhD Thesis, University of London, 1991.

[6] Kopilovich, *On Perfect Binary Arrays,* Electron. Lett. 1988; 24(9):566-567.

[7] J.Jedwab, C.J.Mitchell, *Infinite Families of Quasiperfect and Doubly Quasiperfect Binary Arrays,* Electron. Lett. 1990; 26(5):294-295.

[8] J.Jedwab, C.J.Mitchell, *Constructing New Perfect Binary Arrays,* Electron. Lett. 1988; 24(11):650-652.

[9] L. Bomer, M. Antweiler, *Perfect Binary Arrays With 36 Elements* , Electron. Lett., 1987, 23(9), pp730-732.

[10] L. Bomer, M. Antweiler, *Two-Dimensional Perfect Binary Arrays With 64 Elements* , IEEE Trans. On Inform. Theory, Vol.36, No.2, pp411-414.

[11] H.D. Luke, L.Bomer, and M.Antweiler, *Perfect Binary Arrays,* Signal Proc. 17(1989) 69-80.

[12] P.Wild, *Infinite Families of Perfect Binary Arrays,* Electron. Lett. 1988; 24(14):845-847.

[13] D. Calabro, J.K.Wolf, *On The Synthesis of Two-Dimensional Arrays With Desirable Correlation Properties,* Information and Control, 1968; 11: 537-560.

[14] Y.X.Yang, *On The Construction of 3-Dimensional Hadamard Matrices,* Proc. of 5-th National Conference on Pattern Recognition and Machine Intelligence, XiAn, China, 1986.

[15] W.K.Chan, M.K.Siu, *Summary Of Perfect $s \times t$ Arrays, $1 \leq s \leq t \leq 100$* , Electron. Lett. 1991; 27(9):709-710.

[16] W.K.Chan, M.K.Siu, and P.Tong, *Two-Dimensional Binary Arrays With Good Autocorrelation,* Inform. and Control, 1979; 42, 125-130.

[17] J.A.Davis, J.Jedwab, *A Summary of Menon Difference Sets,* Congresses Numerantium, 93(1993), pp203-207.

[18] J.A.Davis, J.Jedwab, *A Unifying Construction for Difference Sets,* HP Laboratories Technical Report, 1996.

[19] P.J. Shlichta, *Three- And Four-Dimensional Hadamard Matrices,* Bull. Amer. Phys. Soc. Ser. 11, Vol.16, pp825-826, 1971.

[20] P.J. Shlichta, *Higher-Dimensional Hadamard Matrices,* IEEE Trans. On Inform. Theory, Vol.IT-25, No.5, pp.566-572, 1979.

[21] Y.X.Yang, *Higher-Dimensional Walsh–Hadamard Transforms,* J. Beijing Univ. of Posts and Telecomm., Vol.11,No.2, pp22-30, 1988.

[22] Y.X. Yang, *On The Perfect Binary Arrays*, J. of Electronics(China), 1990; 7(2): 175-181.

[23] J.Jedwab, *Perfect Arrays, Barker Arrays And Difference Sets*, PhD Thesis, University of London, 1991.

[24] Kopilovich, *On Perfect Binary Arrays*, Electron. Lett. 1988; 24(9):566-567.

[25] J.Jedwab, C.J.Mitchell, *Infinite Families of Quasiperfect and Doubly Quasiperfect Binary Arrays*, Electron. Lett. 1990; 26(5):294-295.

[26] J.Jedwab, C.J.Mitchell, *Constructing New Perfect Binary Arrays*, Electron. Lett. 1988; 24(11):650-652.

[27] L. Bomer, M. Antweiler, *Perfect Binary Arrays With 36 Elements* , Electron. Lett., 1987, 23(9), pp730-732.

[28] L. Bomer, M. Antweiler, *Two-Dimensional Perfect Binary Arrays With 64 Elements* , IEEE Trans. On Inform. Theory, Vol.36, No.2, pp411-414.

[29] H.D. Luke, L.Bomer, and M.Antweiler, *Perfect Binary Arrays*, Signal Proc. 17(1989) 69-80.

[30] P.Wild, *Infinite Families of Perfect Binary Arrays*, Electron. Lett. 1988; 24(14):845-847.

[31] D. Calabro, J.K.Wolf, *On The Synthesis of Two-Dimensional Arrays With Desirable Correlation Properties*, Information and Control, 1968; 11: 537-560.

[32] Y.X.Yang, *On The Construction of 3-Dimensional Hadamard Matrices*, Proc. of 5-th National Conference on Pattern Recognition and Machine Intelligence, XiAn, China, 1986.

[33] W.K.Chan, M.K.Siu, *Summary Of Perfect $s \times t$ Arrays, $1 \leq s \leq t \leq 100$* , Electron. Lett. 1991; 27(9):709-710.

[34] W.K.Chan, M.K.Siu, and P.Tong, *Two-Dimensional Binary Arrays With Good Autocorrelation*, Inform. and Control, 1979; 42, 125-130.

[35] J.A.Davis, J.Jedwab, *A Summary of Menon Difference Sets*, Congressus Numerantium, 93(1993), pp203-207.

[36] J.A.Davis, J.Jedwab, *A Unifying Construction for Difference Sets*, HP Laboratories Technical Report, 1996.

Chapter 4
Multi-Dimensional Walsh–Hadamard Transforms

We have known from Part I of this book that orthogonal transforms based on 2-dimensional Walsh and Hadamard matrices are very useful in (1-dimensional) signal processing. In engineering practice, sometimes we have to process higher-dimensional digital signals (e.g., the 2-dimensional image signals and 3-dimensional seismic waves, etc.). Thus this chapter concentrates on the Walsh–Hadamard transforms used for higher-dimensional digital signals.

4.1 Conventional 2-Dimensional Walsh–Hadamard Transforms

4.1.1 2-Dimensional Walsh–Hadamard Transforms

For more details of this subsection the readers are recommended to [1].

An image signal can be represented by a 2-dimensional light intensity function $f(x, y)$, where x and y denote spatial coordinates and the value $f(x, y)$ at any point (x, y) is proportional to the brightness (or grey level) of the image at that point. In order to be in a form suitable for computer processing, an image function $f(x, y)$ must be digitized both spatially and in amplitude. Digitization of the spatial coordinates (x, y) is called image sampling, while amplitude digitization is called grey-level quantization. After digitization, a continuous image $f(x, y)$ is approximated by equally spaced samples arranged in the form of an $N \times N$ matrix:

$$f(x,y) \simeq \begin{bmatrix} f(0,0) & f(0,1) & \cdots & f(0, N-1) \\ f(1,0) & f(1,1) & \cdots & f(1, N-1) \\ \vdots & \vdots & \vdots & \vdots \\ f(N-1,0) & f(N-1,1) & \cdots & f(N-1, N-1) \end{bmatrix}, \qquad (4.1)$$

where each element of this matrix is a discrete quantity, which is called an image element, picture element, pixel, or pel. The right hand side of this equation represents what is commonly called a digital image. The conventional 2-dimensional Walsh–Hadamard transforms are used for the digital image processing.

Let $N = 2^n$, and W_N be the Walsh–Hadamard matrix of size $N \times N$ defined by $W_1 = [1]$ and $W_{2^{n+1}} = \begin{bmatrix} W_{2^n} & W_{2^n} \\ W_{2^n} & -W_{2^n} \end{bmatrix}$. The conventional 2-dimensional Walsh–Hadamard forward transform of the digital image $f = [f(i,j)]$, $0 \le i, j \le 2^n - 1$, is the matrix $F = [F(u,v)]$, $0 \le u, v \le 2^n - 1$, of the same size defined by[1]

$$F = W_N f W_N. \tag{4.2}$$

Because of the known identity $W_N^{-1} = 1/N W_N$, it is easy to see, from Equation (4.2), that the inverse transform is of the form

$$f = \frac{1}{N^2} W_N F W_N. \tag{4.3}$$

For example, the conventional 2-dimensional Walsh–Hadamard forward transform of the basic digital image

$$f = [f(i,j)] = \begin{bmatrix} + & + & + & + \\ - & - & + & + \\ - & - & + & + \\ + & + & - & - \end{bmatrix}$$

is

$$
\begin{aligned}
F &= [F(u,v)] \\
&= W_4 f W_4 \\
&= \begin{bmatrix} + & + & + & + \\ + & + & - & - \\ + & - & - & + \\ + & - & + & - \end{bmatrix}
\begin{bmatrix} + & + & + & + \\ - & - & + & + \\ - & - & + & + \\ + & + & - & - \end{bmatrix}
\begin{bmatrix} + & + & + & + \\ + & + & - & - \\ + & - & - & + \\ + & - & + & - \end{bmatrix} \\
&= \begin{bmatrix} 0 & 0 & 0 & 0 \\ 0 & 0 & 0 & 0 \\ 0 & 16 & 0 & 0 \\ 0 & 0 & 0 & 0 \end{bmatrix}.
\end{aligned}
$$

Recall that another equivalent definition of the Walsh–Hadamard matrix $W_{2^n} = [W(i, j)]$, $0 \le i, j \le 2^n - 1$, is

$$W(i, j) = (-1)^{ij'}, \tag{4.4}$$

where $ij' := \sum_{k=0}^{N-1} i_k j_k$ is the inner product of the two vectors $i = (i_0, \ldots, i_{n-1})$ and $j = (j_0, \ldots, j_{n-1})$, which are the binary expended vectors of the integers $i = \sum_{k=0}^{n-1} i_k 2^k$ and $j = \sum_{k=0}^{n-1} j_k 2^k$. Thus by Equations (4.4), (4.2), (4.3), the conventional 2-dimensional Walsh–Hadamard forward transform and its inverse are equivalently defined by

$$F(u, v) = \sum_{i=0}^{n-1} \sum_{j=0}^{n-1} f(i, j)(-1)^{iu' + jv'} \tag{4.5}$$

and

$$f(i, j) = \frac{1}{N^2} \sum_{u=0}^{n-1} \sum_{v=0}^{n-1} F(u, v)(-1)^{iu' + jv'}, \tag{4.6}$$

where $0 \le i, j, u, v \le 2^n - 1$.

The conventional 2-dimensional Walsh–Hadamard transforms keep most of the properties satisfied by the 1-dimensional cases, e.g., the following Parseval theorem is true

$$\sum_{u=0}^{n-1} \sum_{v=0}^{n-1} [F(u, v)]^2 = \sum_{i=0}^{n-1} \sum_{j=0}^{n-1} [f(i, j)]^2. \tag{4.7}$$

The fast algorithm for the conventional 2-dimensional Walsh–Hadamard transforms can be completed by using the fast ones for some suitable 1-dimensional signals. In fact, Equation (4.5) is equivalent to

$$F(u, v) = \sum_{i=0}^{n-1} (-1)^{iu'} \sum_{j=0}^{n-1} f(i, j)(-1)^{jv'}. \tag{4.8}$$

The inner summation in Equation (4.8) is

$$\sum_{j=0}^{2^n - 1} f(i, j)(-1)^{jv'}$$

$$= f(i, 0)(-1)^{0v'} + f(i, 1)(-1)^{1v'} + \ldots + f(i, N - 1)(-1)^{(N-1)v'}$$

$$=: B(i, v), \tag{4.9}$$

which is clearly the known 1-dimensional Walsh–Hadamard transform of the vector $(f(i,0), f(i,1), \ldots, f(i, 2^n - 1)$, the i-th row of the image $f = [f(i,j)]$. Thus the $B(i,v)$ in Equation (4.9)can be calculated by using the fast Walsh–Hadamard transforms introduced in Part I of the book. After N times of 1-dimensional fast Walsh–Hadamard algorithms, we get the following $N \times N$ matrix

$$B = [B(i,v)] = \begin{bmatrix} B(0,0) & B(0,1) & \ldots & B(0, 2^n - 1) \\ B(1,0) & B(1,1) & \ldots & B(1, 2^n - 1) \\ \vdots & \vdots & \vdots & \vdots \\ B(2^n - 1, 0) & B(2^n - 1, 1) & \ldots & B(2^n - 1, 2^n - 1) \end{bmatrix}.$$

$$(4.10)$$

Substituting of Equation (4.9)into Equation (4.8), we have

$$F(u,v) = \sum_{i=0}^{2^n - 1} B(i,v)(-1)^{iu'}, \qquad (4.11)$$

which is the 1-dimensional Walsh–Hadamard transform for the vector $(B(0,v), B(1,v), \ldots, B(2^n - 1, v))$, the v-th column of the matrix $B = [B(i,v)]$ defined in Equation (4.10).

Hence we have known that a fast algorithm for the conventional 2-dimensional Walsh–Hadamard transform can be completed by using N^2 fast algorithms for the 1-dimensional cases.

4.1.2   Definitions of 4-Dimensional Hadamard Matrices ([1])

It will be clear that the conventional 2-dimensional Walsh–Hadamard transforms introduced in the last subsection are, in fact, special cases of Walsh–Hadamard transforms based on 4-dimensional Hadamard matrices $H = [H(i,j,k,l)] = [(-1)^{ij'+kl'}]$. Thus in this preliminary subsection we state the definitions of 4- and higher-dimensional Hadamard matrices.

A 2-dimensional Hadamard matrix is defined by a binary matrix such that all of its parallel $(2 - 1)$-dimensional layers are orthogonal to each other. A 3-dimensional Hadamard matrix is defined by a binary matrix such that all of its parallel $(3 - 1)$-dimensional layers are orthogonal to each other. Thus it is natural to define a 4-dimensional Hadamard matrix as such a binary matrix that all of its parallel $(4 - 1)$-dimensional layers are mutually orthogonal ([1], [3]). In other words, $H = [H(i,j,k,l)]$, $0 \leq i,j,k,l \leq m-1$, is a four-dimensional Hadamard matrix iff $H(i,j,k,l) = \pm 1$ and

$$\sum_{p=0}^{m-1}\sum_{q=0}^{m-1}\sum_{r=0}^{m-1} H(p,q,r,a)H(p,q,r,b) \tag{4.12}$$

$$= \sum_{p=0}^{m-1}\sum_{q=0}^{m-1}\sum_{s=0}^{m-1} H(p,q,a,s)H(p,q,b,s)$$

$$= \sum_{p=0}^{m-1}\sum_{r=0}^{m-1}\sum_{s=0}^{m-1} H(p,a,r,s)H(p,b,r,s)$$

$$= \sum_{q=0}^{m-1}\sum_{r=0}^{m-1}\sum_{s=0}^{m-1} H(a,q,r,s)H(b,q,r,s)$$

$$= m^3 \delta_{ab}. \tag{4.13}$$

A four-dimensional Hadamard matrix is called 'absolutely proper', if all of its two-dimensional sections, in all possible directions (and therefore all three-dimensional sections), are themselves two-dimensional Hadamard matrices. Besides the extreme case 'absolutely proper', there are several other kinds of intermediate propriety. For example, a four-dimensional Hadamard matrix is called 'three-dimensionally proper', if all of its three-dimensional sections are proper or improper Hadamard matrices. A four-dimensional Hadamard matrix is called 'proper in two directions', if all of its three-dimensional sections in some two directions are proper or improper Hadamard matrices. ([1]).

The definition of direct multiplications can also be extended to four-dimensional cases. In fact, it will be proved that ([1], [3])

1. The direct multiplication of two four-dimensional Hadamard matrices is also a four-dimensional Hadamard matrix. Thus a four-dimensional Hadamard matrix of order 2^t can be generated by direct multiplication of the ones of order 2;

2. A four-dimensional Hadamard matrix is called 'absolutely improper' if none of its two- or three-dimensional section is a Hadamard matrix. A four-dimensional absolutely improper Hadamard matrix of order m^2 can be generated by the successive direct multiplication of four two-dimensional Hadamard matrices, of order m, which are in a set

of mutually perpendicular orientations in which each axis appears twice (e.g., the wx, xy, yz, and zw planes);

3. A four-dimensional matrix is Hadamardian whenever

   - all three-dimensional sections in one direction are three-dimensional Hadamard matrices which are mutually orthogonal, or
   - all three-dimensional sections in two directions are three-dimensional Hadamard matrices.

The first requirement can be satisfied by constructing an m^4 matrix from a set of m^3 Hadamard matrices formed by the successive cyclic m^2-layer permutations of a single m^3 Hadamard matrix. Since the latter can be formed, as shown in the last chapter, from a single m^2 Hadamard matrix, it follows that every two-dimensional Hadamard matrix implies a three-dimensional and a four-dimensional one with at least partial propriety.

The four-dimensional Hadamard matrices can be generalized, in further, to the following general higher-dimensional cases ([1], [3]):

An n-dimensional Hadamard matrix $H = [H(i_1, i_2, \ldots, i_n)]$ of order m is a (± 1)-valued matrix in which all parallel $(n-1)$-dimensional sections are mutually orthogonal; that is

$$\sum_{i_1=0}^{m-1} \cdots \sum_{i_{n-1}=0}^{m-1} H(i_1, i_2, \ldots, i_{n-1}, a) H(i_1, i_2, \ldots, i_{n-1}, b) = 2^{(n-1)} \delta_{ab}$$

$$\sum_{i_1=0}^{m-1} \cdots \sum_{i_{n-2}=0}^{1} \sum_{i_n=0}^{1} H(i_1, i_2, \ldots, i_{n-2}, a, i_n) H(i_1, i_2, \ldots, i_{n-2}, b, i_n) = 2^{(n-1)} \delta_{ab}$$

$$\vdots$$

$$\sum_{i_2=0}^{m-1} \cdots \sum_{i_n=0}^{m-1} H(a, i_2, i_3, \ldots, i_n) H(b, i_2, i_3, \ldots, i_n) = 2^{(n-1)} \delta_{ab}.$$

4.2 Algebraical Theory of Higher-Dimensional Matrices

An n-dimensional matrix of size $m_1 \times \ldots \times m_n$ can be denoted by $A = [A(i_1, \ldots, i_n)]$, where each $A(i_1, \ldots, i_n)$, $0 \le i_1 \le m_1 - 1$, $\ldots$, $0 \le i_n \le$

$m_n - 1$, is called an element of this matrix. Let r be a number. Then rA refers to the matrix

$$rA := [rA(i_1, \ldots, i_n)],$$

i.e., the product of each element with the number r.

Two matrices A and B are called 'equal to each other', denoted by $A = B$, if they have the same size and

$$A(i_1, \ldots, i_n) = B(i_1, \ldots, i_n)$$

for all possible $0 \le i_1 \le m_1 - 1, \ldots, 0 \le i_n \le m_n - 1$.

Let A and B be two n-dimensional matrices having the same size. Then their summation $C = A + B$ is another matrix defined by

$$C := A + B := [A(i_1, \ldots, i_n) + B(i_1, \ldots, i_n)],$$

i.e., the matrix formed by element-wise summation of their parent matrices A and B. Similarly, $A - B$ is defined as the matrix formed by element-wise minus of their parent matrices.

One of the most useful definitions for the higher-dimensional matrices is, possibly, the following operation called the multiplication.

Definition 4.2.1 *Let $A = [A(i_1, \ldots, i_n)]$ and $B = [B(j_1, \ldots, j_n)]$ be two n-dimensional matrices of sizes $a_1 \times a_2 \times a_n$ and $b_1 \times b_2 \times b_n$, respectively.*

1. *If $n = 2s$ is even and $(a_{s+1}, \ldots, a_n) = (b_1, \ldots, b_s)$, then the multiplication between A and B is the following n-dimensional matrix $C = [C(k_1, \ldots, k_n)] := AB$ of size $a_1 \times \ldots \times a_s \times b_{s+1} \times \ldots \times b_n$ defined by*

$$C(k_1, \ldots, k_n) = \sum_{e(1)=0}^{b_1-1} \sum_{e(2)=0}^{b_2-1} \cdots \sum_{e(s)=0}^{b_s-1} A(k_1, \ldots, k_s, e(1), \ldots e(s)).$$

$$B(e(1), \ldots, e(s), k_{s+1}, \ldots, k_n).$$

Clearly, this definition is the natural generalization of the regular multiplication between two 2-dimensional matrices.

2. *If $n = 2s + 1$ is odd and $(a_{s+1}, \ldots, a_{2s}) = (b_1, \ldots, b_s)$ and $a_n = b_n$, then the multiplication between A and B is the following n-dimensional matrix $D = [D(d_1, \ldots, d_n)] := AB$ of size $a_1 \times \ldots \times$*

$a_s \times b_{s+1} \times \ldots \times b_n$ *defined by*

$$D(d_1,\ldots,d_n) = \sum_{e(1)=0}^{b_1-1} \sum_{e(2)=0}^{b_2-1} \cdots \sum_{e(s)=0}^{b_s-1} A(d_1,\ldots,d_s,e(1),\ldots e(s),d_n).$$

$$B(e(1),\ldots,e(s),d_{s+1},\ldots,d_n).$$

With these definitions in mind, it is easy to verify the following theorem.

Theorem 4.2.1 *Let k and s be two integers. And let A, B, and C be three n-dimensional matrices such that the following operations are well defined. Then*

$$A(B+C) = AB + AC;$$
$$(B+C)A = BA + CA;$$
$$(k+s)A = kA + sA;$$
$$k(A+B) = kA + kB;$$
$$k(sA) = (ks)A;$$
$$1A = A;$$
$$k(AB) = (kA)B = A(kB).$$

Theorem 4.2.2 *Let A, B, and C be three n-dimensional matrices of suitable sizes such that the following multiplication is well defined. Then*

$$A(BC) = (AB)C.$$

Proof. We prove only the case of even $n = 2s$, because the odd n case can be proved by the same way.

Let $A = [A(i_1,\ldots,i_n)]$, $B = [B(j_1,\ldots,j_n)]$, and $C = [C(k_1,\ldots,k_n)]$. Assume that

$$V := A(BC) = [V(v_1,\ldots,v_n)] \text{ and } U := (AB)C = [U(u_1,\ldots,u_n)].$$

When the above multiplications have been well defined, the sizes of U and V are clearly equivalent to each other. By Definition 4.2.1, we have, on the one hand,

$$V(v_1,\ldots,v_n) = \sum_{e(1),\ldots,e(s)} A(v_1,\ldots,v_s,e(1),\ldots,e(s))\Big\{ \sum_{f(1),\ldots,f(s)}$$

$$B(e(1),\ldots,e(s),f(1),\ldots,f(s))C(f(1),\ldots,f(s),$$
$$v_{s+1},\ldots,v_n)\Big\}$$

$$= \sum_{e(1),\ldots,e(s),f(1),\ldots,f(s)} A(v_1,\ldots,v_s,e(1),\ldots,e(s)).$$

$$B(e(1),\ldots,e(s),f(1),\ldots,f(s)).$$

$$C(f(1),\ldots,f(s),v_{s+1},\ldots,v_n). \tag{4.14}$$

On the other hand,

$$U(u_1,\ldots,u_n) = \sum_{f(1),\ldots,f(s)} \Big\{ \sum_{e(1),\ldots,e(s)} A(u_1,\ldots,u_s,e(1),\ldots,e(s)).$$

$$B(e(1),\ldots,e(s),f(1),\ldots,f(s))\Big\}.C(f(1),\ldots,$$
$$f(s),u_{s+1},\ldots,u_n)$$

$$= \sum_{e(1),\ldots,e(s),f(1),\ldots,f(s)} A(u_1,\ldots,u_s,e(1),\ldots,e(s)).$$

$$B(e(1),\ldots,e(s),f(1),\ldots,f(s)).$$

$$C(f(1),\ldots,f(s),u_{s+1},\ldots,u_n). \tag{4.15}$$

The proof is finished by comparing the right hand sides of Equation (4.14)and Equation (4.15).   **Q.E.D.**

Definition 4.2.2 *Let* $A = [A(i_1,\ldots,i_n)]$ *be an* n-*dimensional matrix.*

1. *If* $n = 2s$ *is an even integer, then the matrix*

$$A' := [A'(j_1,\ldots,j_n)] = [A(j_{s+1},\ldots,j_n,j_1,\ldots,j_s)]$$

is called the transpose matrix of the mother matrix A;

2. *If $n = 2s + 1$ is an odd integer, then the matrix*

$$A' := [A'(j_1,\ldots,j_n)] = [A(j_{s+1},\ldots,j_{2s},j_1,\ldots,j_s,j_n)]$$

is called the transpose matrix of the mother matrix A.

Theorem 4.2.3 *Let A and B be two n-dimensional matrices, and k a number. Then the transpose operations defined by Definition 4.2.2 satisfy the following identities:*

$$(A + B)' = A' + B',$$
$$(A')' = A,$$
$$(AB)' = B'A',$$
$$(kA)' = kA',$$

where it has been assumed that all of these operations are Well defined.

Proof. We prove only the third identity $(AB)' = B'A'$, because the other identities are trivial.

Without loss of the generality, we consider the case of $n = 2s$ even (the case $n = 2s + 1$ odd can be proved by the same way).

Let $A = [A(i_1,\ldots,i_n)]$ and $B = [B(j_1,\ldots,j_n)]$ be of the sizes $a_1 \times \ldots \times a_n$ and $b_1 \times \ldots \times b_n$, respectively. First, it is clear that the two matrices:

$$(AB)' := U = [U(u_1,\ldots,u_n)] \text{ and } B'A' := V = [V(v_1,\ldots,v_n)]$$

have the same size of $b_{s+1} \times \ldots \times b_{2s} \times a_1 \times \ldots \times a_s$.

In addition, by Definition 4.2.1, we have

$$U(u_1,\ldots,u_n) = \sum_{e(1),\ldots,e(s)} A(u_{s+1},\ldots,u_n,e(1),\ldots,e(s))$$

$$\cdot B(e(1),\ldots,e(s),u_1,\ldots,u_s) \tag{4.16}$$

and

$$V(v_1,\ldots,v_n) = \sum_{e(1),\ldots,e(s)} B(e(1),\ldots,e(s),v_1,\ldots,v_s)$$

$$\cdot A(v_{s+1},\ldots,v_n,e(1),\ldots,e(s)). \tag{4.17}$$

The proof is finished by comparing Equation (4.16) with Equation (4.16). **Q.E.D.**

Definition 4.2.3 *An n-dimensional matrix I of size $I_1 \times \ldots \times I_n$ is called a unit matrix iff for every n-dimensional matrix A, the following identity is satisfied*

$$IA = AI = A$$

provided that AI and IA are well defined.

Remark: The requirement $AI = IA$ in Definition 4.2.3 implies that:

1. If $n = 2s$ is even, then the size of I should satisfy $(I_1, \ldots, I_s) = (I_{s+1}, \ldots, I_n)$;

2. If $n = 2s + 1$ is odd, then the size of I should satisfy $(I_1, \ldots, I_s) = (I_{s+1}, \ldots, I_{n-1})$.

Theorem 4.2.4 *For every given integer n and the size $I_1 \times \ldots \times I_n$ satisfying the conditions stated in the previous remark, there exists one and only one unit matrix. In fact,*

1. *If $n = 2s$ is even, then the unit matrix $I = [I(i_1, \ldots, i_n)]$ of size $I_1 \times \ldots I_s \times I_1 \times \ldots I_s$ is defined by*

$$I(i_1, \ldots, i_n) = \begin{cases} 1 \ if \ (i_1, \ldots, i_s) = (i_{s+1}, \ldots, i_n) \\ 0 \ otherwise, \end{cases}$$

2. *If $n = 2s + 1$ is odd, then the unit matrix $J = [J(j_1, \ldots, j_n)]$ of size $J_1 \times \ldots J_s \times J_1 \times \ldots J_s \times J_n$ is defined by*

$$J(j_1, \ldots, j_n) = \begin{cases} 1 \ if \ (j_1, \ldots, j_s) = (j_{s+1}, \ldots, j_{n-1}) \\ 0 \ otherwise. \end{cases}$$

This theorem can be proved by the direct verification.

For example, the matrix

$$I = [I(i, j)] = \begin{bmatrix} 1 & 0 \\ 0 & 1 \end{bmatrix}$$

is the 2-dimensional unit matrix of size 2×2. The 3-dimensional unit matrix $I = [I(i, j, k)]$ of size $2 \times 2 \times 2$ is described by

$$[I(i, j, 0)] = \begin{bmatrix} 1 & 0 \\ 0 & 1 \end{bmatrix} \ and \ [I(i, j, 1)] = \begin{bmatrix} 1 & 0 \\ 0 & 1 \end{bmatrix}$$

The 3-dimensional unit matrix $I = [I(i, j, k)]$ of size $2 \times 2 \times 3$ is described by

$$[I(i,j,0)] = \begin{bmatrix} 1 & 0 \\ 0 & 1 \end{bmatrix}, \quad [I(i,j,1)] = \begin{bmatrix} 1 & 0 \\ 0 & 1 \end{bmatrix}, \quad \text{and } [I(i,j,2)] = \begin{bmatrix} 1 & 0 \\ 0 & 1 \end{bmatrix}.$$

Definition 4.2.4 *A matrix B is called the inverse matrix of A, denoted by A^{-1}, if $BA = AB = I$, the unit matrix. An invertible matrix is called non-singular.*

It is not difficult to prove the following statements:

1. If the matrix A is non-singular, then it has a unique inverse matrix;

2. If both A and B are non-singular, then so is their multiplication matrix AB. Moreover, $(AB)^{-1} = B^{-1}A^{-1}$. (Of course, it has been assumed that these operations are well defined.)

The following theorem makes it reasonable to concentrate on the even-dimensional matrices.

Theorem 4.2.5 *Let $n = 2s + 1$ be odd, and $A = [A(i_1, \ldots, i_n)]$ a matrix of size $a_1 \times \ldots \times a_n$. Then A is non-singular if and only if its $(n-1)$-dimensional section $[B(j_1, \ldots, j_{n-1})] := [A(j_1, \ldots, j_{n-1}, i_n)]$ is non-singular for each i_n, $0 \le i_n \le a_n - 1$.*

Proof. $\Longrightarrow$: Let $C = [C(i_1, \ldots, i_n)] = A^{-1}$. Then it is easy to verify that for each $0 \le k \le a_n - 1$

$$[A(i_1, \ldots, i_{n-1}, k)]^{-1} = [C(i_1, \ldots, i_{n-1}, k)].$$

$\Longleftarrow$: Let

$$V^{(k)} = [V^{(k)}(i_1, \ldots, i_{2s})] := [A(i_1, \ldots, i_{2s}, k)]^{-1}, \quad 0 \le k \le a_n - 1.$$

Then it can be verified that $A^{-1} = V^{(a_n - 1)}$. **Q.E.D.**

Theorem 4.2.6 *Let $n = 2s > 2$ be even. The n-dimensional matrix $A = [A(i_1, \ldots, i_s, \ldots, i_n)]$ of size $a_1 \times \ldots \times a_s \times a_1 \times \ldots \times a_s$ is non-singular if the following two conditions are satisfied:*

1. *If $i_s \ne i_n$, then $A(i_1, \ldots, i_s, \ldots, i_{2s}) = 0$;*

2. *If $i_s = i_n = k$, then the following $(n-2)$-dimensional section B is non-singular, where*

$$B = [B(i_1, \ldots, i_{s-1}, i_{s+1}, \ldots, i_{n-1})]$$
$$:= [A(i_1, \ldots, i_{s-1}, k, i_{s+1}, \ldots, i_{n-1}, k)].$$

This theorem can be generalized as the following theorem. In fact, Theorem 4.2.6 is the special case of the following Theorem 4.2.7 by letting $m = 2$ and V be the 2-dimensional unit matrix.

Theorem 4.2.7 *Let $n = 2s$ and $m = 2r$ be even, $U = [U(i_1, \ldots, i_n)]$ and $V = [V(j_1, \ldots, j_m)]$ two non-singular matrices of sizes $a_1 \times \ldots \times a_s \times a_1 \times \ldots \times a_s$ and $b_1 \times \ldots \times b_r \times b_1 \times \ldots \times b_r$, respectively. Then the following $(m+n)$-dimensional matrix $H = [H(h_1, \ldots, h_{m+n})]$ of size*

$$a_1 \times \ldots \times a_s \times b_1 \times \ldots \times b_r \times a_1 \times \ldots \times a_s \times b_1 \times \ldots \times b_r$$

is non-singular too, where

$$H(h_1, \ldots, h_{m+n}) = U(h_1, \ldots, h_s, h_{s+r+1}, \ldots, h_{2s+r})$$
$$V(h_{s+1}, \ldots, h_{s+r}, h_{2s+r+1}, \ldots, h_{m+n}).$$

Proof. Let $A = [A(i_1, \ldots, i_n)] = U^{-1}$ and $B = [B(j_1, \ldots, j_m)] = V^{-1}$ be the inverse matrices of the matrices U and V, respectively. And let $F = [F(h_1, \ldots, h_{m+n})]$ be the $(m+n)$-dimensional matrix defined by

$$F(h_1, \ldots, h_{m+n}) = A(h_1, \ldots, h_s, h_{s+r+1}, \ldots, h_{2s+r})$$
$$B(h_{s+1}, \ldots, h_{s+r}, h_{2s+r+1}, \ldots, h_{m+n}).$$

Assume $HF = C = [C(c_1, \ldots, c_{m+n})]$. Then by Definition 4.2.1, we have

$$C(c_1, \ldots, c_{m+n}) = \sum_{e(1), \ldots, e(r+s)} H(c_1, \ldots, c_{r+s}, e(1), \ldots, e(r+s))$$

$$\cdot F(e(1), \ldots, e(r+s), c_{r+s+1}, \ldots, c_{m+n})$$

$$= \sum_{e(1), \ldots, e(r+s)} U(c_1, \ldots, c_s, e(1), \ldots, e(s))$$

$$\cdot V(c_{s+1}, \ldots, c_{s+r}, e(s+1), \ldots, e(s+r))$$

$$\cdot A(e(1), \ldots, e(s), c_{r+s+1}, \ldots, c_{r+2s}).$$

$$B(e(s+1), \ldots, e(s+r), c_{2s+r+1}, \ldots, c_{2s+2r})$$

$$= \left[\sum_{e(1), \ldots, e(s)} U(c_1, \ldots, c_s, e(1), \ldots, e(s)) \right.$$

$$\left. \cdot A(e(1), \ldots, e(s), c_{r+s+1}, \ldots, c_{r+2s}) \right].$$

$$\left[\sum_{e(s+1), \ldots, e(r+s)} V(c_{s+1}, \ldots, c_{s+r}, e(s+1), \ldots, e(s+r)) \right.$$

$$\left. \cdot B(e(s+1), \ldots, e(s+r), c_{2s+r+1}, \ldots, c_{2s+2r}) \right]$$

$$= \delta[(c_1, \ldots, c_s) - (c_{r+s+1}, \ldots, c_{r+2s})]$$

$$\cdot \delta[(c_{s+1}, \ldots, c_{s+r}) - (c_{r+2s+1}, \ldots, c_{2r+2s})]$$

$$= \delta[(c_1, \ldots, c_{r+s}) - (c_{r+s+1}, \ldots, c_{2r+2s})]$$

$$= \begin{cases} 1 \text{ if } (c_1, \ldots, c_{r+s}) = (c_{r+s+1}, \ldots, c_{2r+2s}) \\ 0 \text{ otherwise.} \end{cases}$$

Thus we have proved that $HF = I$, the unit matrix.

By the same way, it can be proved that $FH = I$, the unit matrix. Thus H is non-singular and its inverse matrix is F.   **Q.E.D.**

The following useful corollary follows from Theorem 4.2.7:

Corollary 4.2.1 *Let $n = 2s$ be even. And let $A_1 = [A_1(i, j)]$, $\ldots$, $A_s = [A_s(i, j)]$ be 2-dimensional non-singular matrices of sizes $n_1 \times n_1$, $\ldots$, $n_s \times n_s$, respectively. Then the following n-dimensional matrix $H = [H(i_1, \ldots, i_n)]$ of size $n_1 \times \ldots \times n_s \times n_1 \times \ldots \times n_s$ is also non-singular, where*

$$H(i_1, \ldots, i_n) := A_1(i_1, i_{s+1}) A_2(i_2, i_{s+2}) \ldots A_s(i_s, i_n).$$

Proof. Let $B_1 = [B_1(i, j)]$, $\ldots$, $B_s = [B_s(i, j)]$ be the inverse matrices of

$A_1 = [A_1(i,j)], \ldots, A_s = [A_s(i,j)]$, respectively. And let

$$F(i_1,\ldots,i_n) := B_1(i_1,i_{s+1})B_2(i_2,i_{s+2})\ldots B_s(i_s,i_n). \qquad (4.18)$$

By the same way as that used in the proof of Theorem 4.2.7 it can be proved that the matrix $F = [F(i_1,\ldots,i_n)]$ defined by Equation (4.18)is the inverse matrix of H.   **Q.E.D.**

The definitions of direct multiplications for 2- and 3-dimensional matrices have been shown to be very useful. Now we generalize this concept to the general higher-dimensional cases.

Definition 4.2.5 *Let $A = [A(i_1,\ldots,i_n)]$, $0 \le i_s \le a_s - 1$, $1 \le s \le n$, and $B = [B(j_1,\ldots,j_n)]$, $0 \le j_s \le b_s - 1$, $1 \le s \le n$, be two n-dimensional matrices of sizes $a_1 \times \ldots \times a_n$ and $b_1 \times \ldots \times b_n$, respectively. The direct multiplication of A and B, denoted by $A \otimes B$, is an n-dimensional matrix, $C = [C(k_1,\ldots,k_n)]$, $0 \le k_s \le (a_s b_s) - 1$, $1 \le s \le n$, of size $(a_1 b_1) \times \ldots \times (a_n b_n)$, defined by*

$$C(k_1,\ldots,k_n) = A\left(\left\lfloor \frac{k_1}{a_1} \right\rfloor, \ldots, \left\lfloor \frac{k_n}{a_n} \right\rfloor\right) B([k_1]_{a_1},\ldots,[k_n]_{a_n}),$$

where $\lfloor x \rfloor$ stands for the largest integer upper-bounded by x, and $[x]_m \equiv x\mathrm{mod}(m)$.

Theorem 4.2.8 *Let $A = [A(i_1,\ldots,i_n]$ and $B = [B(j_1,\ldots,j_n]$ be two n-dimensional matrices of sizes $a_1 \times \ldots \times a_n$ and $b_1 \times \ldots \times b_n$, respectively.*

1. *If both A and B are non-singular, then so is their direct multiplication. Precisely, $(B \otimes A)^{-1} = B^{-1} \otimes A^{-1}$;*

2. *$(B \otimes A)' = B' \otimes A'$.*

Proof. We prove only the first statement.

If $n = 2s$ is even, then the sizes of A^{-1} and B^{-1} are $a_{s+1} \times \ldots \times a_n \times a_1 \times \ldots \times a_s$ and $b_{s+1} \times \ldots \times b_n \times b_1 \times \ldots \times b_s$, respectively. Write

$$A^{-1} = [A^{-1}(i_{s+1},\ldots,i_n,i_1,\ldots,i_s)];$$

and

$$B^{-1} = [B^{-1}(j_{s+1},\ldots,j_n,j_1,\ldots,j_s)].$$

Thus $B^{-1} \otimes A^{-1} := [D(i_{s+1}, \ldots, i_n, i_1, \ldots, i_s)]$ is determined by

$$D\left(i_{s+1}, \ldots, i_n, i_1, \ldots, i_s\right) = A^{-1}\left([i_{s+1}]_{a_{s+1}}, \ldots, [i_n]_{a_n}, [i_1]_{a_1}, \ldots, [i_s]_{a_s}\right).$$

$$B^{-1}\left(\left\lfloor \frac{i_{s+1}}{a_{s+1}} \right\rfloor, \ldots, \left\lfloor \frac{i_n}{a_n} \right\rfloor, \left\lfloor \frac{i_1}{a_1} \right\rfloor, \ldots, \left\lfloor \frac{i_s}{a_s} \right\rfloor\right).$$

By Definition 4.2.5, we have

$$B \otimes A := [C(k_1, \ldots, k_n)] = \left[A\left([k_1]_{a_1}, \ldots, [k_n]_{a_n}\right) B\left(\left\lfloor \frac{k_1}{a_1} \right\rfloor, \ldots, \left\lfloor \frac{k_n}{a_n} \right\rfloor\right)\right].$$

It can be proved that the matrix $E = [E(e_1, \ldots, e_n)] = (B \otimes A)(B^{-1} \otimes A$
is a unit matrix. In fact,

$$E(e_1, \ldots, e_n) = \sum_{c(s+1), \ldots, c(n)} C(e_1, \ldots, e_s, c(s+1), \ldots, c(n))$$

$$\cdot D(c(s+1), \ldots, c(n), e_{s+1}, \ldots, e_n)$$

$$= \sum_{c(s+1), \ldots, c(n)} A\left([e_1]_{a_1}, \ldots, [e_s]_{a_s}, [c(s+1)]_{a_{s+1}}, \ldots, \right.$$

$$[c(n)]_{a_n}\right) B\left(\left\lfloor \frac{e_1}{a_1} \right\rfloor, \ldots, \left\lfloor \frac{e_s}{a_s} \right\rfloor, \left\lfloor \frac{c(s+1)}{a_{s+1}} \right\rfloor, \ldots, \left\lfloor \frac{c(n)}{a_n} \right\rfloor\right)$$

$$\cdot A^{-1}\left([c(s+1)]_{a_{s+1}}, \ldots, [c(n)]_{a_n}, [e_{s+1}]_{a_1}, \ldots, [e_n]_{a_s}\right)$$

$$\cdot B^{-1}\left(\left\lfloor \frac{c(s+1)}{a_{s+1}} \right\rfloor, \ldots, \left\lfloor \frac{c(n)}{a_n} \right\rfloor, \left\lfloor \frac{e_{s+1}}{a_1} \right\rfloor, \ldots, \left\lfloor \frac{e_n}{a_s} \right\rfloor\right)$$

$$= \left\{ \sum_{a(s+1), \ldots, a(n)} A\left([e_1]_{a_1}, \ldots, [e_s]_{a_s}, a(s+1), \ldots, a(n)\right) \right.$$

$$\left. \cdot A^{-1}\left(a(s+1), \ldots, a(n), [e_{s+1}]_{a_1}, \ldots, [e_n]_{a_s}\right)\right\}.$$

$$\left\{ \sum_{b(s+1), \ldots, b(n)} B\left(\left\lfloor \frac{e_1}{a_1} \right\rfloor, \ldots, \left\lfloor \frac{e_s}{a_s} \right\rfloor, b(s+1), \ldots, b(n)\right) \right.$$

$$\left. \cdot B^{-1}\left(b(s+1), \ldots, b(n), \left\lfloor \frac{e_{s+1}}{a_1} \right\rfloor, \ldots, \left\lfloor \frac{e_n}{a_s} \right\rfloor\right)\right\}$$

$$= \delta \left\{ \left([e_1]_{a_1}, \ldots, [e_s]_{a_s} \right) - \left([e_{s+1}]_{a_1}, \ldots, [e_n]_{a_s} \right) \right\}$$

$$\cdot \delta \left\{ \left(\left\lfloor \frac{e_1}{a_1} \right\rfloor, \ldots, \left\lfloor \frac{e_s}{a_s} \right\rfloor \right) - \left(\left\lfloor \frac{e_{s+1}}{a_1} \right\rfloor, \ldots, \left\lfloor \frac{e_n}{a_s} \right\rfloor \right) \right\}$$

$$= \delta \left((e_1, \ldots, e_s) - (e_{s+1}, \ldots, e_n) \right)$$

$$= \begin{cases} 1 \text{ if } (e_1, \ldots, e_s) = (e_{s+1}, \ldots, e_n) \\ 0 \text{ otherwise.} \end{cases} \tag{4.19}$$

Therefore the matrix E is a unit matrix.

By the same way, it can be proved that the matrix $(B^{-1} \otimes A^{-1})(B \otimes A)$ is also a unit matrix. Hence it is proved that $B^{-1} \otimes A^{-1} = (B \otimes A)^{-1}$.

Similarly, it can be proved that if $n = 2s + 1$ is odd, then the equation $B^{-1} \otimes A^{-1} = (B \otimes A)^{-1}$ is also satisfied.   **Q.E.D.**

Theorem 4.2.9 *The following equations are true for the direct multiplication operation:*

1. $(A \otimes B) \otimes C = A \otimes (B \otimes C)$;

2. $(A + B) \otimes (C + D) = A \otimes C + A \otimes D + B \otimes C + B \otimes D$;

3. $(A \otimes B)(C \otimes D) = (AC) \otimes (BD)$;

4. $(A_1 B_1) \otimes (A_2 B_2) \otimes \ldots \otimes (A_n B_n) = (A_1 \otimes \ldots \otimes A_n)(B_1 \otimes \ldots \otimes B_n)$;

5. $(AB)^{(k)} = A^{(k)} B^{(k)}$, *where* $X^{(k)} := \overbrace{X \otimes \ldots \otimes X}^{k}$.

Proof. We prove only the third identity, because the fourth and fifth identities can be implied by the third one, and the first two identities can be proved by direct verification.

Let

$$A = [A(a(1)), \ldots, a(n))], \quad 0 \leq a(i) \leq a_i - 1, \ 1 \leq i \leq n$$

$$B = [B(b(1)), \ldots, b(n))], \quad 0 \leq b(i) \leq b_i - 1, \ 1 \leq i \leq n$$

$$C = [C\left(c\left(1\right),\ldots,c\left(n\right)\right)], \quad 0 \le c\left(i\right) \le c_i - 1, \ 1 \le i \le n$$

and

$$D = [D\left(d\left(1\right),\ldots,d\left(n\right)\right)], \quad 0 \le d\left(i\right) \le d_i - 1, \ 1 \le i \le n.$$

If $n = 2s$ is even, then AC and BD are well defined only if $(a_{s+1},\ldots,a_n) = (c_1,\ldots,c_s)$ and $(b_{s+1},\ldots,b_n) = (d_1,\ldots,d_s)$. For $0 \le e\left(i\right) \le a_i b_i - 1$, $0 \le f\left(i\right) \le c_i d_i - 1, 1 \le i \le n$. Let

$$A \otimes B = E = [E\left(e\left(1\right),\ldots,e\left(n\right)\right)]$$

$$= \left[B\left([e\left(1\right)]_{b_1},\ldots,[e\left(n\right)]_{b_n}\right) A\left(\left\lfloor \frac{e\left(1\right)}{b_1} \right\rfloor,\ldots,\left\lfloor \frac{e\left(n\right)}{b_n} \right\rfloor\right) \right]$$

and

$$C \otimes D = F = [F\left(f\left(1\right),\ldots,f\left(n\right)\right)]$$

$$= \left[D\left([f\left(1\right)]_{d_1},\ldots,[f\left(n\right)]_{d_n}\right) C\left(\left\lfloor \frac{f\left(1\right)}{d_1} \right\rfloor,\ldots,\left\lfloor \frac{f\left(n\right)}{d_n} \right\rfloor\right) \right].$$

Hence the matrix $(A \otimes B)(C \otimes D) := G = [G\left(g\left(1\right),\ldots,g\left(n\right)\right)]$ has the size $(a_1 b_1) \times \ldots \times (a_s b_s) \times (c_{s+1} d_{s+1}) \times \ldots \times (c_n d_n)$ and its general term represented by

$$G(g(1), \quad \ldots, g(n)) = \sum_{0 \le f(i) \le c_i d_i - 1, 1 \le i \le s} E\left(g\left(1\right),\ldots,g\left(s\right),f\left(1\right),\ldots,f\left(s\right)\right)$$

$$\cdot F\left(f\left(1\right),\ldots,f\left(s\right),g\left(s+1\right),\ldots,f\left(n\right)\right)$$

$$= \sum_{0 \le f(i) \le c_i d_i - 1, 1 \le i \le s} B\left([g\left(1\right)]_{b_1},\ldots,[g\left(s\right)]_{b_s},[f\left(1\right)]_{b_{s+1}},\right.$$

$$\left.\ldots,[f\left(s\right)]_{b_n}\right) A\left(\left\lfloor \frac{g\left(1\right)}{b_1} \right\rfloor,\ldots,\left\lfloor \frac{g\left(s\right)}{b_s} \right\rfloor,\left\lfloor \frac{f\left(1\right)}{b_{s+1}} \right\rfloor,\ldots,\left\lfloor \frac{f\left(s\right)}{b_n} \right\rfloor\right)$$

$$\cdot D\left([f\left(1\right)]_{d_1},\ldots,[f\left(s\right)]_{d_s},[g\left(s+1\right)]_{d_{s+1}},\ldots,[g\left(n\right)]_{d_n}\right)$$

$$\cdot C\left(\left\lfloor \frac{f\left(1\right)}{d_1} \right\rfloor,\ldots,\left\lfloor \frac{f\left(s\right)}{d_s} \right\rfloor,\left\lfloor \frac{g\left(s+1\right)}{d_{s+1}} \right\rfloor,\ldots,\left\lfloor \frac{g\left(n\right)}{d_n} \right\rfloor\right)$$

$$= \left\{ \sum_{0 \le c(i) \le c_i-1, 1 \le i \le s} A\left(\left\lfloor \frac{g(1)}{b_1} \right\rfloor, \ldots, \left\lfloor \frac{g(s)}{b_s} \right\rfloor, c(1), \ldots, c(s) \right) \right.$$

$$\cdot C\left(c(1), \ldots, c(s), \left\lfloor \frac{g(s+1)}{d_{s+1}} \right\rfloor, \ldots, \left\lfloor \frac{g(n)}{d_n} \right\rfloor \right) \right\}$$

$$\cdot \left\{ \sum_{0 \le d(i) \le d_i-1, \ 1 \le i \le s} B\left([g(1)]_{b_1}, \ldots, [g(s)]_{b_s}, d(1), \ldots, d(s) \right) \right.$$

$$\cdot D\left(d(1), \ldots, d(s), [g(s+1)]_{d_{s+1}}, \ldots, [g(n)]_{d_n} \right) \right\}$$

Because the size of the matrix BD is $b_1 \times \ldots \times b_s \times d_{s+1} \times \ldots \times d_n$, and then by the definition of direct multiplication we have proved that $(AC) \otimes (BD) = [G(g(1), \ldots, g(n))]$. In other word, the equation $(AC) \otimes (BD) = (A \otimes B)(C \otimes D)$ has been proved.

By the same way it can be proved that the equation $(AC) \otimes (BD) = (A \otimes B)(C \otimes D)$ is also true when $n = 2s + 1$ is odd.   **Q.E.D.**

Theorem 4.2.10 *Let I and J be two unit matrices. Then $I \otimes J$ is also a unit matrix.*

Corollary 4.2.2 $I_m \otimes I_n = I_{mn}$, *where I_k stands for the n-dimensional unit matrix of size $k \times \ldots \times k$.*

The following three theorems are very useful for the fast algorithms of higher-dimensional Walsh–Hadamard transforms.

Theorem 4.2.11 *Let A be an n-dimensional matrix of size $p \times \ldots \times p$, and*

$$B = A^{(m)} := \overbrace{A \otimes \ldots \otimes A}^{m}.$$

Then the matrix B can be decomposed as the multiplication of the following m sparse matrices, precisely,

$$B = \prod_{s=0}^{m-1} \left(I_{p^s} \otimes A \otimes I_{p^{m-1-s}} \right),$$

where I_k refers to the n-dimensional unit matrix of size $k \times \ldots \times k$.

Proof. It can be proved by using induction on the integer m.

If $m = 2$, the theorem follows by applying the third identity of Theorem 4.2.9 for $B = C = I_p$ and $A = D$.

Assume that the theorem is true for $m = M - 1$.

Now it is sufficient to prove that the theorem is true for $m = M$.

$$B = \left(\overbrace{A \otimes \ldots \otimes A}^{M-1} \right) \otimes A$$

$$= \left[\prod_{s=0}^{M-2} \left(I_{p^s} \otimes A \otimes I_{p^{M-2-s}} \right) \right] \otimes A$$

(by the assumption on the case of $n = M - 1$)

$$= \left\{ \left[\prod_{s=0}^{M-2} \left(I_{p^s} \otimes A \otimes I_{p^{M-2-s}} \right) \right] \otimes I_p \right\} \left(I_{p^{M-1}} \otimes A \right)$$

(by the third identity of Theorem 4.2.9)

$$= \left[\prod_{s=0}^{M-2} \left(I_{p^s} \otimes A \otimes I_{p^{M-2-s}} \otimes I_p \right) \right] \left[I_{p^{M-1}} \otimes A \right]$$

(by the equation $(AB) \otimes I = (A \otimes I)(B \otimes I)$)

$$= \prod_{s=0}^{M-1} \left(I_{p^s} \otimes A \otimes I_{p^{M-1-s}} \right)$$

(by Corollary 4.2.2).

Thus the theorem works for the case of $m = M$.   **Q.E.D.**

Theorem 4.2.12 Let $H = [H(i,j,k,l)] = \left[(-1)^{B(i,j,k)+C(j,k,l)} \right]$, $0 \le i,j,$ $k,l \le 1$, be a 4-dimensional (± 1)-valued matrix of size $2 \times 2 \times 2 \times 2$. Then the H can be decomposed as the multiplication of two sparse matrices, i.e., $H = UV$, where $U = [U(i,j,k,l)] = \left[\delta(j,l)(-1)^{B(i,j,k)} \right]$ and $V = [V(i,j,k,l)] = \left[\delta(i,k)(-1)^{C(j,k,l)} \right]$.

Theorem 4.2.13 The following decompositions hold:

1. Let $H = [H(i,j,k,l,m,n)] = \left[(-1)^{B(i,j,k,l)+C(j,k,l,m)+D(k,l,m,n)} \right]$, $0 \le i,j,k,l,m,n \le 1$, be a 6-dimensional (± 1)-valued matrix of size $2 \times 2 \times 2 \times 2 \times 2 \times 2$. Then this matrix can be decomposed into

the multiplication of three sparse matrices, i.e., $H = UVW$, *where*

$$U = [U(i,j,k,l,m,n)] = \left[\delta((j,k) - (m,n))(-1)^{B(i,j,k,l)}\right],$$
$$V = [V(i,j,k,l,m,n)] = \left[\delta((i,k) - (l,n))(-1)^{C(j,k,l,m)}\right],$$
$$W = [W(i,j,k,l,m,n)] = \left[\delta((i,j) - (l,m))(-1)^{D(k,l,m,n)}\right].$$

2. *Let* $H = [H(i,j,k,l,m,n)] = \left[(-1)^{B(i,j,k,l,m)+C(k,l,m,n)}\right]$, $0 \leq i,j,k,l,$
   $m,n \leq 1$, *be a 6-dimensional* (± 1)*-valued matrix of size* $2 \times 2 \times 2 \times 2 \times 2 \times 2$. *Then this matrix can be decomposed into the multiplication of two sparse matrices, i.e.,* $H = UV$, *where*

$$U = [U(i,j,k,l,m,n)] = \left[\delta(k,n)(-1)^{B(i,j,k,l,m)}\right],$$
$$V = [V(i,j,k,l,m,n)] = \left[\delta((i,j) - (l,m))(-1)^{C(k,l,m,n)}\right].$$

The proofs for Theorems 4.2.12 and 4.2.12 can be finished by direct verifications. Now we are ready for the introduction of higher-dimensional Walsh–Hadamard transforms and their fast algorithms.

4.3 Multi-Dimensional Walsh–Hadamard Transforms

This section will introduce many new higher-dimensional orthogonal transforms used for the processing of 2-dimensional (e.g., images) and/or 3-dimensional (e.g. seismic waves) digital signals. These new transforms share the following advantages:

1. Their transform matrices are higher-dimensional (± 1)-valued matrices. Thus only plus and minus are required. In other words, it is easier to implement these new transforms than some other known orthogonal transforms that employ productions;

2. The forward transforms are almost the same as their inverse transforms. Thus the hardware implementation is also much easy;

3. Fast algorithms have been found;

4. The known 2-dimensional Walsh–Hadamard transforms introduced in the first section of this chapter are special cases of these new transforms;

5. The Parseval theorems and some other properties are also retained.

4.3.1   Transforms Based on 3-Dimensional Hadamard Matrices

Let $W = [W(i, j, k)]$, $0 \leq i, j, k \leq 2^n - 1$, be a 3-dimensional ($\pm 1$)-valued matrix of size $2^n \times 2^n \times 2^n$ such that

$$W^{-1} = \frac{1}{2^n} W, \qquad (4.20)$$

or equivalently, $WW = 2^n I$. A 2-dimensional digital signal $f = [f(i, j)]$, $0 \leq i, j \leq 2^n - 1$, can be treated as a 3-dimensional matrix of size $2^n \times 1 \times 2^n$. Thus

$$F = Wf \qquad (4.21)$$

and

$$f = \frac{1}{2^n} WF \qquad (4.22)$$

are a pair of orthogonal forward and inverse transforms.

Equation (4.20) can be satisfied by the following matrices:

$$W = [W(i, j, k)] = [(-1)^{ij' + ik' + jk' + an + b(ii' + jj') + ckk'}], \qquad (4.23)$$

where $0 \leq i, j, k \leq 2^n - 1$, $a, b, c \in \{0, 1\}$. Here and henceforth, we use ij' to stand for the dot product of the vectors $i = (i_0, \ldots, i_{n-1})$ and $j = (j_0, \ldots, j_{n-1})$, which are the binary expanded vectors of the integers $i = \sum_{k=0}^{n-1} i_k 2^k$ and $j = \sum_{k=0}^{n-1} j_k 2^k$, respectively.

Theorem 4.3.1 *Let $W = [W(i, j, k)]$ be the matrix in Equation 4.23, $f = [f(i, j)]$, $0 \leq i, j \leq 2^n - 1$, an image signal. Then the transforms defined by Equations (4.21) and (4.21) satisfy*

1. $F = [F(i, j)] = Wf$ *and* $f = \frac{1}{2^n} WF$;

2. *The fast algorithm is*

$$F = Wf = \prod_{s=0}^{n-1} (I_{2^s} \otimes A \otimes I_{2^{n-1-s}}), f$$

where $A = [A(p, q, r)]$, $0 \leq p, q, r \leq 1$, is a 3-dimensional matrix of size $2 \times 2 \times 2$ defined by

$$A(p, q, r) = (-1)^{pq + pr + qr + a + b(p+q) + cr}.$$

This fast algorithm can be finished by $n \times 4^n$ plus and/or minus operations;

3. *(Parseval's Theorem) If $F = [F(i,j)] = Wf$ then*

$$\sum_{i,k=0}^{2^n-1} F(i,k)^2 = 2^n \sum_{i,k=0}^{2^n-1} f(i,k)^2.$$

4. *If all elements in $f = [f(i,j)]$ are integers then the elements in $F = [F(i,j)]$ are also integers, and the elements in each column of F have the same parity.*

4.3.2 Transforms Based on 4-Dimensional Hadamard Matrices

Let $W = [W(i,j,k,l)]$, $0 \le i,j,k,l \le 2^n-1$, be a 4-dimensional ($\pm$)-valued matrix of size $2^n \times 2^n \times 2^n \times 2^n$ such that

$$W^{-1} = \frac{1}{4^n}W, \qquad (4.24)$$

or equivalently, $WW = 4^n I$. A 2-dimensional digital signal $f = [f(i,j)]$, $0 \le i,j \le 2^n - 1$, can be treated as a 4-dimensional matrix of size $2^n \times 2^n \times 1 \times 1$. Thus $F = Wf$ and $f = (1/4^n)WF$ are a pair of orthogonal forward and inverse transforms. For different transform matrices W we have the following transforms and their fast algorithms:

Class 1: The transform matrix $W = [W(i,j,k,l)]$, $0 \le i,j,k,l \le 2^n - 1$, is defined by

$$W(i,j,k,l) = (-1)^{ik'+jl'+an+b(ii'+kk')+c(jj'+ll')},$$

where $a,b,c \in \{0,1\}$;

The forward and inverse transforms are: $F = Wf$ and $f = \frac{1}{4^n}WF$, respectively.

The fast transform is:

$$F = \prod_{s=0}^{n-1}(I_{2^s} \otimes A_1 \otimes I_{2^{n-1-s}}),(I_{2^s} \otimes B_1 \otimes I_{2^{n-1-s}})f$$

where $A_1 = [A_1(p,q,r,s)]$, $B_1 = [B_1(p,q,r,s)]$, $0 \le p,q,r,s \le 1$, are defined by

$$A_1(p,q,r,s) = \delta(q,s)(-1)^{pr+a+b(p+r)}$$

and
$$B_1(p, q, r, s) = \delta(p, r)(-1)^{qs+c(q+s)}.$$

This fast algorithm needs $n.2^{2n+1}$ plus and/or minus operations.

Remark: If $a = b = c = 0$, then this transform reduces to the conventional 2-dimensional Walsh–Hadamard transform introduced in the first section of this chapter.

Class 2: The transform matrix $W = [W(i, j, k, l)]$, $0 \leq i, j, k, l \leq 2^n - 1$, is defined by

$$W(i, j, k, l) = (-1)^{ij'+ik'+jl'+kl'+an+b(ii'+kk')+c(jj'+ll')},$$

where $a, b, c \in \{0, 1\}$.

The forward and inverse transforms are: $F = Wf$ and $f = \frac{1}{4^n}WF$, respectively.

The fast transform is

$$F = \prod_{s=0}^{n-1}(I_{2^s} \otimes A_2 \otimes I_{2^{n-1-s}}), (I_{2^s} \otimes B_2 \otimes I_{2^{n-1-s}})f$$

where $A_2 = [A_2(p, q, r, s)]$, $B_2 = [B_2(p, q, r, s)]$, $0 \leq p, q, r, s \leq 1$, are defined by

$$A_2(p, q, r, s) = \delta(q, s)(-1)^{pq+pr+a+b(p+r)}$$

and

$$B_2(p, q, r, s) = \delta(p, r)(-1)^{rs+qs+c(q+s)}.$$

This fast algorithm needs $n \cdot 2^{2n+1}$ plus and/or minus operations.

Class 3: The transform matrix $W = [W(i, j, k, l)]$, $0 \leq i, j, k, l \leq 2^n - 1$, is defined by

$$W(i, j, k, l) = (-1)^{il'+kj'+an+b(ii'+kk')+c(jj'+ll')},$$

where $a, b, c \in \{0, 1\}$.

The forward and inverse transforms are: $F = Wf$ and $f = \frac{1}{4^n}WF$, respectively.

The fast transform is

$$F = \prod_{s=0}^{n-1}(I_{2^s} \otimes A_3 \otimes I_{2^{n-1-s}}), f$$

where $A_3 = [A_3(p, q, r, s)]$, $0 \leq p, q, r, s \leq 1$, is defined by

$$A_3(p, q, r, s) = (-1)^{ps+qr+a+b(p+r)+c(q+s)}.$$

This fast algorithm needs $3n \cdot 2^{2n}$ plus and/or minus operations.

Class 4: The transform matrix $W = [W(i, j, k, l)]$, $0 \leq i, j, k, l \leq 2^n - 1$, is defined by

$$W(i, j, k, l) = (-1)^{ij'+il'+kj'+kl'+an+b(ii'+kk')+c(jj'+ll')},$$

where $a, b, c \in \{0, 1\}$;

The forward and inverse transforms are: $F = Wf$ and $f = \frac{1}{4^n} WF$, respectively.

The fast transform is

$$F = \prod_{s=0}^{n-1} (I_{2^s} \otimes A_4 \otimes I_{2^{n-1-s}}), f$$

where $A_4 = [A_4(p, q, r, s)]$, $0 \leq p, q, r, s \leq 1$, is defined by

$$A_4(p, q, r, s) = (-1)^{pq+ps+qr+qs+a+b(p+r)+c(q+s)}.$$

This fast algorithm needs $3n \cdot 2^{2n}$ plus and/or minus operations.

Class 5: The transform matrix $W = [W(i, j, k, l)]$, $0 \leq i, j, k, l \leq 2^n - 1$, is defined by

$$W(i, j, k, l) = (-1)^{ik'+il'+kj'+an+b(ii'+kk')+c(jj'+ll')},$$

where $a, b, c \in \{0, 1\}$.

The forward and inverse transforms are: $F = Wf$ and $f = \frac{1}{4^n} WF$, respectively.

The fast transform is

$$F = \prod_{s=0}^{n-1} (I_{2^s} \otimes A_5 \otimes I_{2^{n-1-s}}), f$$

where $A_5 = [A_5(p, q, r, s)]$, $0 \leq p, q, r, s \leq 1$, is defined by

$$A_5(p, q, r, s) = (-1)^{pr+ps+qr+a+b(p+r)+c(q+s)}.$$

This fast algorithm needs $3n \cdot 2^{2n}$ plus and/or minus operations.

Class 6: The transform matrix $W = [W(i,j,k,l)]$, $0 \le i,j,k,l \le 2^n - 1$, is defined by

$$W(i,j,k,l) = (-1)^{ij'+ik'+il'+jk'+kl'+an+b(ii'+kk')+c(jj'+ll')},$$

where $a, b, c \in \{0, 1\}$.

The forward and inverse transforms are: $F = Wf$ and $f = \frac{1}{4^n}WF$, respectively.

The fast transform is

$$F = \prod_{s=0}^{n-1}(I_{2^s} \otimes A_6 \otimes I_{2^{n-1-s}}), f$$

where $A_6 = [A_6(p,q,r,s)]$, $0 \le p,q,r,s \le 1$, is defined by

$$A_6(p,q,r,s) = (-1)^{pq+pr+ps+qr+qs+a+b(p+r)+c(q+s)}.$$

This fast algorithm needs $3n \cdot 2^{2n}$ plus and/or minus operations.

Class 7: The transform matrix $W = [W(i,j,k,l)]$, $0 \le i,j,k,l \le 2^n - 1$, is defined by

$$W(i,j,k,l) = (-1)^{il'+kj'+jl'+an+b(ii'+kk')+c(jj'+ll')},$$

where $a, b, c \in \{0, 1\}$.

The forward and inverse transforms are: $F = Wf$ and $f = \frac{1}{4^n}WF$, respectively.

The fast transform is

$$F = \prod_{s=0}^{n-1}(I_{2^s} \otimes A_7 \otimes I_{2^{n-1-s}})f,$$

where $A_7 = [A_7(p,q,r,s)]$, $0 \le p,q,r,s \le 1$, is defined by

$$A_7(p,q,r,s) = (-1)^{ps+qr+qs+a+b(p+r)+c(q+s)}.$$

This fast algorithm needs $3n \cdot 2^{2n}$ plus and/or minus operations.

Class 8: The transform matrix $W = [W(i,j,k,l)]$, $0 \le i,j,k,l \le 2^n - 1$, is defined by

$$W(i,j,k,l) = (-1)^{ij'+il'+kj'+jl'+kl'+an+b(ii'+kk')+c(jj'+ll')},$$

where $a, b, c \in \{0, 1\}$.

The forward and inverse transforms are: $F = Wf$ and $f = \frac{1}{4^n}WF$, respectively.

The fast transform is

$$F = \prod_{s=0}^{n-1}(I_{2^s} \otimes A_8 \otimes I_{2^{n-1-s}}), f$$

where $A_8 = [A_8(p, q, r, s)]$, $0 \leq p, q, r, s \leq 1$, is defined by

$$A_8(p, q, r, s) = (-1)^{pq+ps+qr+qs+rs+a+b(p+r)+c(q+s)}.$$

This fast algorithm needs $3n \cdot 2^{2n}$ plus and/or minus operations.

Theorem 4.3.2 *All of the transforms described in this subsection satisfy the following properties:*

1. *(Parseval's Theorem) If $F = [F(i, j)] = Wf$, then*

$$\sum_{i,j=0}^{2^n-1} F(i, j)^2 = 4^n \sum_{i,k=0}^{2^n-1} f(i, k)^2.$$

2. *If all elements in $f = [f(i, j)]$ are integers, then the elements in $F = [F(i, j)]$ have the same parity.*

4.3.3 Transforms Based on 6-Dimensional Hadamard Matrices

Let $W = [W(i, j, k, p, q, r)]$, $0 \leq i, j, k, p, q, r \leq 2^n - 1$, be a 6-dimensional Hadamard matrix of size $2^n \times 2^n \times 2^n \times 2^n \times 2^n \times 2^n$ such that

$$W^{-1} = \frac{1}{8^n}W, \tag{4.25}$$

or equivalently, $WW = 8^n I$. A 3-dimensional digital signal $f = [f(i, j, k)]$, $0 \leq i, j, k \leq 2^n - 1$, can be treated as a 6-dimensional matrix of size $2^n \times 2^n \times 2^n \times 1 \times 1 \times 1$. Thus $F = Wf$ and $f = \frac{1}{8^n}WF$ produce a pair of orthogonal forward and inverse transforms. For different transform matrices W, we have the following transforms and their fast algorithms:

First Class: The transform matrix $W = [W(i, j, k, p, q, r)]$, $0 \leq i, j, k, p, q, r \leq 2^n - 1$, is defined by

$$W(i, j, k, p, q, r) = (-1)^{g(i,j,k,p,q,r)},$$

where $g(i,j,k,p,q,r) = a_1ij' + ip' + a_2jk' + jq' + kr'a_1pq' + a_2qr' + a_3n + a_4(ii' + pp') + a_5(jj' + qq') + a_6(kk' + rr')$, $a_1, a_2, a_3, a_4, a_5, a_6 \in \{0,1\}$.

The forward and inverse transforms are: $F = Wf$ and $f = \frac{1}{8^n}WF$, respectively.

The fast transform is

$$F = \prod_{s=0}^{n-1}(I_{2^s} \otimes A_1 \otimes I_{2^{n-1-s}})(I_{2^s} \otimes B_1 \otimes I_{2^{n-1-s}})(I_{2^s} \otimes C_1 \otimes I_{2^{n-1-s}})f,$$

where $A_1 = [A_1(i,j,k,p,q,r)]$, $B_1 = [B_1(i,j,k,p,q,r)]$, $C_1 = [C_1(i,j,k,p,q,r)]$, $0 \leq i,j,k,p,q,r \leq 1$, are defined by

$$A_1(i,j,k,p,q,r) = \delta((j,k) - (q,r))(-1)^{a_1ij+ip+a_3+a_4(i+p)},$$

$$B_1(i,j,k,p,q,r) = \delta((i,k) - (p,r))(-1)^{a_2jk+jq+a_5(j+q)},$$

and

$$C_1(i,j,k,p,q,r) = \delta((i,j) - (p,q))(-1)^{kr+a_1pq+a_2qr+a_5(k+r)}.$$

This fast algorithm needs $3n \cdot 2^{3n}$ plus and/or minus operations.

Second Class: The transform matrix $W = [W(i,j,k,p,q,r)]$, $0 \leq i,j,k,p,q,r \leq 2^n - 1$, is defined by

$$W(i,j,k,p,q,r) = (-1)^{g(i,j,k,p,q,r)},$$

where $g(i,j,k,p,q,r) = a_1ij' + a_2ik' + iq' + a_3jk' + jp' + kr' + a_1pq' + a_2pr' + a_3qr' + a_4n + a_5(ii' + pp') + a_6(jj' + qq') + a_7(kk' + rr')$, $a_1, a_2, a_3, a_4, a_5, a_6, a_7 \in \{0,1\}$.

The forward and inverse transforms are: $F = Wf$ and $f = \frac{1}{8^n}WF$, respectively.

The fast transform is

$$F = \prod_{s=0}^{n-1}(I_{2^s} \otimes A_2 \otimes I_{2^{n-1-s}})(I_{2^s} \otimes B_2 \otimes I_{2^{n-1-s}})f,$$

where $A_2 = [A_2(i,j,k,p,q,r)]$, $B_2 = [B_2(i,j,k,p,q,r)]$, $0 \leq i,j,k,p,q,r \leq 1$, are defined by

$$A_2(i,j,k,p,q,r) = \delta(k,r)(-1)^{a_1ij+a_2ik+iq+a_3jk+jp+a_4+a_5(i+p)+a_6(j+q)}$$

and

$$B_2(i,j,k,p,q,r) = \delta((i,j)-(p,q))(-1)^{kr+a_1pq+a_2pr+a_3qr+a_7(k+r)}.$$

This fast algorithm needs $4n \cdot 2^{3n}$ plus and/or minus operations.

Third Class: The transform matrix $W = [W(i,j,k,p,q,r)]$, $0 \le i,j,k,\ p,q,r \le 2^n - 1$, is defined by

$$W(i,j,k,p,q,r) = (-1)^{g(i,j,k,p,q,r)},$$

where $g(i,j,k,p,q,r) = a_1ij' + a_2ik + ip' + iq' + a_3jk' + jp' + kr' + a_1pq' + a_2pr' + a_3qr + a_4n + a_5(ii' + pp') + a_6(jj' + qq') + a_7(kk' + rr')$, $a_1, a_2, a_3, a_4, a_5, a_6, a_7 \in \{0,1\}$.

The forward and inverse transforms are: $F = Wf$ and $f = \frac{1}{8^n}WF$, respectively.

The fast transform is

$$F = \prod_{s=0}^{n-1}(I_{2^s} \otimes A_3 \otimes I_{2^{n-1-s}})(I_{2^s} \otimes B_3 \otimes I_{2^{n-1-s}})f,$$

where $A_3 = [A_3(i,j,k,p,q,r)]$ and $B_3 = [B_3(i,j,k,p,q,r)]$ are defined by

$$A_3(i,j,k,p,q,r)$$
$$= \delta(k,r)(-1)^{a_1ij+a_2ik+ip+iq+a_3jk+jp+a_4+a_5(i+p)+a_6(j+q)}$$

and

$$B_3(i,j,k,p,q,r) = \delta((i,j)-(p,q))(-1)^{kr+a_1pq+a_2pr+a_3qr+a_7(k+r)}.$$

This fast algorithm needs $4n \cdot 2^{3n}$ plus and/or minus operations.

Fourth Class: The transform matrix $W = [W(i,j,k,p,q,r)]$, $0 \le i,j,k,\ p,q,r \le 2^n - 1$, is defined by

$$W(i,j,k,p,q,r) = (-1)^{g(i,j,k,p,q,r)},$$

where $g(i,j,k,p,q,r)$ is one of the following 17 functions:

1. $g(i,j,k,p,q,r) = a_1ij' + a_2ik' + ir' + a_3jk' + jq' + kp' + a_1pq' + a_2pr' + a_3qr' + a_4n + a_5(ii' + pp') + a_6(jj' + qq') + a_7(kk' + rr');$

2. $g(i,j,k,p,q,r) = a_1ij' + a_2ik' + ip' + a_3jk' + jr' + kq' + a_1pq' + a_2pr' + a_3qr' + a_4n + a_5(ii' + pp') + a_6(jj' + qq') + a_7(kk' + rr');$

3. $g(i,j,k,p,q,r) = a_1ij' + a_2ik' + ip' + iq' + a_3jk' + jp' + jq' + jr' + kq' + a_1pq' + a_2pr' + a_3qr' + a_4n + a_5(ii' + pp') + a_6(jj' + qq') + a_7(kk' + rr');$

4. $g(i,j,k,p,q,r) = a_1ij' + a_2ik' + ip' + iq' + a_3jk' + jp' + jr' + kq' + a_1pq' + a_2pr' + a_3qr' + a_4n + a_5(ii' + pp') + a_6(jj' + qq') + a_7(kk' + rr');$

5. $g(i,j,k,p,q,r) = a_1ij' + a_2ik' + ip' + iq' + a_3jk' + jp' + jq' + jr' + kq' + kr' + a_1pq' + a_2pr' + a_3qr' + a_4n + a_5(ii' + pp') + a_6(jj' + qq') + a_7(kk' + rr');$

6. $g(i,j,k,p,q,r) = a_1ij' + a_2ik' + iq' + ir' + a_3jk' + jp' + jq' + jr' + kp' + kq' + a_1pq' + a_2pr' + a_3qr' + a_4n + a_5(ii' + pp') + a_6(jj' + qq') + a_7(kk' + rr');$

7. $g(i,j,k,p,q,r) = a_1ij' + a_2ik' + iq' + ir' + a_3jk' + jp' + jq' + kp' + a_1pq' + a_2pr' + a_3qr' + a_4n + a_5(ii' + pp') + a_6(jj' + qq') + a_7(kk' + rr');$

8. $g(i,j,k,p,q,r) = a_1ij' + a_2ik' + iq' + ir' + a_3jk' + jp' + jr' + kp' + kq' + kr' + a_1pq' + a_2pr' + a_3qr' + a_4n + a_5(ii' + pp') + a_6(jj' + qq') + a_7(kk' + rr');$

9. $g(i,j,k,p,q,r) = a_1ij' + a_2ik' + iq' + ir' + a_3jk' + jp' + kp' + kr' + a_1pq' + a_2pr' + a_3qr' + a_4n + a_5(ii' + pp') + a_6(jj' + qq') + a_7(kk' + rr');$

10. $g(i,j,k,p,q,r) = a_1ij' + a_2ik' + ip' + ir' + a_3jk' + jq' + jr' + kp' + kq' + kr' + a_1pq' + a_2pr' + a_3qr' + a_4n + a_5(ii' + pp') + a_6(jj' + qq') + a_7(kk' + rr');$

11. $g(i,j,k,p,q,r) = a_1ij' + a_2ik' + ip' + ir' + a_3jk' + jq' + kq' + a_1pq' + a_2pr' + a_3qr' + a_4n + a_5(ii' + pp') + a_6(jj' + qq') + a_7(kk' + rr');$

12. $g(i,j,k,p,q,r) = a_1ij' + a_2ik' + ip' + ir' + a_3jk' + jr' + kp' + kq' + kr' + a_1pq' + a_2pr' + a_3qr' + a_4n + a_5(ii' + pp') + a_6(jj' + qq') + a_7(kk' + rr');$

13. $g(i,j,k,p,q,r) = a_1ij' + a_2ik' + ip' + ir' + a_3jk' + jr' + kp' + kq' + a_1pq' + a_2pr' + a_3qr' + a_4n + a_5(ii' + pp') + a_6(jj' + qq') + a_7(kk' + rr');$

14. $g(i,j,k,p,q,r) = a_1ij' + a_2ik' + ip' + iq' + ir' + a_3jk' + jp' + jq' + kp' + kr' + a_1pq' + a_2pr' + a_3qr' + a_4n + a_5(ii' + pp') + a_6(jj' + qq') + a_7(kk' + rr');$

15. $g(i,j,k,p,q,r) = a_1ij' + a_2ik' + ip' + iq' + ir' + a_3jk' + jp' + jq' + kp' + a_1pq' + a_2pr' + a_3qr' + a_4n + a_5(ii' + pp') + a_6(jj' + qq') + a_7(kk' + rr');$

16. $g(i,j,k,p,q,r) = a_1ij' + a_2ik' + ip' + iq' + ir' + a_3jk' + jp' + jr' +$
$kp' + kq' + a_1pq' + a_2pr' + a_3qr' + a_4n + a_5(ii' + pp') + a_6(jj' + qq') +$
$a_7(kk' + rr');$

17. $g(i,j,k,p,q,r) = a_1ij' + a_2ik' + ip' + iq' + ir' + a_3jk' + jp' + kp' + kr' +$
$a_1pq' + a_2pr' + a_3qr' + a_4n + a_5(ii' + pp') + a_6(jj' + qq') + a_7(kk' + rr');$

The forward and inverse transforms are $F = Wf$ and $f = \frac{1}{8^n}WF$, respectively.

The fast transform is

$$F = \prod_{s=0}^{n-1}(I_{2^s} \otimes A \otimes I_{2^{n-1-s}})f,$$

where the matrix A is the corresponding transform matrix of size $2 \times 2 \times 2 \times 2 \times 2 \times 2$. This fast algorithm needs $7n.2^{3n}$ plus and/or minus operations.

For more details of three-dimensional Hadamard matrices the readers are recommended to the papers [4-8].

Bibliography

[1] N.Ahmed and K.R.Rao, *Orthogonal Transforms for Digital Signal Processing*, Springer-Verlag, 1975.

[2] P.J. Shlichta, *Higher-Dimensional Hadamard Matrices*, IEEE Trans. On Inform. Theory, Vol.IT-25, No.5, pp.566-572, 1979.

[3] P.J. Shlichta, *Three- And Four-Dimensional Hadamard Matrices*, Bull. Amer. Phys. Soc. Ser. 11, Vol.16, pp825-826, 1971.

[4] Y.X.Yang, *Operations and Applications of Higher Dimensional Matrices*, J. of Chengdu Institute of Radio Engineering, Vol.16, No.2, pp191-199, 1987.

[5] Y.X.Yang, *Higher-Dimensional Walsh–Hadamard Transforms*, J. of Beijing Univ. of Posts and Telecomm., Vol.11, No.2, pp22-30, 1988.

[6] D.F.Elliott and K.R.Rao, *Fast Transforms, Algorithms, Analyses, Applications*, New York, 1982.

[7] H.F.Harmuth, *Transmission of Information by Orthogonal Functions*, 2nd ed. New York: Springer-Verlag, 1972.

[8] R.C.Gonzalez and P.Wintz, *Digital Image Processing (Second Edition)*, Addison-Wesley Publishing Company, 1987.

Part III
General Higher-Dimensional Cases

Chapter 5
n-Dimensional Hadamard Matrices
of Order 2

The simplest higher-dimensional Hadamard matrices are those of order 2. Informally, an n-dimensional Hadamard matrix of order 2 (in short, a 2^n Hadamard matrix) is a binary n-cube of order 2 in which all parallel $(n - 1)$-dimensional sections are mutually orthogonal. This chapter will concentrate on the constructions, enumeration and applications of 2^n Hadamard matrices.

5.1   Constructions of 2^n Hadamard Matrices

The number of n-dimensional Hadamard matrices of order 2 is probably a substantial fraction of the number of possible binary 2^n matrices, but the latter are so numerous that for $n > 4$, exhaustive search routines are impractical. Thus the first question to be answered should be: how can one construct as many 2^n Hadamard matrices as possible?

Here are four intuitive constructions of 2^n Hadamard matrices ([1]) :

Minimal ([1]): If the population of black positions has the minimum value of $1/4$, then the only requirement for an Hadamard matrix is that black–black correlation be absent. For a 2^n matrix this means merely that black–black nearest neighbors must be avoided. Here and henceforth we treat a binary 2^n matrix as an n-cube of order 2 with the element '1' being replaced by the black position and '-1' by the white position.

Petrie Polygon ([1]): These are n-cubes in which the Petrie polygon is colored black and the rest of the matrix white, and all of the latter are the nearest neighbors to at least one black position. This method may be extendable to higher-dimensions by using a Petrie-polygon-colored matrix as a trial solution for a computer search.

Antipodal $(n-2)$-Dimensional Sections ([1]): This construction can be finished by the following two steps:

1. A pair of antipodal $(n-2)$-dimensional sections are selected; one is colored black and the other white.

2. The remaining positions are regarded as chains between the two $(n-2)$-dimensional sections; alternate ones are colored black and white.

Double Proximity Shells ([1]): A position in a 2^n matrix may be considered to be surrounded by a shell of nearest neighbors (having one intervening tie line), which in turn is surrounded by successive proximity shells until the ultimate shell, a single antipodal position, is reached. (The population of these shells is given by the binomial coefficients of order n.) Colors are assigned by giving the next two proximity shells the opposite color, and so on. This scheme results in one coloring pattern when n is even and two coloring patterns when n is odd. All of these are completely proper 2^n Hadamard matrices. Moreover, all of the lower-dimensional sections of these matrices are members of the same family.

Clearly the above four constructions cannot exhaust all 2^n Hadamard matrices. In order to produce more 2^n Hadamard matrices, we introduce, in the following subsection, the concept of H–Boolean function, which is one of the equivalent forms of 2^n Hadamard matrix.

5.1.1 Equivalence Between 2^n Hadamard Matrices and H–Boolean Functions

Definition 5.1.1 *A Boolean function $f(x_1, x_2, \ldots, x_n)$ of n variables is called an H–Boolean function if and only if the Hamming weights of the following n Boolean functions $g_1(.), g_2(.), \ldots, g_n(.)$ of $(n-1)$ variables are 2^{n-2}, where*

$$g_1(x_1, x_2, \ldots, x_{n-1}) = f(x_1, x_2, \ldots, x_{n-1}, 0) + f(x_1, x_2, \ldots, x_{n-1}, 1)$$

$$g_2(x_1, \ldots, x_{n-2}, x_n) = f(x_1, \ldots, x_{n-2}, 0, x_n) + f(x_1, \ldots, x_{n-2}, 1, x_n)$$

$$\vdots$$

$$g_n(x_2, x_3, \ldots, x_n) = f(0, x_2, \ldots, x_n) + f(1, x_2, \ldots, x_n).$$

Remark. The Hamming weight of a Boolean function means the number of '1's in the truth table of this Boolean function. The Boolean functions $g_1(.), g_2(.), \ldots, g_n(.)$ in Definition 5.1.1 must be treated as Boolean functions of $(n-1)$ but not of n variables.

On the other hand, the 2^n Hadamard matrix is defined by:

Definition 5.1.2 *An n-dimensional Hadamard matrix $H = [H(i_1, i_2, \ldots, i_n)]$ of order 2 is a binary matrix in which all parallel $(n-1)$-dimensional sections are mutually orthogonal; that is, all $H(i_1, i_2, \ldots, i_n) = -1$ or 1, $i_1, i_2, \ldots, i_n = 0$ or 1, and*

$$\sum_{i_1=0}^{1} \sum_{i_2=0}^{1} \cdots \sum_{i_{n-1}=0}^{1} H(i_1, i_2, \ldots, i_{n-1}, a) H(i_1, i_2, \ldots, i_{n-1}, b) = 2^{(n-1)} \delta_{ab}$$

$$\sum_{i_1=0}^{1} \cdots \sum_{i_{n-2}=0}^{1} \sum_{i_n=0}^{1} H(i_1, i_2, \ldots, i_{n-2}, a, i_n) H(i_1, i_2, \ldots, i_{n-2}, b, i_n) = 2^{(n-1)} \delta_{ab}$$

$$\vdots$$

$$\sum_{i_2=0}^{1} \sum_{i_3=0}^{1} \cdots \sum_{i_n=0}^{1} H(a, i_2, i_3, \ldots, i_n) H(b, i_2, i_3, \ldots, i_n) = 2^{(n-1)} \delta_{ab}.$$

Comparing the above Definition 5.1.1 and Definition 5.1.2, the equivalent relationship between H–Boolean functions and 2^n Hadamard matrices becomes straightforward, i.e.:

Theorem 5.1.1 *A Boolean function $f(x_1, x_2, \ldots, x_n)$ is an H–Boolean function of n variables if and only if the n-dimensional binary matrix $[H(i_1, i_2, \ldots, i_n)]$, $i_1, i_2, \ldots, i_n = 0$ or 1, defined by*

$$H(i_1, i_2, \ldots, i_n) = (-1)^{f(i_1, i_2, \ldots, i_n)}$$

is a 2^n Hadamard matrix.

It is Theorem 5.1.1 that motivates the study of H–Boolean functions in the following subsections.

5.1.2   Existence of H–Boolean Functions

This subsection states some basic properties and necessary and/or sufficient conditions of H–Boolean functions.

At first, it is not difficult to verify that the following Boolean functions
are all H–Boolean:

$$f_1(x_1,\ldots,x_n) = \sum_{1\leq i<j\leq n} x_i x_j,$$

$$f_2(x_1,\ldots,x_n) = x_1 \sum_{k=2}^{n} x_k + \prod_{k=m}^{n} x_k, \quad (2\leq m\leq n)$$

$$f_3(x_1,\ldots,x_n) = x_1 \sum_{k=2}^{n} x_k + x_1 x_2 \ldots x_{n-1} + x_2 x_3 \ldots x_n,$$

$$f_4(x_1,\ldots,x_n) = \sum_{1\leq i<j\leq n} x_i x_j + x_2 x_3 \ldots x_n,$$

$$f_5(x_1,\ldots,x_n) = x_1(x_2 + \ldots + x_n) + x_2(x_3 + \ldots + x_n) + x_1 x_3 x_4 \ldots x_n,$$

$$f_6(x_1,\ldots,x_n) = x_1(x_2 + \ldots + x_n) + x_3 \ldots x_n + x_2 \ldots x_n + x_1 x_3 x_4 \ldots x_n.$$

These examples imply that for each n, there exists at least one n-
dimensional Hadamard matrix of order 2.

Definition 5.1.3 *Two Boolean functions $f(x_1, x_2, \ldots, x_n)$ and $g(x_1, x_2,$
$\ldots, x_n)$ are said to be equivalent to each other if and only if there is a permutation $\tau(.)$ of the set $\{1, 2, \ldots, n\}$ and a binary vector $(a_1, a_2, \ldots, a_n)$,
$a_i = 0$ or 1 for $1 \leq i \leq n$, such that*

$$f(x_1, x_2, \ldots, x_n) = g(\tau(x_1 + a_1, x_2 + a_2, \ldots, x_n + a_n)),$$

where $\tau(y_1, y_2, \ldots, y_n) = (y_{\tau(1)}, y_{\tau(2)}, \ldots, y_{\tau(n)})$.

From the definition of H–Boolean functions and Definition 5.1.3, one
can prove the following theorem.

Theorem 5.1.2 *The Boolean function equivalent to an H–Boolean function is also an H–Boolean function.*

Proof. Equivalent Boolean functions have the same Hamming weight.
Q.E.D.

Lemma 5.1.1 *Let $f(x_1, x_2, \ldots, x_n)$ and $g(x_1, x_2, \ldots, x_n)$ be two Boolean functions. Then*

$$w(f(.) + g(.)) = w(f(.)) + w(g(.)) - 2w(f(.)g(.)),$$

where $w(h(.))$ refers to the Hamming weight of the Boolean function $h(x_1, x_2, \ldots, x_n)$.

Proof. This lemma is, in fact, a direct corollary of the well known identity $(a + b) \mod 2 = a + b - 2ab$, for $a, b = 0$ or 1.   **Q.E.D.**

Lemma 5.1.2 *The following are true:*

1. *When all of the following Boolean functions are treated as of n variables, we have*

$$w(x_1) = 2^{n-1};$$
$$w(x_1 x_2) = 2^{n-2};$$
$$\vdots$$
$$w(x_1 x_2 \ldots x_k) = 2^{n-k};$$
$$\vdots$$
$$w(x_1 x_2 \ldots x_n) = 1.$$

2. *If $f(x_1, x_2, \ldots, x_n)$ is a Boolean function of odd Hamming weight, then there is a Boolean function $g(x_1, x_2, \ldots, x_n)$ of even Hamming weight such that*

$$f(x_1, x_2, \ldots, x_n) = g(x_1, x_2, \ldots, x_n) + x_1 x_2 \ldots x_n.$$

This lemma can be directly verified by the definition of Hamming weight.

Theorem 5.1.3 *The Hamming weight of every H–Boolean function $f(x_1, x_2, \ldots, x_n)$, $n \geq 3$, is even.*

Proof. If, on the contrary, $f(x_1, x_2, \ldots, x_n)$ is an H–Boolean function of odd Hamming weight, then there is a Boolean function $g(x_1, x_2, \ldots, x_n)$ of even Hamming weight such that

$$f(x_1, x_2, \ldots, x_n) = g(x_1, x_2, \ldots, x_n) + x_1 x_2 \ldots x_n.$$

Thus

$$g_1(x_1, x_2, \ldots, x_{n-1}) = f(x_1, x_2, \ldots, x_{n-1}, 0) + f(x_1, x_2, \ldots, x_{n-1}, 1)$$
$$= [g(x_1, x_2, \ldots, x_{n-1}, 0) + g(x_1, x_2, \ldots, x_{n-1}, 1)]$$
$$+ x_1 x_2 \ldots x_{n-1}$$
$$= r(x_1, x_2, \ldots, x_{n-1}) + x_1 x_2 \ldots x_{n-1}.$$

By Lemma 5.1.2, the Hamming weight of this Boolean function of $(n-1)$ variables must be odd, which is different from 2^{n-2}, $n \geq 3$. In other words, $f(x_1, x_2, \ldots, x_n)$ cannot be H–Boolean. **Q.E.D.**

Every Boolean function is a linear combination of the functions x_1; $x_1 x_2$; $\ldots$; $x_1 x_2 \ldots x_k$; $\ldots$; $x_1 x_2 \ldots x_n$ and their equivalent forms. Formally, every Boolean function $f(x_1, x_2, \ldots, x_n)$ can be denoted by its polynomial form as

$$f(x_1, x_2, \ldots, x_n) = a + \sum_{i=1}^{n} b(i)x_i + \sum_{1 \leq i < j \leq n} c(i,j)x_i x_j$$
$$+ \sum_{1 \leq i < j < k \leq n} d(i,j,k)x_i x_j x_k + \cdots,$$

in which every term 'x_i' is said to be of degree one, '$x_i x_j$' of degree two, and '$x_i x_j x_k$' of degree three, and so on, where $b(i)$, $c(i,j)$, and $d(i,j,k)$, etc., are zero or one.

Lemma 5.1.3 *Let $f(x_1, x_2, \ldots, x_n)$ be a Boolean function consisting of the sum of K terms of degree $n-1$. Then its Hamming weight $w(f)$ is*

$$w(f) = 2\lfloor (K+1)/2 \rfloor,$$

where $\lfloor x \rfloor$ refers to the floor function, i.e., the largest integer up to x, e.g., $\lfloor 1/2 \rfloor = 0$ and $\lfloor 3/2 \rfloor = 1$.

Proof. It can be proved by using induction on the integer K.

Case $K = 1$: The Hamming weight of the single term

$$f(x_1, x_2, \ldots, x_n) = x_1 x_2 \ldots x_{n-1}$$

is equal to 2, i.e., the lemma is right if $K = 1$.

Case $K = 2$: Let

$$f(x_1, x_2, \ldots, x_n) = g(x_1, x_2, \ldots, x_n) + h(x_1, x_2, \ldots, x_n)$$

be the sum of two terms of degree $n-1$. Then by Lemma 5.1.1 and Lemma 5.1.2, the Hamming weight of $f(.)$ is $w(f) = w(g) + w(h) - 2w(g.h) = 2 + 2 - 2 \times 1 = 2$. In other words, the lemma is right if $K = 2$.

In general, suppose that the lemma is correct for the cases of $K = 2m - 1$ and $K = 2m$. It is sufficient to prove the correctness of the cases of $K = 2m + 1$ and $K = 2m + 2$.

For $K = 2m + 1$ let

$$f(x_1, x_2, \ldots, x_n) = g_1(x_1, x_2, \ldots, x_n) + g_2(x_1, x_2, \ldots, x_n) \ldots$$
$$+ g_{2m}(x_1, x_2, \ldots, x_n) + g_{2m+1}(x_1, x_2, \ldots, x_n),$$

where $g_i(x_1, x_2, \ldots, x_n)$, $1 \le i \le 2m + 1$, are different single-terms of degree $n - 1$. Hence, for each $i \ne j$,

$$g_i(x_1, x_2, \ldots, x_n)g_j(x_1, x_2, \ldots, x_n) = x_1 x_2 \ldots x_n.$$

Therefore,

$$
\begin{aligned}
w(f) &= w([g_1 + \ldots + g_{2m}] + g_{2m+1}) \\
&= w(g_1 + \ldots + g_{2m}) + w(g_{2m+1}) - 2w([g_1 + \ldots + g_{2m}]g_{2m+1}) \\
&= 2\lfloor (2m + 1)/2 \rfloor + 2 \\
&= 2m + 2 \\
&= 2\lfloor [(2m + 1) + 1]/2 \rfloor,
\end{aligned}
$$

where the second equation is due to the facts of $[g_1 + \ldots + g_{2m}]g_{2m+1} = 0$ and the assumption about the case of $K = 2m$. Up to now, it is clear that the lemma is correct for the case of $K = 2m + 1$.

For $K = 2m + 2$, let

$$f(x_1, x_2, \ldots, x_n) = g_1(x_1, x_2, \ldots, x_n) + g_2(x_1, x_2, \ldots, x_n) \ldots$$
$$+ g_{2m+1}(x_1, x_2, \ldots, x_n) + g_{2m+2}(x_1, x_2, \ldots, x_n),$$

where $g_i(x_1, x_2, \ldots, x_n)$, $1 \le i \le 2m + 2$, are different single-terms of degree $n - 1$. Therefore,

$$w(f) = w([g_1 + \ldots + g_{2m+1}] + g_{2m+2})$$

$$= w(g_1 + \ldots + g_{2m+1}) + w(g_{2m+2}) - 2w([g_1 + \ldots + g_{2m+1}]g_{2m+2})$$
$$= 2(m+1) + 2 - 2$$
$$= 2(m+1)$$
$$= 2\lfloor [(2m+2)+1]/2 \rfloor,$$

where the second equation is owed to the property $[g_1 + \ldots + g_{2m+1}]g_{2m+2} = x_1 x_2 \ldots x_n$ and the assumption about the case of $K = 2m+1$. It is clear that the lemma is also correct for the case of $K = 2m + 2$. **Q.E.D.**

One can easily verify that $f(x_1, x_2, x_3) = x_1 x_2 + x_1 x_3$ is an H–Boolean function of 3-variables consisting of 2 terms of degree 2; and that

$$g(x_1, x_2, x_3, x_4) = x_1 x_2 x_3 + x_1 x_2 x_4 + x_1 x_3 x_4 + x_2 x_3 x_4$$

is an H–Boolean function of 4-variables consisting of 3 terms of degree 3. Motivated by these two examples, it is naturally to ask that, except for these two examples, are there any other H–Boolean functions of n-variable ($n > 4$) consisting of terms of degree just $n - 1$? The following theorem gives us a negative answer to this problem.

Theorem 5.1.4 *No H–Boolean functions of n-variable consist of terms of degree $n - 1$ if $n > 4$.*

Proof. On the contrary, if $f(x_1, x_2, \ldots, x_n)$ is an H–Boolean function consisting of k terms of degree $(n-1)$, then the following Boolean function

$$g(x_1, x_2, \ldots, x_{n-1}) = f(x_1, x_2, \ldots, x_{n-1}, 0) + f(x_1, x_2, \ldots, x_{n-1}, 1)$$

of $(n-1)$-variable consists of no more than k terms of degree $(n-2)$. Hence, by Lemma 5.1.3 the Hamming weight of $g(x_1, x_2, \ldots, x_{n-1})$ is bounded above by

$$w(g) = 2\lfloor (k+1)/2 \rfloor \le 2\lfloor (n+1)/2 \rfloor < 2^{n-2}, \quad if \quad n > 4.$$

Therefore, by Definition 5.1.1, $f(x_1, x_2, \ldots, x_n)$ is not an H–Boolean function. **Q.E.D.**

In order to introduce more non-existence results, we prove the following popular lemma.

Lemma 5.1.4 *If* $f(x_1, x_2, \ldots, x_n)$ *and* $g(x_1, x_2, \ldots, x_n)$ *are two different Boolean functions, then*

$$w(f(x_1, x_2, \ldots, x_n) + g(x_1, x_2, \ldots, x_n)) \leq w(f(x_1, x_2, \ldots, x_n))$$
$$+ w(g(x_1, x_2, \ldots, x_n)).$$

The equation is satisfied if and only if

$$f(x_1, x_2, \ldots, x_n)g(x_1, x_2, \ldots, x_n) = 0.$$

Theorem 5.1.5 *For* $n \geq 9$ *no H–Boolean functions of n-variable consist of terms of degrees* $(n-1)$ *or* $(n-2)$.

Proof. Let $f(x_1, x_2, \ldots, x_n)$ be a Boolean function consisting of terms of degrees $(n-1)$ or $(n-2)$. Consider the following Boolean function:

$$g(x_1, x_2, \ldots, x_{n-1}) = f(x_1, x_2, \ldots, x_{n-1}, 0) + f(x_1, x_2, \ldots, x_{n-1}, 1),$$

which consists of terms of degrees $(n-2)$ or $(n-3)$.

Because of that, among the set of $(n-1)$-variable's single-terms, there are at most $(n-1)$ terms of degree $(n-2)$ and $(n-1)(n-2)/2$ terms of degree $(n-3)$. On the other hand, the Hamming weights of each $(n-1)$-variable's single-term of degree $(n-1)$ and $(n-2)$ are 2 and 4, respectively. Therefore, by Lemma 5.1.4 the Hamming weight of $g(x_1, x_2, \ldots, x_{n-1})$ is upper-bounded by

$$4 \times (n-1)(n-2)/2 + 2 \times (n-1) < 2^{n-2}, \quad for \ \ n \geq 9.$$

Thus by Definition 5.1.1, the function $f(x_1, x_2, \ldots, x_n)$ is not H–Boolean function. **Q.E.D.**

By the same approach as that used in Theorem 5.1.5, we have the following theorems (their proofs are all omitted for saving space) :

Theorem 5.1.6 *For* $n \geq 13$, *no H–Boolean functions of n-variable consist of terms of degrees just* $(n-1)$, $(n-2)$, *or* $(n-3)$.

Theorem 5.1.7 *For* $n \geq 18$, *no H–Boolean functions of n-variable consist of terms of degrees just* $(n-1)$, $(n-2)$, $(n-3)$ *or* $(n-4)$.

Definition 5.1.4 *The minimum degree of a Boolean function is defined by the minimum value of the degrees of the single-terms contained in this Boolean function. For example, the minimum degree of*

$$f(x_1, x_2, \ldots, x_n) = x_1 + x_1 x_2$$

is 1, while the minimum degree of

$$f(x_1, x_2, \ldots, x_n) = x_1 x_3 + x_2 x_4 + x_5 x_6 x_7$$

is 2.

The previous theorems 5.1.4, 5.1.5, 5.1.6, and 5.1.7 can be generalized as:

Theorem 5.1.8 *For every given integer k, there exists an integer N such that no H–Boolean functions of n-variable are of minimum degree $(n-k)$ for $n \geq N$*

Proof. From Lemma 5.1.4, we know that the Hamming weights of n-variable's Boolean functions of minimum degree $(n-k)$ are upper bounded by $\sum_{i=1}^{k} 2^i n!/[i!(n-i)!]$, whilst, on the other hand,

$$\sum_{i=1}^{k} 2^i n!/[i!(n-i)!] \leq 2^{n-1}$$

for sufficiently large n. The theorem follows by Definition 5.1.1.   **Q.E.D.**

Definition 5.1.5 *A Boolean function $f(x_1, x_2, \ldots, x_n)$ is said to be independent of some variable, say x_i, if and only if the equation*

$$f(x_1, \ldots, x_{i-1}, 0, x_{i+1}, \ldots, x_n) = f(x_1, \ldots, x_{i-1}, 1, x_{i+1}, \ldots, x_n)$$

is satisfied by all $x_1, \ldots, x_{i-1}, x_{i+1}, \ldots, x_n = 0$ or 1, otherwise the function is said to be dependent on x_i.

A Boolean function $f(x_1, x_2, \ldots, x_n)$ is said to be linear in some variable, say x_i, if and only if the function $f(x_1, x_2, \ldots, x_n) + x_i$ is independent of x_i.

Theorem 5.1.9 *The following two results are true:*

1. *H–Boolean functions of n-variable are dependent on every variable.*

2. *No H–Boolean functions are linear in any variable.*

Proof. If the Boolean function $f(x_1, x_2, \ldots, x_n)$ is independent of x_i, then

$$f(x_1, x_2, \ldots, x_{i-1}, 0, x_{i+1}, \ldots, x_n) + f(x_1, x_2, \ldots, x_{i-1}, 1, x_{i+1}, \ldots, x_n) = 0$$

i.e., the function is not H–Boolean.

The second statement can be proved by the same way.   **Q.E.D.**

The other equivalent definition of H–Boolean function is stated by

Theorem 5.1.10 *The necessary and sufficient conditions for $f(x_1, x_2, \ldots, x_n)$ being an H–Boolean function of n-variable is that*

$$2^{n-2} + 2w(f(x_1, x_2, \ldots, x_{i-1}, 0, x_{i+1}, \ldots, x_n)$$
$$f(x_1, x_2, \ldots, x_{i-1}, 1, x_{i+1}, \ldots, x_n))$$
$$= w(f(x_1, x_2, \ldots, x_n)),$$

for each $1 \leq i \leq n$.

Proof. The proof is finished by the following equivalent equations:

$$w(f\ (x_1, x_2, \ldots, x_{i-1}, 0, x_{i+1}, \ldots, x_n)$$
$$+ f(x_1, x_2, \ldots, x_{i-1}, 1, x_{i+1}, \ldots, x_n)) = 2^{n-2}$$

is equivalent to

$$2^{n-2} = w(f(x_1, x_2, \ldots, x_{i-1}, 0, x_{i+1}, \ldots, x_n))$$
$$+ w(f(x_1, x_2, \ldots, x_{i-1}, 1, x_{i+1}, \ldots, x_n))$$
$$- 2w(f(x_1, x_2, \ldots, x_{i-1}, 0, x_{i+1}, \ldots,$$
$$x_n)f(x_1, x_2, \ldots, x_{i-1}, 1, x_{i+1}, \ldots, x_n))$$
$$= w(f(x_1, x_2, \ldots, x_n))$$
$$- 2w(f(x_1, x_2, \ldots, x_{i-1}, 0, x_{i+1}, \ldots,$$
$$x_n)f(x_1, x_2, \ldots, x_{i-1}, 1, x_{i+1}, \ldots, x_n)).$$

The last equation is due to the identity

$$w(f(x_1, x_2, \ldots, x_{i-1}, 0, x_{i+1}, \ldots, x_n)) + w(f(x_1, x_2, \ldots, x_{i-1}, 1,$$
$$x_{i+1}, \ldots, x_n)) = w(f(x_1, x_2, \ldots, x_n)).$$

Q.E.D.

Theorem 5.1.11 *Let*

$$f(x_1, x_2, \ldots, x_n) = x_{i_1} x_{i_2} \ldots x_{i_r} + x_{j_1} x_{j_2} \ldots x_{j_s}$$

be a Boolean function consisting of two terms. Then the following statements are true:

- *$f(.)$ is not H–Boolean if $n = 1$ or $n = 2$;*

- *For $n = 3$, $f(.)$ is H–Boolean if and only if $f(x_1, x_2, x_3)$ is equivalent to $x_1 x_2 + x_1 x_3$;*

- *For $n = 4$, $f(.)$ is H–Boolean if and only if $f(x_1, x_2, x_3, x_4)$ is equivalent to $x_1 x_2 + x_3 x_4$;*

- *For $n \geq 5$, $f(.)$ cannot be H–Boolean.*

Proof. The statements (1), (2) and (3) are immediately from Theorems 5.1.2, 5.1.9, and 5.1.16. The proof for the fourth statement is divided into the following three steps:

Step 1: First, we prove that a necessary condition of $f(.)$ being H–Boolean is $r \geq 3$.

In fact, if $r = 1$, by Theorems 5.1.2 and 5.1.9 it is reasonable to assume that

$$f(x_1, x_2, \ldots, x_n) = x_1 + x_1^a x_2 x_3 \ldots x_n, \tag{5.1}$$

where $a = 0$ or 1 and $x_1^1 := x_1$ and $x_1^0 := 1$.

By Theorem 5.1.9, the function $f(.)$ in Equation (5.1) is not H–Boolean for $a = 0$. Moreover, by Theorem 5.1.16, this $f(.)$ is not H–Boolean for $a = 1$ and $n \geq 5$ too. Thus $r > 1$.

If $r = 2$,

$$f(x_1, x_2, \ldots, x_n) = x_1 x_2 + x_1^a x_2^b x_3 \ldots x_n. \tag{5.2}$$

By the same way as the case of $r = 1$, it can be proved that the $f(.)$ in Equation (5.2) is not H–Boolean for $r = 2$.

Therefore we have $r \geq 3$.

Step 2: Then, in the same way as in Step 1, it can be proved that $s \geq 3$.

Step 3: Finally, we prove that $f(.)$ is not an H–Boolean function even if $r \geq 3$ and $s \geq 3$.

In fact, by Theorem 5.1.9, we obtain the following necessary condition for $f(.)$ being H–Boolean

$$x_{i_1} x_{i_2} \ldots x_{i_r} x_{j_1} x_{j_2} \ldots x_{j_s} = x_1 x_2 \ldots x_n.$$

Thus the Hamming weight of $f(.)$ is

$$w(f(.)) = 2^{n-r} + 2^{n-s} - 2.$$

By Theorem 5.1.10, another necessary condition for for $f(.)$ being H–Boolean is that

$$2w(f(0, x_2, \ldots, x_n)f(1, x_2, \ldots, x_n)) = 2^{n-r} + 2^{n-s} - 2 - 2^{n-2}.$$

Because the conditions $r \geq 3$ and $s \geq 3$, we have

$$2^{n-r} + 2^{n-s} - 2 - 2^{n-2} < 0$$

or equivalently

$$w(f(0, x_2, \ldots, x_n)f(1, x_2, \ldots, x_n)) < 0,$$

which is clearly impossible.

In one word, for $n \geq 5$, the function $f(.)$ is not H–Boolean. **Q.E.D.**

An n-dimensional Hadamard matrix of order m is called absolutely improper if and only if none of its subsections is an Hadamard matrix of lower-dimension. In the coming chapter many absolutely improper Hadamard matrices will be constructed for different n and m. The following theorem proves that no 2^n-Hadamard matrix is absolutely improper when $n \geq 3$.

Theorem 5.1.12 *There exist no absolutely improper n-dimensional Hadamard matrix of order 2, if $n \geq 3$.*

Proof. Every binary n-dimensional matrix of order 2 can be denoted by $A = [A(x_1, x_2, \ldots, x_n)]$, where

$$A(x_1, x_2, \ldots, x_n) = (-1)^{f(x_1, x_2, \ldots, x_n)}$$

$$0 \leq x_1, x_2, \ldots, x_n \leq 1.$$

Case 1: If the function $f(x_1, x_2, \ldots, x_n)$ is linear, then, by Theorem 5.1.9, the matrix A is not Hadamard.

Case 2: If the function $f(.)$ is not linear, without loss of generality,

$$x_1 x_2 x_{i_1} \ldots x_{i_k},$$

say, is a term of this function, then

$$f(x_1, x_2, d_3, d_4, \ldots, d_n) = x_1 x_2 + ax_1 + bx_2 + c, \qquad (5.3)$$

where $a, b, c = 0$ or 1, and $d_j = 1$ iff $j \in \{i_1, i_2, \ldots, i_k\}$, otherwise $d_j = 0$.

The function in Equation (5.3) is clearly an H–Boolean function of 2-variables, i.e., at least one 2-dimensional subsection of the matrix A is Hadamard, which is equivalent to saying that A is not an absolutely improper n-dimensional Hadamard matrix of order 2.

The proof is completed by the above two cases.   **Q.E.D.**

Opposite to the concept of absolutely improper, an n-dimensional Hadamard matrix of order m is called absolutely proper if and only if all of its subsections are Hadamard matrices of lower-dimensional ones. The following theorem lists all of the possible n-dimensional absolutely proper Hadamard matrices of order 2.

Theorem 5.1.13 *Let* $A = [A(x_1, x_2, \ldots, x_n)]$, $0 \le x_i \le 1$, $A(x_1, x_2, \ldots, x_n) = 1$ *or* -1. *Then the matrix* A *is an* n-*dimensional absolutely proper Hadamard matrix of order 2 if and only if there is a Boolean function* $f(x_1, x_2, \ldots, x_n)$ *of the following form*

$$f(x_1, x_2, \ldots, x_n) = \sum_{1 \le i < j \le n} x_i x_j + a_0 + \sum_{i=1}^{n} a_i x_i$$

satisfying

$$A(x_1, x_2, \ldots, x_n) = (-1)^{f(x_1, x_2, \ldots, x_n)},$$

where $a_i = 0$ *or* 1.

Proof. $\Longleftarrow$: The matrix defined by

$$A(x_1, x_2, \ldots, x_n) = (-1)^{f(x_1, x_2, \ldots, x_n)}$$

with

$$f(x_1, x_2, \ldots, x_n) = \sum_{1 \le i < j \le n} x_i x_j + a_0 + \sum_{i=1}^{n} a_i x_i$$

is clearly an n-dimensional absolutely proper Hadamard matrix of order 2. In fact, each of its lower subsection is of the form $[(-1)^{h(x_1,x_2,\ldots,x_m)}]$, $1 \leq m \leq n$, with

$$h(x_1, x_2, \ldots, x_m) = \sum_{1 \leq i < j \leq m} x_i x_j + b_0 + \sum_{i=1}^{m} b_i x_I,$$

which is clearly an H–Boolean function.

$\Longrightarrow$: Assume that the Boolean function $g(x_1, x_2, \ldots, x_n)$ satisfies

$$A(x_1, x_2, \ldots, x_n) = (-1)^{g(x_1,x_2,\ldots,x_n)}.$$

Now we try to prove that if $[A]$ is an n-dimensional absolutely proper Hadamard matrix of order 2, then the function $g(.)$ must be of the form of

$$g(x_1, x_2, \ldots, x_n) = \sum_{1 \leq i < j \leq n} x_i x_j + a_0 + \sum_{i=1}^{n} a_i x_i.$$

For any prefixed $1 \leq i < j \leq n$ and x_k, $(k \neq i, k \neq j)$, the function $g(.)$ can be divided into

$$\begin{aligned}
g(x_1, x_2, \ldots, x_n) = {} & x_i E(x_1, \ldots, x_{i-1}, x_{i+1}, \ldots, x_{j-1}, x_{j+1}, \ldots, x_n) \\
& + x_j B(x_1, \ldots, x_{i-1}, x_{i+1}, \ldots, x_{j-1}, x_{j+1}, \ldots, x_n) \\
& + x_i x_j C(x_1, \ldots, x_{i-1}, x_{i+1}, \ldots, x_{j-1}, x_{j+1}, \ldots, x_n) \\
& + D(x_1, \ldots, x_{i-1}, x_{i+1}, \ldots, x_{j-1}, x_{j+1}, \ldots, x_n),
\end{aligned}$$

where $E(.)$, $B(.)$, $C(.)$ and $D(.)$ are Boolean functions of $(n-2)$-variable independent of x_i and x_j.

Let $r(x_i, x_j) = g(x_1, \ldots, x_i, \ldots, x_j, \ldots, x_n)$, which is a Boolean function of 2-variable x_i and x_j.

Because the Hadamard matrix A is absolutely proper, the matrix

$$[F(x_i, x_j)] = [(-1)^{r(x_i,x_j)}]$$

should be a 2-dimensional Hadamard matrix of order 2. Thus

$$\sum_{x_j=0}^{1} (-1)^{r(0,x_j)+r(1,x_j)} = 0,$$

i.e., $r(0,0) + r(0,1) + r(1,0) + r(1,1) \equiv 1\bmod 2$. Hence

$$C(x_1, \ldots, x_{i-1}, x_{i+1}, \ldots, x_{j-1}, x_{j+1}, \ldots, x_n) = 1.$$

Therefore $g(.)$ is of the form

$$
\begin{aligned}
g(x_1, x_2, \ldots, x_n) = {} & x_i E(x_1, \ldots, x_{i-1}, x_{i+1}, \ldots, x_{j-1}, x_{j+1}, \ldots, x_n) \\
& + x_j B(x_1, \ldots, x_{i-1}, x_{i+1}, \ldots, x_{j-1}, x_{j+1}, \ldots, x_n) \\
& + x_i x_j \\
& + D(x_1, \ldots, x_{i-1}, x_{i+1}, \ldots, x_{j-1}, x_{j+1}, \ldots, x_n).
\end{aligned}
$$

Because of that the above equation is true for any $1 \le i < j \le n$, the function $g(.)$ must be the form of

$$g(x_1, x_2, \ldots, x_n) = \sum_{1 \le i < j \le n} x_i x_j + a_0 + \sum_{i=1}^{n} a_i x_i.$$

Q.E.D.

5.1.3 Constructions of H–Boolean Functions

We have known that the construction of H–Boolean functions is in fact the construction of 2^n-Hadamard matrices. This subsection shows some powerful such constructions.

First, it is easy to verify that the following result is true:

Theorem 5.1.14 *If $f(x_1, x_2, \ldots, x_n)$ is an H–Boolean function, then so is*

$$f(x_1, x_2, \ldots, x_n) + a_0 + \sum_{i=1}^{n} a_i x_i.$$

Because of this theorem and Theorem 5.1.4, we have:

Corollary 5.1.1 *No H–Boolean functions are of the form*

$$f(x_1, x_2, \ldots, x_n) = a + \sum_{i=1}^{n} a_i x_i + \sum_{k=1}^{n} b_k x_1 x_2 \ldots x_{k-1} x_{k+1} \ldots x_n$$

if $n > 4$.

The following construction makes it possible to produce many H–Boolean functions of large m-variable from those of small m-variable ones, which can be found by computer search.

Theorem 5.1.15 *If $f(x_1, x_2, \ldots, x_n)$ and $g(y_1, y_2, \ldots, y_m)$ are H–Boolean functions of variables n and m, respectively, then*

$$r(x_1, x_2, \ldots, x_n, y_1, y_2, \ldots, y_m) = f(x_1, x_2, \ldots, x_n) + g(y_1, y_2, \ldots, y_m)$$

is an H–Boolean function of $(m + n)$ variables.

Proof. It can be proved directly by using the definition of H–Boolean functions.   **Q.E.D.**

Theorem 5.1.16 *Let $f(x_1, x_2, \ldots, x_n, y_1, y_2, \ldots, y_m) = h(x_1, x_2, \ldots, x_n) \cdot r(y_1, y_2, \ldots, y_m)$, where $h(.)$ and $r(.)$ are Boolean functions of n- and m-variable, respectively. Then $f(.)$ is H–Boolean function of $(m + n)$-variable if and only if $h(x_1, x_2, \ldots, x_n) = a + x_1 + x_2 + \ldots + x_n$ and $r(y_1, y_2, \ldots, y_m) = b + y_1 + y_2 + \ldots + y_m$.*

Proof. $\Longleftarrow$: It is easy to verify that the function

$$[a + x_1 + x_2 + \ldots + x_n][b + y_1 + y_2 + \ldots + y_m]$$

is indeed H–Boolean.

$\Longrightarrow$: At first we prove that both $h(.)$ and $r(.)$ should be linear functions, if $f(.)$ is H–Boolean.

In fact, one necessary condition for $f(.)$ being H–Boolean is that

$$\begin{aligned}
2^{m+n-2} &= w(f(0, x_2, \ldots, x_n, y_1, y_2, \ldots, y_m) \\
&\quad + f(1, x_2, \ldots, x_n, y_1, y_2, \ldots, y_m)) \\
&= w(r(.))w(h(0, x_2, \ldots, x_n) + h(1, x_2, \ldots, x_n)).
\end{aligned}$$

Therefore there should exist integers, say p and q, such that

$$w(r(y_1, y_2, \ldots, y_m)) = 2^p$$

and

$$w(h(0, x_2, \ldots, x_n) + h(1, x_2, \ldots, x_n)) = 2^q.$$

Because no H–Boolean functions are linear, we have $p < m$. And because of that H–Boolean functions are dependent on each of their variables, we have $q < n$. Hence the identity $m + n - 2 = p + q$ implies $q = n - 1$, in other words, $h(0, x_2, \ldots, x_n) + h(1, x_2, \ldots, x_n) \equiv 1$, i.e., the function $h(.)$ is linear in variable x_1.

By the same way it can be proved that $h(.)$ (and $r(.)$) should be linear in every variable.

Then because H–Boolean functions should dependent on every variable, $h(.)$ and $r(.)$ are of the forms:

$$h(x_1, x_2, \ldots, x_n) = a + x_1 + x_2 + \ldots + x_n$$

and

$$r(y_1, y_2, \ldots, y_m) = b + y_1 + y_2 + \ldots + y_m.$$

Q.E.D.

This theorem provides us numerous H–Boolean functions consisting of terms of degree 2.

Theorem 5.1.17 ([1], [2], [3]) *A Boolean function $f(x_1, x_2, \ldots, x_n)$ is H–Boolean if the following three conditions are satisfied:*

C1: $f(x_1, x_2, \ldots, x_{n-2}, 1, 1) = 1$ *and* $f(x_1, x_2, \ldots, x_{n-2}, 0, 0) = 0$;

C2: *The Hamming weights of*

$$h_1(x_1, x_2, \ldots, x_{n-2}) := f(x_1, x_2, \ldots, x_{n-2}, 0, 1)$$

and

$$h_2(x_1, x_2, \ldots, x_{n-2}) := f(x_1, x_2, \ldots, x_{n-2}, 1, 0)$$

are 2^{n-3}.

C3: *For every $(a_1, a_2, \ldots, a_{n-2})$, if $h_i(a_1, a_2, \ldots, a_{n-2}) = 1$ (or resp. 0), $i=1$ or 2, then*

$$h_i(1 - a_1, a_2, \ldots, a_{n-2}) = h_i(a_1, 1 - a_2, \ldots, a_{n-2})$$

$$= \ldots$$

$$= h_i(a_1, a_2, \ldots, 1 - a_{n-2})$$

$$= 0 \quad \text{(or, resp., 1).}$$

Proof. The proof is finished by the following steps:

Step 1: At first we prove that the Hamming weight of

$$A(x_2, x_3, \ldots, x_n) := f(0, x_2, x_3, \ldots, x_n) + f(1, x_2, x_3, \ldots, x_n)$$

is 2^{n-2}. In fact the known conditions imply that

$$A(x_2, \ldots, x_{n-2}, 1, 1) = f(0, x_2, \ldots, x_{n-2}, 1, 1) + f(1, x_2, \ldots, x_{n-2}, 1, 1)$$
$$= 0 + 0 = 0,$$

$$A(x_2, \ldots, x_{n-2}, 0, 0) = f(0, x_2, \ldots, x_{n-2}, 0, 0) + f(1, x_2, \ldots, x_{n-2}, 0, 0)$$
$$= 1 + 1 = 0,$$

$$A(x_2, \ldots, x_{n-2}, 0, 1) = f(0, x_2, \ldots, x_{n-2}, 0, 1) + f(1, x_2, \ldots, x_{n-2}, 0, 1)$$
$$= h_1(0, x_2, \ldots, x_{n-2}) + h_1(1, x_2, \ldots, x_{n-2})$$
$$= 0 + 1 \text{ (or } 1 + 0)$$
$$= 1,$$

$$A(x_2, \ldots, x_{n-2}, 1, 0) = f(0, x_2, \ldots, x_{n-2}, 1, 0) + f(1, x_2, \ldots, x_{n-2}, 1, 0)$$
$$= h_2(0, x_2, \ldots, x_{n-2}) + h_2(1, x_2, \ldots, x_{n-2})$$
$$= 0 + 1 \text{ (or } 1 + 0)$$
$$= 1.$$

The above four equations imply that the Hamming weight of

$$A(x_2, x_3, \ldots, x_n)$$

is indeed 2^{n-2}. In fact, $A(x_2, x_3, \ldots, x_n) = 1$ is satisfied by the points of the form $(x_2, x_3, \ldots, x_{n-2}, 0, 1)$ or $(x_2, x_3, \ldots, x_{n-2}, 1, 0)$.

Because the variables $x_1, x_2, \ldots, x_{n-2}$ are replaceable by each other, the function $A(x_2, x_3, \ldots, x_n)$ has the same Hamming weight as those of the following $n - 2$ Boolean functions of $(n - 1)$-variable:

$$f(x_1, 0, x_3, \ldots, x_n) + f(x_1, 1, x_3, \ldots, x_n)$$

$$f(x_1, x_2, 0, x_4, \ldots, x_n) + f(x_1, x_2, 1, x_4, \ldots, x_n)$$

$$\vdots$$

$$f(x_1, x_2, \ldots, x_{n-3}, 0, x_{n-1}, x_n) + f(x_1, x_2, \ldots, x_{n-3}, 1, x_{n-1}, x_n).$$

Step 2: Then, we try to prove that the Hamming weight of

$$B(x_1, \ldots, x_{n-2}, x_n) := f(x_1, \ldots, x_{n-2}, 0, x_n) + f(x_1, \ldots, x_{n-2}, 1, x_n)$$

is also 2^{n-2}.

In fact, for the Hamming weight of $h_2(x_1, x_2, \ldots, x_{n-2})$ is 2^{n-3}, the function

$$B(x_1, \ldots, x_{n-2}, 0) = f(x_1, \ldots, x_{n-2}, 0, 0) + f(x_1, \ldots, x_{n-2}, 1, 0)$$
$$= 1 + h_2(x_1, x_2, \ldots, x_{n-2})$$

is equal to 1 at 2^{n-3} points of the form $(x_1, \ldots, x_{n-2}, 0)$.

Similarly, because the Hamming weight of $h_1(x_1, x_2, \ldots, x_{n-2})$ is 2^{n-3}, the function

$$B(x_1, \ldots, x_{n-2}, 1) = f(x_1, \ldots, x_{n-2}, 0, 1) + f(x_1, \ldots, x_{n-2}, 1, 1)$$
$$= h_1(x_1, x_2, \ldots, x_{n-2}) + 0$$
$$= h_1(x_1, x_2, \ldots, x_{n-2})$$

is equal to 1 at 2^{n-3} points of the form $(x_1, \ldots, x_{n-2}, 1)$.

Therefore, the Hamming weight of $B(x_1, \ldots, x_{n-2}, x_n)$ is proved to be $2^{n-3} + 2^{n-3} = 2^{n-2}$.

Step 3: In the same way as in Step 2, it can be proved that the Hamming weight of

$$C(x_1, \ldots, x_{n-1}) := f(x_1, \ldots, x_{n-1}, 0) + f(x_1, \ldots, x_{n-1}, 1)$$

is also 2^{n-2}.

By the definition of H–Boolean functions and the above three steps, it is clear that $f(x_1, x_2 \ldots, x_n)$ is an H–Boolean function of n-variables. **Q.E.D.**

Remark. The conditions in Theorem 5.1.17 can be stated in many equivalent forms. For example, they can be replaced by the following three new conditions:

C'1: $f(1, 0, x_3, x_4, \ldots, x_n) = 1$ and $f(1, 1, x_3, x_4, \ldots, x_n) = 0$;

C'2: The Hamming weights of

$$R_1(x_3, x_4, \ldots, x_n) := f(0, 0, x_3, x_4, \ldots, x_n)$$

and

$$R_2(x_3, x_4, \ldots, x_n) := f(0, 1, x_3, x_4, \ldots, x_n)$$

are 2^{n-3};

C'3: For every $(a_3, a_4, \ldots, a_n)$, if $R_i(a_3, a_4, \ldots, a_n) = 1$ (or, resp., 0), i=1 or 2, then

$$R_i(1 - a_3, a_4, \ldots, a_n) = R_i(a_3, 1 - a_4, \ldots, a_n)$$
$$= \ldots$$
$$= R_i(a_3, a_4, \ldots, 1 - a_n)$$
$$= 0 \quad \text{(or, resp., 1)}.$$

Theorem 5.1.18 ([1], [2], [3]) *A Boolean function $f(x_1, x_2, \ldots, x_n)$ is H–Boolean if the following conditions are satisfied:*

CC1: *The Hamming weight of $f(x_1, x_2, \ldots, x_n)$ itself is 2^{n-2};*

CC2: *For every $(a_1, a_2, \ldots, a_n)$ if $f(a_1, a_2, \ldots, a_n) = 1$, then*

$$f(1 - a_1, a_2, \ldots, a_n) = f(a_1, 1 - a_2, \ldots, a_n)$$
$$= \ldots$$
$$= f(a_1, a_2, \ldots, 1 - a_n).$$

Proof. It is sufficient to prove that the Hamming weight of

$$g(x_1, x_2, \ldots, x_{n-1}) := f(x_1, x_2, \ldots, x_{n-1}, 0) + f(x_1, x_2, \ldots, x_{n-1}, 1)$$

is 2^{n-2}.

In fact, group the binary vectors of length n into 2^n groups, say $X_1, X_2,$ $\ldots, X_{2^{n-1}}$, such that each X_i consists of just two vectors and $(a_1, a_2, \ldots, a_{n-1}, a_n)$ and $(b_1, b_2, \ldots, b_{n-1}, b_n)$ belong to the same group if and only if $(a_1, a_2, \ldots, a_{n-1}) = (b_1, b_2, \ldots, b_{n-1})$.

From the condition CC2, at most one of the two points in each X_i satisfies $f(.) = 1$. While, on the other hand, the condition CC1 ensures that just one of the two points in each X_i satisfies $f(.) = 1$.

If $f(a_1, a_2, \ldots, a_n) = 1$, then, by the condition CC2, $f(a_1, a_2, \ldots, 1 - a_n) = 0$. Thus

$$g(a_1, a_2, \ldots, a_{n-1}) = f(a_1, a_2, \ldots, 1 - a_n) + f(a_1, a_2, \ldots, a_n)$$
$$= 1 + 0$$
$$= 1.$$

In other words, the Hamming weight of $g(x_1, x_2, \ldots, x_{n-1})$ is larger than or equal to that of $f(x_1, x_2, \ldots, x_n)$.

If $g(a_1, a_2, \ldots, a_{n-1}) = 1$, i.e.,

$$f(a_1, \ldots, a_{n-1}, 1) + f(a_1, \ldots, a_{n-1}, 0) = 1,$$

then one and only one of $f(a_1, \ldots, a_{n-1}, 1)$ and $f(a_1, \ldots, a_{n-1}, 0)$ is "1". Therefore, the Hamming weight of $g(x_1, \ldots, x_{n-1})$ is less than or equal to that of $f(x_1, \ldots, x_n)$.

It has been proved that the Hamming weight of $g(x_1, x_2, \ldots, x_{n-1})$ is equal to 2^{n-2}, the Hamming weight of $f(x_1, x_2, \ldots, x_n)$.   **Q.E.D.**

5.2 Enumeration of 2^n Hadamard Matrices

The enumeration for n-dimensional Hadamard matrices of order 2 is a very difficult problem, which is still open up to now. This section will show only some enumeration about small n and special H–Boolean functions.

5.2.1 Classification of 2^4 Hadamard Matrices

It is easy to see that the numbers of 2^2 and 2^3 Hadamard matrices are 8 and 64, respectively.

According to the definition of H–Boolean function, a 4-variable's Boolean function $B(i, j, k, l)$ is H–Boolean if and only if the following four conditions are simultaneously satisfied

C1: $w(B(0, j, k, l) + B(1, j, k, l)) = 4;$

C2: $w(B(i, 0, k, l) + B(i, 1, k, l)) = 4;$

C3: $w(B(i, j, 0, l) + B(i, j, 1, l)) = 4;$

C4: $w(B(i, j, k, 0) + B(i, j, k, 1)) = 4;$

where $w(g(.))$ refers to the Hamming weight of $g(.)$.

In order to enumerate the 2^4 Hadamard matrices we divide the Boolean functions of the form

$$f(i,j,k,l) = b_1 ij + b_2 ik + b_3 il + b_4 kl + b_5 jk + b_6 jl, \quad b_h = 0 \text{ or } 1,$$

into ten equivalence classes with their representative functions being:

1. $f_1(i,j,k,l) = ij;$

2. $f_2(i,j,k,l) = ij + ik;$

3. $f_3(i,j,k,l) = ij + kl;$

4. $f_4(i,j,k,l) = ij + ik + il;$

5. $f_5(i,j,k,l) = ij + ik + kj;$

6. $f_6(i,j,k,l) = ij + ik + jl;$

7. $f_7(i,j,k,l) = ij + ik + il + jk;$

8. $f_8(i,j,k,l) = ij + ik + jl + lk + ik;$

9. $f_9(i,j,k,l) = ij + ik + il + jk + jl;$

10. $f_{10}(i,j,k,l) = ij + ik + jl + jk + il + kl;$

respectively, where two functions $f(i,j,k,l)$ and $g(i,j,k,l)$ belong to the same equivalence class if and only if there exists a permutation of 4 elements, say $\tau(.)$, such that $f(i,j,k,l) = g(\tau(i,j,k,l))$.

The general form of 4-variable's Boolean functions without linear terms is

$$B(i,j,k,l) = b_1 ij + b_2 ik + b_3 il + b_4 jk + b_5 jl + b_6 kl$$
$$+ c_1 ijk + c_2 jkl + c_3 ikl + c_4 jkl + d_1 ijkl.$$

If the number of H–Boolean functions of this form is N, then there are $2^5 N$ H–Boolean functions of 4-variables, because there are 2^5 4-variable's linear terms at all and $B(i,j,k,l)$ is H–Boolean iff so is $B(i,j,k,l) + a_0 + a_1 i + a_2 j + a_3 k + a_4 l$.

Based on the above representative functions of equivalence classes, we find that every 4-variable's H–Boolean function without linear terms is equivalent to one of the following 11 types:

Type 1: $B_1(i,j,k,l) = c_1ijk + c_2jkl + c_3ikl + c_4jkl + d_1ijkl;$

Type 2: $B_2(i,j,k,l) = f_1(i,j,k,l) + B_1(i,j,k,l);$

Type 3: $B_3(i,j,k,l) = f_2(i,j,k,l) + B_1(i,j,k,l);$

Type 4: $B_4(i,j,k,l) = f_3(i,j,k,l) + B_1(i,j,k,l);$

Type 5: $B_5(i,j,k,l) = f_4(i,j,k,l) + B_1(i,j,k,l);$

Type 6: $B_6(i,j,k,l) = f_5(i,j,k,l) + B_1(i,j,k,l);$

Type 7: $B_7(i,j,k,l) = f_6(i,j,k,l) + B_1(i,j,k,l);$

Type 8: $B_8(i,j,k,l) = f_7(i,j,k,l) + B_1(i,j,k,l);$

Type 9: $B_9(i,j,k,l) = f_8(i,j,k,l) + B_1(i,j,k,l);$

Type 10: $B_{10}(i,j,k,l) = f_9(i,j,k,l) + B_1(i,j,k,l);$

Type 11: $B_{11}(i,j,k,l) = f_{11}(i,j,k,l) + B_1(i,j,k,l).$

Now we try to list those non-equivalent H–Boolean functions of different types.

About Type 1: A Boolean function of Type 1 is H–Boolean iff

$$\begin{cases} w(c_3kl + c_1jk + c_2jl + d_1jkl) = 4 & \ldots\ldots(A) \\ w(c_4kl + c_1ik + c_2il + d_1ikl) = 4 & \ldots\ldots(B) \\ w(c_1ij + c_3il + c_4jl + d_1ijl) = 4 & \ldots\ldots(C) \\ w(c_2ij + c_3ik + c_4jk + d_1ijk) = 4 & \ldots\ldots(D). \end{cases}$$

The left side of the equation (A) is

$$2c_3 + w(c_1jk + c_2jl + d_1jkl) - 2c_3(c_1 + c_2 + d_1)\mathrm{mod}2.$$

The left side of the equation (B) is

$$2c_4 + w(c_1ik + c_2il + d_1ikl) - 2c_4(c_1 + c_2 + d_1)\mathrm{mod}2.$$

$(A) - (B)$ implies

$$(c_3 - c_4)[1 - (c_1 + c_2 + d_1)\mathrm{mod}2] = 0 \quad \ldots\ldots(E)$$

and similarly, $(C) - (D)$ implies

$$(c_1 - c_2)[1 - (c_3 + c_4 + d_1)\bmod 2] = 0 \quad \dots\dots (F).$$

The solution of equations (A) to (F) is $c_1 = c_2 = c_3 = c_4 = 1$ and $d_1 = 0$. Hence there is only one H–Boolean function in Type 1 which is denoted by

$$A_1(i, j, k, l) = ijk + ijl + jkl + ikl.$$

To simplify the mathematical expressions, we introduce the following notations:

$$_1B_m(j, k, l) = B_m(0, j, k, l) + B_m(1, j, k, l);$$
$$_2B_m(i, k, l) = B_m(i, 0, k, l) + B_m(i, 1, k, l);$$
$$_3B_m(i, j, l) = B_m(i, j, 0, l) + B_m(i, j, 1, l);$$
$$_4B_m(i, k, k) = B_m(i, j, k, 0) + B_m(i, j, k, 1).$$

About Type 2: Because the simultaneous equations

$$\begin{cases} w[_1B_2(j, k, l)] = 4 \\ w[_2B_2(i, k, l)] = 4 \\ w[_3B_2(i, j, l)] = 4 \\ w[_4B_2(i, j, k)] = 4 \end{cases}$$

have no solutions, no H–Boolean functions are of the form of Type 2.

About Type 3: Because the simultaneous equations

$$\begin{cases} w[_3B_3(i, j, l)] = 4 \\ w[_1B_3(j, k, l)] = 4 \\ w[_2B_3(i, k, l)] = 4 \\ w[_4B_3(i, j, k)] = 4 \end{cases}$$

have no solutions, H–Boolean functions are of the form of Type 3.

About Type 4: Because the simultaneous equations

$$\begin{cases} w[_1B_4(j, k, l)] = 4 \\ w[_2B_4(i, k, l)] = 4 \\ w[_3B_4(i, j, l)] = 4 \\ w[_4B_4(i, j, k)] = 4 \end{cases}$$

have only one solution $c_1 = c_2 = c_3 = c_4 = d_1 = 0$, there is only one H–Boolean function in Type 4. This H–Boolean is

$$A_2(i, j, k, l) = ij + kl.$$

About Type 5: Because the simultaneous equations

$$\begin{cases} w[_1B_5(j, k, l)] = 4 \\ w[_2B_5(i, k, l)] = 4 \\ w[_3B_5(i, j, l)] = 4 \\ w[_4B_5(i, j, k)] = 4 \end{cases}$$

have five solutions

1. $c_1 = c_2 = c_3 = c_4 = d_1 = 0$,

2. $c_1 = c_2 = c_3 = 0$, $c_4 = 1$, $d_1 = 0$,

3. $c_1 = 1$, $c_2 = c_3 = 0$, $c_4 = 1$, $d_1 = 0$,

4. $c_1 = 0$, $c_2 = 1$, $c_3 = 0$, $c_4 = 1$, $d_1 = 0$,

5. $c_1 = c_2 = 0$, $c_3 = c_4 = 1$, $d_1 = 0$,

while the H–Boolean functions produced by the last three solutions are equivalent to each other, there are three non-equivalent H–Boolean functions in Type 5. They are:

$$A_3(i, j, k, l) = ij + ik + il,$$
$$A_4(i, j, k, l) = ij + ik + il + jkl,$$
$$A_5(i, j, k, l) = ij + ik + il + ijk + jkl.$$

About Type 6: Since the simultaneous equations

$$\begin{cases} w[_1B_6(j, k, l)] = 4 \\ w[_2B_6(i, k, l)] = 4 \\ w[_3B_6(i, j, l)] = 4 \\ w[_4B_6(i, j, k)] = 4 \end{cases}$$

have one solution $c_1 = c_2 = c_3 = c_4 = 1$ and $d_1 = 0$, there is one H–Boolean function of Type 6 which is:

$$A_6(i, j, k, l) = ij + ik + kj + ijk + ijl + ikl + jkl.$$

About Type 7: The simultaneous equations

$$\begin{cases} w[_1 B_7(j, k, l)] = 4 \\ w[_2 B_7(i, k, l)] = 4 \\ w[_3 B_7(i, j, l)] = 4 \\ w[_4 B_7(i, j, k)] = 4 \end{cases}$$

have two solutions: (1) $c_1 = c_2 = c_3 = c_4 = d_1 = 0$; (2) $c_1 = c_2 = 0$, $c_3 = c_4 = 1$, $d_1 = 0$. They correspond to the following two non-equivalent H–Boolean functions:

$$A_7(i, j, k, l) = ij + ik + jl,$$
$$A_8(i, j, k, l) = ij + ik + jl + ikl + jkl.$$

About Type 8: The simultaneous equations

$$\begin{cases} w[_1 B_8(j, k, l)] = 4 \\ w[_2 B_8(i, k, l)] = 4 \\ w[_3 B_8(i, j, l)] = 4 \\ w[_4 B_8(i, j, k)] = 4 \end{cases}$$

have three solutions:

1. $c_1 = 1$, $c_2 = c_3 = 0$, $c_4 = 1$, and $d_1 = 0$,

2. $c_1 = c_2 = c_3 = c_4 = d_1 = 0$,

3. $c_1 = c_2 = c_3 = 0$, $c_4 = 1$, $d_1 = 0$,

which correspond to the following three non-equivalent H–Boolean functions:

$$A_9(i, j, k, l) = ij + ik + il + jk + jkl,$$
$$A_{10}(i, j, k, l) = ij + ik + il + jk + ijk + jkl,$$
$$A_{11}(i, j, k, l) = ij + ik + il + ik.$$

About Type 9: The simultaneous equations

$$\begin{cases} w[_1B_9(j,k,l)] = 4 \\ w[_2B_9(i,k,l)] = 4 \\ w[_3B_9(i,j,l)] = 4 \\ w[_4B_9(i,j,k)] = 4 \end{cases}$$

have six solutions:

1. $c_1 = c_2 = c_3 = c_4 = d_1 = 0$,

2. $c_1 = c_2 = c_3 = c_4 = 1, d_1 = 0$,

3. $c_1 = 0, c_2 = 1, c_3 = 0, c_4 = 1, d_1 = 0$,

4. $c_1 = 1, c_2 = 0, c_3 = 1, c_4 = d_1 = 0$,

5. $c_1 = c_2 = 0, c_3 = c_4 = 1, d_1 = 0$,

6. $c_1 = c_2 = 1, c_3 = c_4 = 0, d_1 = 0$,

while the Boolean functions corresponding to the last four solutions are equivalent to each other. There are three non-equivalent H–Boolean functions in Type 9. They are

$$A_{12}(i,j,k,l) = ij + ik + jl + kl,$$
$$A_{13}(i,j,k,l) = ij + ik + jl + kl + ijk + ijl + jkl + ikl,$$
$$A_{14}(i,j,k,l) = ij + ik + jl + kl + ijl + jkl.$$

About Type 10: Since the simultaneous equations

$$\begin{cases} w[_1B_{10}(j,k,l)] = 4 \\ w[_2B_{10}(i,k,l)] = 4 \\ w[_3B_{10}(i,j,l)] = 4 \\ w[_4B_{10}(i,j,k)] = 4 \end{cases}$$

have three solutions:

1. $c_1 = c_2 = c_3 = c_4 = d_1 = 0$,

2. $c_1 = c_2 = c_3 = 0, c_4 = 1, d_1 = 0$,

3. $c_1 = c_2 = 0, c_3 = 1, c_4 = d_1 = 0$,

the Boolean functions corresponding to the last two solutions are equivalent to each other, there are two non-equivalent H–Boolean functions in Type 10. They are

$$A_{15}(i, j, k, l) = ij + ik + il + jl + jk,$$
$$A_{16}(i, j, k, l) = ij + ik + il + jk + jl + ikl.$$

About Type 11: Since the simultaneous equations

$$\begin{cases} w[_1 B_{11}(j, k, l)] = 4 \\ w[_2 B_{11}(i, k, l)] = 4 \\ w[_3 B_{11}(i, j, l)] = 4 \\ w[_4 B_{11}(i, j, k)] = 4 \end{cases}$$

have five solutions:

1. $c_1 = c_2 = c_3 = c_4 = d_1 = 0$,

2. $c_1 = c_2 = c_3 = 0$, $c_4 = 1$, $d_1 = 0$,

3. $c_1 = c_2 = 0$, $c_3 = 1$, $c_4 = d_1 = 0$,

4. $c_1 = 0$, $c_2 = 1$, $c_3 = c_4 = d_1 = 0$,

5. $c_1 = 1$, $c_2 = c_3 = c_4 = d_1 = 0$,

the Boolean functions corresponding to the last four solutions are equivalent to each other, there are two non-equivalent H–Boolean functions in Type 11. They are

$$A_{17}(i, j, k, l) = ij + ik + il + jl + jk + kl,$$
$$A_{18}(i, j, k, l) = ij + ik + il + jk + jl + kl + jkl.$$

In a word, there are in total 18 non-equivalent H–Boolean functions from Type 1 to Type 11 denoted by $A_m(i, j, k, l)$, $(1 \leq m \leq 18)$. Therefore all the H–Boolean functions of 4-variable without linear terms can be divided into 18 equivalent classes X_1, X_2, ..., X_{18} such that X_m, $1 \leq m \leq 18$, consists of all functions equivalent to $A_m(i, j, k, l)$.

When the linear terms are considered, all 4-variable's H–Boolean functions are divided into 18 classes, say Y_1, Y_2, ..., Y_{18}, with

$$Y_m = \{f(i, j, k, l) + g(i, j, k, l) : f(.) \in X_m, \ g(.) \text{ linear}\}.$$

The construction of 2^4 Hadamard matrices is clear: every 2^4 Hadamard matrix corresponds to a 4-variable's H–Boolean function belonging to one of Y_m, $(1 \leq m \leq 18)$.

Finally, we turn to enumerating the 2^4 Hadamard matrices. Let N be the number of different Boolean functions produced by all $A_m(i, j, k, l)$, $1 \leq m \leq 18$, produced by permutations of $\{1, 2, 3, 4\}$. Then the number of 2^4 Hadamard matrices should be $2^5 N = 32N$.

The number N is found by the following 18 steps:

Step 1: $A_1(i, j, k, l)$ keeps unchanged under any permutations among the variables i, j, k, and l.

Step 2: After permuting the four variables i, j, k, and l, $A_2(i, j, k, l)$ results in three different H–Boolean functions: $ij + kl$; $ik + jl$; and $il + jk$.

Step 3: After permuting the four variables i, j, k, and l, $A_3(i, j, k, l)$ results in four different H–Boolean functions:

$$ij + ik + il; ij + jk + kl,$$
$$ik + jk + kl; il + jl + kl.$$

Step 4: After permuting the four variables i, j, k, and l, $A_4(i, j, k, l)$ results in four different H–Boolean functions:

$$ij + ik + il + jkl; ij + jk + jl + ikl,$$
$$ik + jk + kl + ijl; il + jl + kl + ijk.$$

Step 5: After permuting the four variables i, j, k, and l, $A_5(i, j, k, l)$ results in 12 different H–Boolean functions:

$$ij + ik + il + ijk + jkl; ij + ik + il + ijl + jkl,$$
$$ij + ik + il + ikl + jkl; ij + jk + jl + ijk + ikl,$$
$$ij + jk + jl + ijl + ikl; ij + jk + jl + jkl + ikl,$$
$$ik + jk + kl + ijk + ijl; ik + jk + kl + ikl + jkl,$$
$$ik + jk + kl + jkl + ijl; il + jl + kl + ijl + ijk,$$
$$il + jl + kl + ikl + ijk; il + jl + kl + jkl + ijk.$$

Step 6: After permuting the four variables i, j, k, and l, $A_6(i, j, k, l)$ results in 4 different H–Boolean functions:

$$ij + ik + jk + ijk + ijl + ikl + jkl; ij + il + jl + ijk + ijl + ikl + jkl,$$
$$ik + il + kl + ijk + ijl + ikl + jkl; jk + jl + kl + ijk + ijl + ikl + jkl.$$

Step 7: After permuting the four variables i, j, k, and l, $A_7(i,j,k,l)$ results in 12 different H–Boolean functions:

$$ij + ik + jl; \; ij + il + jk; \; ik + il + jk; \; ij + ik + kl,$$
$$il + ik + jl; \; il + ij + lk; \; jk + ji + kl; \; jk + jl + ik,$$
$$ij + jl + kl; \; jl + jk + il; \; kl + ik + jl; \; kl + kj + jl.$$

Step 8: After permuting the four variables i, j, k, and l, $A_8(i,j,k,l)$ results in 12 different H–Boolean functions:

$$ij + ik + jl + ikl + jkl; \; ij + il + jk + ikl + jkl,$$
$$ik + il + jk + ijl + jkl; \; ij + ik + kl + ijl + jkl,$$
$$il + ik + jl + ijk + jkl; \; il + ij + lk + ijk + jkl,$$
$$jk + ji + kl + ijl + ikl; \; jk + jl + ik + jkl + ikl,$$
$$ij + jl + kl + ijk + ikl; \; jl + jk + il + ijk + ikl,$$
$$kl + ik + jl + ijk + ijl; \; kl + kj + il + ijk + ijl.$$

Step 9: After permuting the four variables i, j, k, and l, $A_9(i,j,k,l)$ results in 12 different H–Boolean functions:

$$ij + ik + il + jk; \; ij + ik + il + jl; \; ij + ik + il + kl,$$
$$ij + jk + jl + ik; \; ij + jk + jl + il; \; ij + jk + jl + kl,$$
$$ik + jk + kl + ij; \; ik + jk + kl + il; \; ik + jk + kl + jl,$$
$$il + jl + kl + ik; \; il + jl + kl + ij; \; il + jl + kl + jk.$$

Step 10: After permuting the four variables i, j, k, and l, $A_{10}(i,j,k,l)$ results in 12 different H–Boolean functions:

$$ij + ik + il + jk + ijk + jkl; \; ij + ik + il + jl + ijl + jkl,$$
$$ij + ik + il + kl + ikl + jkl; \; ij + jk + jl + ik + ijk + ikl,$$
$$ij + jk + jl + il + ijl + ikl; \; ij + jk + jl + kl + jkl + ikl,$$
$$ik + jk + kl + ij + ijk + ijl; \; ik + jk + kl + il + ikl + ijl,$$
$$ik + jk + kl + jl + jkl + ijl; \; il + jl + kl + ik + ikl + ijk,$$
$$il + jl + kl + ij + ijl + ijk; \; il + jl + kl + jk + ijk + jkl.$$

Step 11: After permuting the four variables i, j, k, and l, $A_{11}(i,j,k,l)$ results in 12 different H–Boolean functions:

$$ij + ik + il + jk + jkl; \; ij + ik + il + jl + jkl; \; ij + ik + il + kl + jkl,$$
$$ij + jk + jl + ik + ikl; \; ij + jk + jl + il + ikl; \; ij + jk + jl + kl + ikl,$$
$$ik + jk + kl + ij + ijl; \; ik + jk + kl + il + ijl; \; ik + jk + kl + jl + ijl,$$
$$il + jl + kl + ik + ijk; \; il + jl + kl + ij + ijk; \; il + jl + kl + jk + ijk.$$

Step 12: After permuting the four variables i, j, k, and l, $A_{12}(i, j, k, l)$ results in 3 different H–Boolean functions: $ij + ik + jl + lk$; $ij + il + jk + lk$; and $ik + il + jk + jl$.

Step 13: After permuting the four variables i, j, k, and l, $A_{13}(i, j, k, l)$ results in 3 different H–Boolean functions:

$$ij + ik + jl + lk + ijk + ijl + ikl + jkl \,,$$
$$ij + il + jk + kl + ijk + ijl + ikl + jkl \,,$$
$$ik + il + jk + jl + ijk + ijl + ikl + jkl \,.$$

Step 14: After permuting the four variables i, j, k, and l, $A_{14}(i, j, k, l)$ results in 12 different H–Boolean functions:

$$ij + ik + jl + lk + ijl + jkl; \; ij + il + jk + kl + ijk + jkl,$$
$$ik + il + jk + jl + ijk + jkl; \; ik + ij + kl + jl + ikl + jkl,$$
$$il + ik + jl + jk + ijl + jkl; \; il + ij + kl + kj + ikl + jkl,$$
$$ij + jk + il + kl + ijl + ikl; \; ij + jl + ki + kl + ijk + ikl,$$
$$jk + jl + ik + il + ijk + ikl; \; jl + jk + il + ik + ijl + ikl,$$
$$ik + kl + ij + jl + ijk + ijl; \; jk + kl + ij + il + ijk + ijl.$$

Step 15: After permuting the four variables i, j, k, and l, $A_{15}(i, j, k, l)$ results in 6 different H–Boolean functions:

$$ij + ik + il + jk + jl; \; ij + ik + il + kl + jk,$$
$$ij + ik + il + lk + jl; \; jk + ij + jl + ik + kl,$$
$$jl + ij + jk + il + kl; \; kl + ki + kj + il + jl.$$

Step 16: After permuting the four variables i, j, k, and l, $A_{16}(i, j, k, l)$ results in 12 different H–Boolean functions:

$$ij + ik + il + jk + jl + ikl; \; ij + ik + il + kl + jk + ijl,$$
$$ij + ik + il + kl + lj + ijk; \; ij + jk + jl + ik + il + jkl,$$
$$ij + jk + jl + ik + kl + ijl; \; ij + jk + jl + li + kl + ijk,$$
$$ik + jk + kl + ij + il + jkl; \; ik + jk + kl + li + lj + ijk,$$
$$il + jl + lk + ij + ik + jkl; \; il + jl + kl + jk + ij + ikl,$$
$$il + jl + kl + kj + ik + ijl; \; ik + jk + kl + ij + jl + ikl.$$

Step 17: The $A_{17}(i, j, k, l)$ keeps unchanged under any permutation of the four variables i, j, k, and l.

Step 18: After permuting the four variables i, j, k, and l, $A_{18}(i, j, k, l)$ results in 4 different H–Boolean functions:

$$A_{17}(i, j, k, l) + jkl; \ A_{17}(i, j, k, l) + ikl,$$
$$A_{17}(i, j, k, l) + ijl; \ A_{17}(i, j, k, l) + ijk.$$

From the above 18 steps, we find that

$$N = 1+3+4+4+12+4+12+12+12+12+12+3+3+12+6+12+1+4 = 129.$$

With the above we have finished the proof of the following theorem.

Theorem 5.2.1 *There are* $32 \times 129 = 4128$ *4-dimensional Hadamard matrices of order 2.*

5.2.2 Enumeration of 2^5 Hadamard Matrices

Based on the classification approach used in the last subsection and with the help of a computer search, we obtain the following theorem.

Theorem 5.2.2 *There are* $12,086,336$ *5-dimensional Hadamard matrices of order 2.*

To save space we omit its proof here.

Besides Theorem 5.2.2, the following partial results are also true.

1. The number of 5-variable's H–Boolean functions of the form

$$a_0 + \sum_{i=1}^{5} a_i x_i + \sum_{1 \leq i < j \leq 5} b_{ij} x_i x_j \quad a_i, \ b_{ij} = 0 \text{ or } 1$$

   is 49152;

2. There exists no 5-variable's H–Boolean functions of the form

$$\sum_{1 \leq i < j < k < l \leq 5} c_{ijkl} x_i x_j x_k x_l, c_{ijkl} = 0 \text{ or } 1;$$

3. The number of 5-variable's H–Boolean functions of the form

$$a_0 + \sum_{i=1}^{5} a_i x_i + \sum_{1 \leq i < j \leq 5} b_{ij} x_i x_j$$
$$+ \sum_{1 \leq i < j < k < l \leq 5} c_{ijkl} x_i x_j x_k x_l$$

   is 60416;

4. The number of 5-variable's H–Boolean functions of the form

$$a_0 + \sum_{i=1}^{5} a_i x_i + \sum_{1 \leq i < j < k \leq 5} b_{ijk} x_i x_j x_k, \quad a_i, \ b_{ijk} = 0 \text{ or } 1$$

is 640;

5. The number of 5-variable's H–Boolean functions of the form

$$a_0 + \sum_{i=1}^{5} a_i x_i + \sum_{1 \leq i < j < k \leq 5} b_{ijk} x_i x_j x_k$$

$$+ \sum_{1 \leq i < j < k < l \leq 5} c_{ijkl} x_i x_j x_k x_l$$

is 6720.

5.2.3 Enumeration of General 2^n Hadamard Matrices

The problem of enumerating the general 2^n Hadamard matrices, or equivalently the H–Boolean functions of n-variable, is still open. This subsection will provide some partial enumeration results.

Theorem 5.2.3 *The number of m-variable's H–Boolean functions of the form*

$$a_0 + \sum_{i=1}^{m} a_i x_i + \sum_{1 \leq i < j \leq m} b_{ij} x_i x_j$$

is equal to

$$2^{m+1}[2^{m(m-1)/2} - \sum_{k=1}^{m} (-1)^k \binom{m}{k} 2^{m(m-1)/2 - k(2m-k-1)/2}].$$

Proof. It is sufficient to prove that the number of m-variable's H–Boolean functions of the form

$$g(x_1, x_2, \ldots, x_m) = \sum_{1 \leq i < j \leq m} b_{ij} x_i x_j$$

is equal to

$$2^{m(m-1)/2} - \sum_{k=1}^{m} (-1)^k \binom{m}{k} 2^{m(m-1)/2 - k(2m-k-1)/2}.$$

The Hamming weight of every non-constant linear Boolean function of $(m-1)$-variable is 2^{m-2}. Thus the necessary condition for

$$w[g(x_1,\ldots,x_{k-1},0,x_{k+1},\ldots,x_m) + g(x_1,\ldots,x_{k-1},1,x_{k+1},\ldots,x_m)]$$

$$= w\left[\sum_{i=1}^{k-1} b_{ik}x_i + \sum_{j=k+1}^{m} b_{kj}x_j\right]$$

$$= 2^{m-2} \qquad (1 \le k \le m)$$

is that $(b_{1,k},\ldots,b_{k-1,k},b_{k,k+1},\ldots,b_{k,m})$ is not the zero-vector for each $1 \le k \le m$.

Therefore the enumeration of the above $g(x_1, x_2, \ldots, x_m)$ is equal to the number of binary symmetric matrices

$$A = \begin{bmatrix} 0 & b_{12} & b_{13} & b_{14} & \ldots & b_{1m} \\ b_{12} & 0 & b_{23} & b_{24} & \ldots & b_{2m} \\ b_{13} & b_{23} & 0 & b_{34} & \ldots & b_{3m} \\ \vdots & \vdots & \vdots & \vdots & \vdots & \vdots \\ b_{1m} & b_{2m} & b_{3m} & b_{4m} & \ldots & 0 \end{bmatrix}$$

satisfying that none of its row is all-zero.

Let $A_k, 1 \le k \le m$, be the set of binary symmetric matrices with its k-th row be zero-vector; and $N(A_k)$ be the number of matrices contained in A_k.

By the well-known Polya's enumeration identity, we have

$$N(A_1 \bigcup \ldots \bigcup A_m) = \sum_{i=1}^{m} N(A_i) - \sum_{1 \le i < j \le m} N(A_i \bigcap A_j)$$

$$+ \sum_{1 \le i < j < k \le m} N(A_i \bigcap A_j \bigcap A_k) + \ldots$$

$$+ (-1)^{m-1} N(A_1 \bigcap \ldots \bigcap A_m).$$

Because of the identity

$$N(A_{i_1} \bigcap \ldots \bigcap A_{i_k}) = 2^{m(m-1)/2 - k(2m-k-1)/2}$$

we have

$$N(A_1 \bigcup \ldots \bigcup A_m) = \sum_{k=1}^{m} (-1)^{k-1} \binom{m}{k} 2^{m(m-1)/2 - k(2m-k-1)/2}.$$

The enumeration of the above Boolean functions $g(x_1, \ldots, x_m)$ is

$$N\left(\overline{A_1 \bigcup \cdots \bigcup A_m}\right) = 2^{m(m-1)/2} - N\left(A_1 \bigcup \cdots \bigcup A_m\right).$$

The theorem follows.   **Q.E.D.**

Theorem 5.2.4 *The number of m-variable's H–Boolean functions of the form*

$$a_0 + \sum_{i=1}^{m} a_i x_i + \sum_{1 \le i < j \le m} b_{ij} x_i x_j + x_{i_1} x_{i_2} \ldots x_{i_k},$$

where $1 \le i_1 < i_2 < \ldots < i_k \le m$ and $3 \le k \le m-1$, is equal to

$$2^{m+1} \times \binom{m}{k} \times 2^{k(k-1)/2} \times \{2^{m(m-1)/2 - k(k-1)/2}$$

$$- \sum_{s=1}^{m} (-1)^{s-1} \sum_{r=0}^{k} \binom{k}{r} \binom{m-k}{s-r}$$

$$\times 2^{(m-k)(k-r)} \times 2^{(m-k)(m-k-1)/2 - [(m-k-1)(s-r) - (s-r)(s-r-1)/2]}\}.$$

Proof. It is sufficient to prove that the number of H–Boolean functions of the form

$$\sum_{1 \le i < j \le m} b_{ij} x_i x_j + x_1 x_2 \ldots x_k$$

is equal to

$$2^{k(k-1)/2} \times \{2^{m(m-1)/2 - k(k-1)/2} - \sum_{s=1}^{m} (-1)^{s-1} \sum_{r=0}^{k} \binom{k}{r} \binom{m-k}{s-r}$$

$$\times 2^{(m-k)(k-r)} \times 2^{(m-k)(m-k-1)/2 - [(m-k-1)(s-r) - (s-r)(s-r-1)/2]}\}.$$

By the definition of H–Boolean function, we know that the function

$$\sum_{1 \le i < j \le m} b_{ij} x_i x_j + x_1 x_2 \ldots x_k$$

is an H–Boolean function of m-variable if and only if the following two identities are satisfied

$$
\begin{cases}
w\left(\displaystyle\sum_{j=1}^{i-1} b_{ji}x_j + \sum_{j=i+1}^{m} b_{ij}x_j + x_1\ldots x_{i-1}x_{i+1}\ldots x_k\right) \\
\qquad = 2^{m-2} & \text{if } 1 \le i \le k \\
w\left(\displaystyle\sum_{j=1}^{i-1} b_{ji}x_j + \sum_{j=i+1}^{m} b_{ij}x_j\right) = 2^{m-2} & \text{if } k+1 \le i \le m.
\end{cases}
$$

The second identity is equivalent to the condition: at least one of b_{1i}, b_{2i}, $\ldots$, b_{i-1i}, b_{ii+1}, $\ldots$, b_{im} is 1, for each i, $k+1 \le i \le m$.

And the first identity is equivalent to the condition: at least one of b_{ik+1}, b_{ik+2}, $\ldots$, b_{im} is 1, for each i, $1 \le i \le k$. In fact, if b_{ik+1}, b_{ik+2}, $\ldots$, b_{im} are all zero, then

$$
w\left(\sum_{j=1}^{i-1} b_{ji}x_j + \sum_{j=i+1}^{k} b_{ij}x_j + x_1\ldots x_{i-1}x_{i+1}\ldots x_k\right)
$$

$$
= w\left(\sum_{j=1}^{i-1} b_{ji}x_j + \sum_{j=i+1}^{k} b_{ij}x_j\right) + w(x_1\ldots x_{i-1}x_{i+1}\ldots x_k)
$$

$$
-2w\left[\left(\sum_{j=1}^{i-1} b_{ji}x_j + \sum_{j=i+1}^{k} b_{ij}x_j\right) x_1\ldots x_{i-1}x_{i+1}\ldots x_k\right]
$$

$$
= 2^{m-2} + 2^{m-k} - 2^{m-k+1}\left[\left(\sum_{j=1}^{i-1} b_{ji} + \sum_{j=i+1}^{k} b_{ij}\right) \bmod 2\right]
$$

$$
\ne 2^{m-2},
$$

where the last to the second equation is due to that the treated Hamming weights are of Boolean functions of $(m-1)$ variables.

Because of the above two equivalent identities, it is sufficient to prove that there are

$$
2^{m(m-1)/2-k(k-1)/2} - \left\{\sum_{s=1}^{m}(-1)^{s-1}\sum_{r=0}^{k}\binom{k}{r}\binom{m-k}{s-r}\right.
$$

$$
\left. \times 2^{(m-k)(k-r)} \times 2^{(m-k)(m-k-1)/2-[(m-k-1)(s-r)-(s-r)(s-r-1)/2]}\right\}
$$

binary (o or 1) matrices of the form

$$
\begin{bmatrix}
0 & 0 & \cdots 0 & b_{1k+1} \cdots & & \cdots & b_{1m} \\
0 & 0 & \cdots 0 & b_{2k+1} \cdots & & \cdots & b_{2m} \\
\vdots & \vdots & \vdots\ \vdots & \vdots & \vdots & \vdots & \vdots \\
0 & 0 & \cdots 0 & b_{kk+1} \cdots & & \cdots & b_{km} \\
b_{1k+1} & b_{2k+1} & \cdots b_{kk+1} & 0 & b_{k+1k+2} \cdots & & b_{k+1m} \\
\vdots & \vdots & \vdots\ \vdots & \vdots & \vdots & \vdots & \vdots \\
b_{1m} & b_{2m} & \cdots b_{km} & b_{k+1m} \cdots & & b_{m-1m} & 0
\end{bmatrix}
$$

without all-zero rows. This enumeration can be finished by the same way as that of Theorem 5.2.3.   **Q.E.D.**

Similarly, it has been proved that

Theorem 5.2.5 *The number of H–Boolean functions of the form*

$$
a_0 + \sum_{i=1}^{m} a_i x_i + \sum_{1 \leq i < j \leq m} b_{ij} x_i x_j + \sum_{1 \leq i < j < k \leq m} c_{ijk} x_i x_j x_k
$$

is bounded above by

$$
2^{n+X} \times \{ 2^{n+Y} - \sum_{i \neq 2^{n-2}} A_i \},
$$

where

$$
A_{2^{n-2} \pm 2^{n-2-h}} = 2^{h(h+1)} \times \frac{(2^{n-1} - 1)(2^{n-2} - 1) \ldots (2^{n-2h} - 1)}{(2^{2k} - 1)(2^{2k-2} - 1) \ldots (2^2 - 1)}
$$

and the other A_is are all zero, and where $1 \leq h \leq \lfloor \frac{n-1}{2} \rfloor$, and

$$
X := \binom{n-1}{2} + \binom{n-1}{3}, \quad \text{and} \quad Y = \binom{n-1}{2}.
$$

In order to introduce more enumerations about H–Boolean function, we need the following definitions.

Definition 5.2.1 *An n-dimensional vector $X = (x_1, x_2, \ldots, x_n)$ is called a characteristic vector of a Boolean function $f(.)$, iff $f(X) = 1$. The matrix produced by arranging all characteristic vectors of $f(.)$ in the dictionary order is called the ordered characteristic matrix of this Boolean function.*

Clearly, Boolean functions and their ordered characteristic matrices are uniquely determined by each other.

Definition 5.2.2 *A binary (0 or 1) matrix of size $2k \times n$ is said to be a matrix of type A_r, $0 \le r \le k$, if and only if*

1. *The rows of this matrix are different from each other;*

2. *The sub-matrices consisting of any $n-1$ columns have and only have r same-row pairs, while the other rows are different from each other.*

For example, $f(x_1, x_2, x_3, x_4) = x_1 x_2 + x_3 x_4$ is an H–Boolean function of 4-variable, and its characteristic matrix is

$$
\begin{bmatrix}
0 & 0 & 1 & 1 \\
0 & 1 & 1 & 1 \\
1 & 0 & 1 & 1 \\
1 & 1 & 0 & 0 \\
1 & 1 & 0 & 1 \\
1 & 1 & 1 & 0
\end{bmatrix}.
$$

This matrix is of type A_1, e.g., its sub-matrix consisting of the first 3 columns has just one pair of same-row (the 4-th and 5-th rows are the same vector $(1,1,0)$) . In fact, this example can be generalized as the following theorem, which indicates another equivalent definition of H–Boolean function.

Theorem 5.2.6 *An n-variable Boolean function of Hamming weight $2k$ is H–Boolean if and only if its ordered characteristic matrix is of type $A_{k-2^{n-3}}$.*

Proof. If $w(f(.)) = 2k$ then the equation

$$
w(f(x_1, \ldots, x_{i-1}, 0, \ldots, x_n) + f(x_1, \ldots, x_{i-1}, 1, \ldots, x_n)) = 2^{n-2}
$$

is equivalent to

$$
w(f(x_1, \ldots, x_{i-1}, 0, \ldots, x_n) f(x_1, \ldots, x_{i-1}, 1, \ldots, x_n)) = k - 2^{n-3},
$$

which infers that the sub-matrix produced by canceling the i-th column of its ordered characteristic matrix has and only has $k - 2^{n-3}$ same-row pairs. The proof is finished by applying $1 \le i \le n$. **Q.E.D.**

A straightforward corollary of Definition 5.2.2 is

Lemma 5.2.1 *A binary matrix is of type A_0 if and only if the Hamming distances between any two rows are larger than or equal to 2.*

Theorem 5.2.7 *There are at least*

$$2 \sum_{s=0}^{2^{n-3}-1} \binom{2^{n-1}}{s}\binom{2^{n-1}-ns}{2^{n-2}-s} + \binom{2^{n-1}}{2^{n-3}}\binom{2^{n-1}-n2^{n-3}}{2^{n-3}}$$

n-variable H–Boolean functions of Hamming weight 2^{n-2}.

Proof. A matrix of type A_0 can be constructed by: (1) choose any s n-dimensional vectors of even Hamming weights as s rows of the matrix; (2) thus there are at least $2^{n-1} - ns$ n-dimensional vectors of odd Hamming weights that have Hamming distances larger than 1 apart from the above chosen s even weight's vectors. Choose the other $2^{n-2} - s$ rows of the matrix from these odd weight ones.   **Q.E.D.**

Theorem 5.2.8 *There are at least*

$$2^{n-1} \sum_{s=0}^{2^{n-3}-\lfloor (n-1)/2 \rfloor} \binom{2^{n-1}-n^2-n}{s}\binom{2^{n-1}-n^2-n-ns}{2^{n-2}-n+1}$$

n-variable H–Boolean functions of Hamming weight $2^{n-2} + 2$

Proof. A matrix of type A_1 can be constructed by: (1) choose $(0, 0, \ldots, 0)$ and $e_i := (0, \ldots, 1, 0, \ldots, 0)$, $1 \le i \le n$, as $n + 1$ rows of the matrix; (2) there are at least $2^n - n^2$ n-dimensional vectors that have Hamming distances larger than 1 apart from the e_i, $1 \le i \le n$. Choose the other $2^{n-2} - n + 1$ rows of the matrix from these vectors.   **Q.E.D.**

5.3   Applications

n-dimensional Hadamard matrices of order 2, or equivalently, the H–Boolean functions, can be widely used in modern cryptography, error-correcting codes, and signal processing. This section concentrates on the close relationships between H–Boolean functions and the strict avalanche criterion, propagation characteristics, Bent functions, and Reed–Muller codes, respectively.

5.3.1   Strict Avalanche Criterion and H–Boolean Functions [4], [5], [6], [7], [8]

The symmetric or private-key cryptosystem is one of the most important kinds of cryptosystems. This kind of cryptosystem, e.g., the famous DES, is constructed by the combination of substitutions and permutations. For the development of symmetric cryptosystem, a significant portion of time has been spent on design or on analysis of the substitution boxes (or called S-boxes in short) . Because the remainder of the cryptosystem algorithm is linear, severe weaknesses in the S-boxes can therefore lead to an insecure cryptosystem.

The Strict Avalanche Criterion (SAC) was introduced by Webster and Tavares [4] in order to combine the ideas of completeness and the avalanche effect of the design of an S-box. A cryptographic transformation is said to be complete if each output bit depends on all of the input bits, and it exhibits the avalanche effect if an average of one-half of the output bits changes whenever a single input bit is changed. Forre [7] extended the notion of SAC by defining higher-order Strict Avalanche Criteria. This subsection will prove that the SAC is, in fact, another equivalent form of H–Boolean. Thus all results about H–Boolean functions can be applied to the study of SAC, and vice versa.

The cryptographic significance of the SAC is highlighted by considering the situation where a cryptographer needs some 'complex' mapping f of n bits onto one bit, although there is no precise mathematical definition for the expression 'complex' . In order to make a more intuitively pleasing meaning of 'complex', we present here an information-theoretic statement: If the conditional entropy

$$H[f(x_1, x_2, \ldots, \overline{x_i}, \ldots, x_n) \mid f(x_1, x_2, \ldots, x_n)]$$

is maximized for all i, $1 \leq i \leq n$, then the Boolean $f(x_1, x_2, \ldots, x_n)$ is said to satisfy the SAC. In other words, little information of $f(x_1, x_2, \ldots,$ $\overline{x_i}, \ldots, x_n)$ will be exposed by a Boolean $f(x_1, x_2, \ldots, x_i, \ldots, x_n)$ satisfying the SAC. The higher-order SAC goes even further, by keeping one or more input bits of $f(.)$ constant, and making the obtained 'sub-functions' complex as well. It is worthwhile pointing out that any function $g(.)$ of $n - 1$ variables will be a relatively bad approximation of $f(.)$ if $f(.)$ satisfies the SAC. Indeed, the output of the best possible $g(.)$ will differ from the output of f with a probability of 1/4. This lack of accuracy of lower-dimensional approximations is a desired property of cryptosystems, because the existence of some lower-dimensional approximation of an encryption could reduce the amount of work for an exhaustive decryption according to the dimension of the domain of the approximation.

The mathematical definition of the Strict Avalanche Criterion (SAC) is (see [4]):

Definition 5.3.1 *A Boolean function $f(x_1, x_2, \ldots, x_n)$ is called "satisfying SAC" if and only if for each unit vector $e_i = (0, \ldots, 0, 1, 0, \ldots, 0)$, the following equation is hold*

$$w[f(x) + f(x + e_i)] = 2^{n-1},$$

where $x = (x_1, x_2, \ldots, x_n)$.

Comparing this definition with the definition of an H–Boolean function, i.e., Definition 5.1.1, it is easy to see that they are the same thing. Hence we have:

Theorem 5.3.1 *A Boolean function $f(x_1, x_2, \ldots, x_n)$ satisfies the SAC if and only if it is an H–Boolean function of n variables.*

Definition 5.3.2 ([8]) *A Boolean function $f(x_1, x_2, \ldots, x_n)$ is said to satisfy a SAC to order k, $0 \le k \le n - 2$, if and only if whenever k variables are fixed arbitrarily, the resulting function of $(n - k)$ variables satisfies the SAC.*

It is easy to see that a SAC of order 0 is the SAC defined in Definition 5.3.1. In addition, if a function satisfies the SAC of order $k > 0$, then it also satisfies the SAC of order j for any $j = 0, 1, \ldots, k - 1$. Thus if $f(x_1, \ldots, x_n)$ is a SAC of order k, then $f(.)$ generates a higher-dimensional Hadamard matrix such that all of its $m(\ge n - k)$-dimensional sections are Hadamard matrices of order 2. In fact, this higher-dimensional Hadamard matrix is defined by $[(-1)^{f(x_1, \ldots, x_n)}]$.

One of the hot topics in SAC criteria is to count the Boolean functions satisfying the SAC criterion. We use the abbreviation SAC(k) for the strict avalanche criterion of order k. Here are a few fundamental results on Boolean functions satisfying SAC(k).

Lemma 5.3.1 ([7], [8]) *If* $f(x_1, x_2, \ldots, x_n)$ *satisfies* SAC(k) *for some* k, $0 \leq k \leq n - 2$, *then so does*

$$f(x_1, x_2, \ldots, x_n) + a + \sum_{i=1}^{n} b_i x_i.$$

Lemma 5.3.2 ([5], [8]) *If* $f(x_1, x_2, \ldots, x_n)$ *satisfies* SAC(k) *for some* k, $0 \leq k \leq n - 3$, *then* $f(.)$ *is a Boolean function of degree up to* $n - k - 1$. *If* $f(x_1, x_2, \ldots, x_n)$ *satisfies* SAC($n - 2$), *then* $f(.)$ *is a Boolean function of degree* 2.

Proof. Recall that every H–Boolean function of n, $n \geq 3$, variables is of the degree less than or equal to $(n - 1)$ (see Theorem 5.1.3) . If $f(x_1, x_2, \ldots, x_n)$ is a Boolean function of degree larger than $n - k - 1$, $0 \leq k \leq n - 3$, say it contains the term $x_1 x_2 \ldots x_d$, $d \geq n - k$, then by fixing $x_{n-k+1} = \ldots = x_n = 0$ we get a function $f(x_1, \ldots, x_{n-k}, 0, \ldots, 0)$, which cannot be an H–Boolean of $(n - k)$ variables, because it is of degree $n - k$. Thus this function $f(x_1, x_2, \ldots, x_n)$ does not satisfy SAC(k).

The fact that a function satisfying SAC($n - 2$) is of degree 2 can be proved by the following observations. On one hand, linear function is not H–Boolean (see Theorem 5.1.9) and does not satisfy SAC($n - 2$). On the other hand, a function satisfying SAC($n - 2$) satisfies SAC($n - 3$) too. Thus, by the first assertion of this lemma, the function has a degree upper bounded by $n - (n - 3) - 1 = 2$.   **Q.E.D.**

Lemma 5.3.3 ([6], [8]) *A quadratic Boolean function of the form*

$$f(x_1, x_2, \ldots, x_n) = \sum_{1 \leq i < j \leq n} a_{ij} x_i x_j, \quad n \geq 2$$

satisfies SAC(k), $0 \leq k \leq n - 2$, *if and only if every variable* x_i *occurs at least* $k + 1$ *times.*

Proof. At first it is easy to prove that a quadratic Boolean function is H–Boolean if and only if this function is dependent on each of its variable. On one hand, if x_1 occurs no more than $k + 1$ times, say,

$$f(x_1, x_2, \ldots, x_n) = x_1(x_2 + x_3 + \ldots + x_d) + \sum_{2 \leq i < j \leq n} a_{ij} x_i x_j, \quad d \leq k + 1,$$

so the function $f(x_1, 0, \ldots, 0, x_{k+2}, \ldots, x_n)$ is independent of x_1 and thus not an H–Boolean. Hence this function does not satisfies SAC(k).

On the other hand, if every variable occurs at least $k + 1$ times, then the function produced by fixing any k variables is a quadratic function dependent on each of its variable. Hence an H–Boolean function.   **Q.E.D.**

Theorem 5.3.2 ([5], [8]) *There are 2^{n+1} Boolean functions of $n \geq 2$ variables which satisfy* SAC$(n - 2)$; *they are exactly the functions of the form*

$$a + \sum_{i=1}^{n} b_i x_i + \sum_{1 \leq i < j \leq n} x_i x_j.$$

Proof. By Lemma 5.3.1 it is sufficient to prove that a quadratic Boolean function of the form $\sum_{1 \leq i < j \leq n} a_{ij} x_i x_j$ satisfies SAC$(n - 2)$ if and only if $a_{ij} = 1$ for all $1 \leq i < j \leq n$ which is, in fact, a direct corollary of Lemma 5.3.3.   **Q.E.D.**

The Lemma 5.3.1, Lemma 5.3.2, and Lemma 5.3.3 can also be used to enumerate the Boolean functions satisfying SAC$(n - 3)$.

Theorem 5.3.3 ([5], [8]) *Define a sequence $\{W_i\}$ of integers by $W_1 = 1$, $W_2 = 2$, and*

$$W_n = W_{n-1} + (n - 1)W_{n-2}, \quad for \quad n \geq 3.$$

Then there are $2^{n+1} W_n$ Boolean functions of $n \geq 3$ variables which satisfy SAC$(n - 3)$.

Proof. By Lemma 5.3.1, it suffices to show that there are W_n Boolean functions of the form $\sum_{1 \leq i < j \leq n} a_{ij} x_i x_j$ satisfying SAC$(n-3)$. By Lemma 5.3.2 and Lemma 5.3.3, any such function is obtained by deleting zero or more terms from the sum $\sum_{1 \leq i < j \leq n} x_i x_j$ in such a way in which the remaining sum has the property that every variable x_i occurs in at least $n - 2$ terms. Thus S is a set of terms which we are allowed to delete if and only if no subscript i occurs in a term $x_i x_j$ in S more than once. It is easy to find a recursion for the number W_n of such sets S: Obviously $W_1 = 1$ (the empty set) and $W_2 = 2$. Clearly, any set of terms $T = \{x_i x_j\}$ which is counted in W_{n-1} is also a set which must be counted in W_n, and this includes all sets which do not contain any term $x_i x_n$. If we have any set T which includes less than or equal to $n - 2$ variables from $x_1, x_2, \ldots, x_{n-1}$, we may add a term $x_k x_n$ to T and get a set to be counted in W_n if and only if x_k does not already occur in a term in T. There are W_{n-2} such sets of T, by our definitions, so we count W_{n-2} sets for each k, $1 \le k \le n-1$. Hence $W_n = W_{n-1} + (n-1)W_{n-2}$ and the theorem follows. **Q.E.D.**

The number W_n appearing in Theorem 5.3.3 is the number of permutations in the symmetric group S_n whose square is the identity. In particular, the following asymptotic formula for W_n has been proved

$$W_n \overset{n\to\infty}{\Longrightarrow} (e^{1/4}\sqrt{2})^{-1} e^{\sqrt{n}} (n/e)^{n/2}.$$

The problem of counting the Boolean functions satisfying the SAC of order $k \le n-4$ is still open. The updated bounds for the number, T_n, of Boolean functions satisfying SAC$(n-4)$ is [8]

$$2^n(n-1)! < T_n < n^5 2^n n!.$$

The cases of $k \le n-4$ are much difficult than the cases of $k = n-2$ and $k = n-3$, because many of such functions are non-quadratic.

Now, we turn to the spectral characterization of Boolean functions satisfying SAC. Every Boolean function $f(x_1, x_2, \ldots, x_n)$ corresponds to a unique function defined by

$$f'(x_1, x_2, \ldots, x_n) := (-1)^{f(x_1, x_2, \ldots, x_n)},$$

which takes values in the range $\{-1, 1\}$. By this notation, Definition 5.3.1 can be equivalently stated as

Lemma 5.3.4 ([7], [9]) *A Boolean function $f(x_1, x_2, \ldots, x_n)$ satisfies* SAC *if and only if its $f'(x_1, x_2, \ldots, x_n)$ satisfies*

$$\sum_{x \in Z_2^n} f'(x)f'(x + e_i) = 0. \tag{5.4}$$

Recall that the Walsh Transform of $f'(x)$, $x := (x_1, x_2, \ldots, x_n)$, is

$$F'(w) = \sum_{x \in Z_2^n} f'(x)(-1)^{x_1 w_1 + x_2 w_2 + \ldots + x_n w_n}, \tag{5.5}$$

where $w = (w_1, w_2, \ldots, w_n)$. The function $f'(x)$ can be recovered from $F'(w)$ by the inverse Walsh transform:

$$f'(x) = 2^{-n} \sum_{w \in Z_2^n} F'(w)(-1)^{x_1 w_1 + x_2 w_2 + \ldots + x_n w_n}. \tag{5.6}$$

From the well-known convolution theorem, which states that

$$h'(x) = \sum_{y \in Z_2^n} f'(y).g'(y+x) \text{ if and only if } H'(w) = F'(w).G'(w), \quad (5.7)$$

we see that the left hand side of Equation (5.4) is also the inverse Walsh transform of $F'(w).F'(w) = [F'(w)]^2$ and by using Equation (5.6) we obtain:

$$\sum_{x \in Z_2^n} f'(x).f'(x+e_i) = 2^{-n} \sum_{w \in Z_2^n} [F'(w)]^2 (-1)^{w_i}, \quad (5.8)$$

where w_i is the i-th coordinate of the vector w. This equation proves the following theorem.

Theorem 5.3.4 ([7], [9]) *A Boolean function $f(x_1, x_2, \ldots, x_n)$ satisfies SAC if and only if its $f'(x)$ has a Walsh transform $F'(w)$ satisfying*

$$\sum_{w \in Z_2^n} [F'(w)]^2 (-1)^{w_i} = 0 \quad (5.9)$$

for all i, $1 \leq i \leq n$.

A geometrical interpretation of Theorem 5.3.4 can be introduced ([7]) if we treat the n-tuples $(w_1, w_2, \ldots, w_n)$ as the corners of an n-dimensional cube with edges of length one. Let's attach to each corner $w = (w_1, w_2, \ldots, w_n)$ a weight $m(w)$ equal to $[F'(w)]^2$. The center of gravity of this n-dimensional body has the coordinates $(\overline{w_1}, \overline{w_2}, \ldots, \overline{w_n})$ with

$$\overline{w_i} = \frac{\sum_{w \in Z_2^n} m(w) w_i}{\sum_{w \in Z_2^n} m(w)} = \frac{\sum_{w \in Z_2^n,\, w_i=1} [F'(w)]^2}{\sum_{w \in Z_2^n} [F'(w)]^2} \quad (5.10)$$

for all $1 \leq i \leq n$. If a Boolean function $f(x)$ satisfies SAC, we know by Theorem 5.3.4 that

$$\sum_{w \in Z_2^n,\, w_i=0} [F'(w)]^2 - \sum_{w \in Z_2^n,\, w_i=1} [F'(w)]^2 = 0. \quad (5.11)$$

Thus

$$\sum_{w \in Z_2^n,\, w_i=0} [F'(w)]^2 = \sum_{w \in Z_2^n,\, w_i=1} [F'(w)]^2. \quad (5.12)$$

And in that case, we have

$$\overline{w_i} = \frac{\sum_{w \in Z_2^n,\, w_i=1} [F'(w)]^2}{\sum_{w \in Z_2^n} [F'(w)]^2} = \frac{\sum_{w \in Z_2^n,\, w_i=0} [F'(w)]^2}{\sum_{w \in Z_2^n} [F'(w)]^2}, \quad (5.13)$$

which shows that the coordinate w_i of the center of gravity of the considered cubic body remains unchanged if all the weights on one 'face' of the cube (the face with $w_i = 0$) are moved to the opposite 'face' (the face with $w_i = 1$) and conversely. Therefore, we can state that a Boolean function $f(x)$ satisfies SAC if and only if the n-cube with weights equal to $[F'(w)]^2$ attached to its corners has a center of gravity which is equidistant from any two opposite 'faces' of the cube, and thus from all the corners of the cube. The center of gravity of the body associated to the Walsh-spectrum of an SAC-fulfilling function therefore has the coordinates $(\frac{1}{2}, \frac{1}{2}, \ldots, \frac{1}{2})$.

The idea that now naturally arises is to use this as a construction for new Boolean functions satisfying SAC from known ones. The pitfall is that $F'(w)$ might be taken as $\pm\sqrt{[F'(w)]^2}$ for each one of the 2^n possible $w's$. For the worst case where all 2^n $w's$ are associated to nonzero values of $[F(w)]^2$, this will yield 2^{2^n} possible choices for the mapping $F(w)$, and some of them have no valid functions $f'(x)$ (i.e., (± 1)-valued) as inverse Walsh transforms. In fact, a function $f'(x)$ is a (± 1)-valued if and only if

$$[f'(x)]^2 = 1, \text{ for all possible } x \in Z_2^n. \tag{5.14}$$

By the convolution theorem, we see that this is equivalent to

Theorem 5.3.5 ([7]) $F'(w)$ *is the Walsh transform of a (± 1)-valued function $f'(x)$ if and only if*

$$\sum_{w \in Z_2^n} F'(w).F'(w+s) = 2^n \delta(s) = \begin{cases} 2^n \text{ for } s = (0, 0, \ldots, 0) \\ 0 \text{ otherwise.} \end{cases}$$

5.3.2   Bent Functions and H–Boolean Functions

Bent functions, defined and first analyzed by Rothaus in 1976, have been the subject of some interest in logic synthesis, digital communications (especially spread-spectrum multiple access communications), coding theory, and cryptography. We will show in this subsection that every Bent function is an H–Boolean that corresponds to a higher-dimensional Hadamard matrix of order 2. The concept of propagation characteristics (PC) and the relationships among PC, Bent, SAC, and H-Boolean will be introduced too.

Definition 5.3.3 ([10]) *A Boolean function $f(x_1, x_2, \ldots, x_n)$ is called a Bent function if its Walsh transform coefficients are all $\pm 2^{n/2}$, i.e.,*

$$F(u_1, u_2, \ldots, u_n) = \sum_{x \in Z_2^n} (-1)^{u.x + f(x)} \qquad (5.15)$$

is a vector with elements $\pm 2^{n/2}$, where $x = (x_1, \ldots, x_n)$, $u = (u_1, \ldots, u_n)$, and $u \cdot x = \sum_{i=1}^n u_i x_i$.

Because the transform coefficients are summations of integers, $\pm 2^{n/2}$ must be integer, i.e., n must be an even integer for $f(x_1, x_2, \ldots, x_n)$ being Bent.

The following three functions are Bent on 6 variables ([10]):

$$f_1 = x_1 x_4 + x_2 x_5 + x_3 x_6 + x_1 x_2 x_3,$$
$$f_2 = x_1 x_2 + x_1 x_4 + x_2 x_6 + x_3 x_5 + x_4 x_5 + x_1 x_2 x_3 + x_2 x_4 x_5,$$
$$f_3 = x_1 x_4 + x_2 x_6 + x_3 x_4 + x_3 x_5 + x_3 x_6 + x_4 x_5 + x_4 x_6$$
$$+ x_1 x_2 x_3 + x_2 x_4 x_5 + x_3 x_4 x_6.$$

Theorem 5.3.6 ([10], [2]) *$f(x_1, x_2, \ldots, x_n)$ is Bent if and only if the $2^n \times 2^n$ matrix H whose (u, v)-th entry is $(1/2^{n/2})F(u + v)$ is a 2-dimensional Hadamard matrix.*

Proof. It is sufficient to prove that if $f(.)$ is Bent, then $(1/2^{n/2})F(u + v)$ is an Hadamard matrix. On one hand, by Definition 5.3.3, this matrix is clearly a (± 1)-matrix. On the other hand,

$$\sum_{u \in Z_2^n} \frac{1}{2^{n/2}} F(u)(1/2^{n/2})F(u + v)$$

$$= \frac{1}{2^n} \sum_{u \in Z_2^n} \sum_{w \in Z_2^n} (-1)^{u.w}(-1)^{f(w)} \sum_{x \in Z_2^n} (-1)^{(u+v).x}(-1)^{f(x)}$$

$$= \frac{1}{2^n} \sum_{w,x \in Z_2^n} (-1)^{v.x}(-1)^{f(w)+f(x)} \sum_{u \in Z_2^n} (-1)^{u.(w+x)}$$

$$= \sum_{w \in Z_2^n} (-1)^{v.w}[(-1)^{f(w)}]^2$$

$$= \sum_{w \in Z_2^n} (-1)^{v.w}$$

$$= 2^n \delta_{v,0}.$$

Thus this matrix is orthogonal.   **Q.E.D.**

Note that if $f(x_1, x_2, \ldots, x_n)$ is Bent, we then may write

$$F(u)/2^{n/2} = (-1)^{g(u)},$$

which defines a Boolean function $g(u)$. It is easy to verify that the Walsh transform coefficients of $g(u)$ are

$$2^{n/2}(-1)^{f(u)} = \pm 2^{n/2}.$$

Therefore $g(u)$ is also a Bent. Thus there is a natural pairing $f \longleftrightarrow g$ of Bent functions. The other straight implications of Theorem 5.3.6 are:

Theorem 5.3.7 ([10], [2]) *Bent functions can also be equivalently defined by:*

1. $f(x_1, x_2, \ldots, x_n)$ *is Bent if and only if the matrix whose (u, v)-th entry is $(-1)^{f(u+v)}$, for $u, v \in Z_2^n$, is a 2-dimensional Hadamard matrix.*

2. $f(x_1, x_2, \ldots, x_n)$ *is Bent if and only if for all $v \neq 0$, $v \in Z_2^n$, the Hamming weight of $g(x) = f(x + v) + f(x)$ is 2^{n-1}.*

3. $f(x_1, x_2, \ldots, x_n)$ *is Bent if and only if the function*

$$f(x_1, x_2, \ldots, x_n) + a + \sum_{i=1}^{n} b_i x_i$$

is Bent.

Theorem 5.3.8 *Every Bent function is H–Boolean.*

Proof. It can be proved by applying the unit vectors $e_i = (0, \ldots, 0, 1, 0, \ldots, 0)$, $1 \leq i \leq n$, to the second statement of Theorem 5.3.7.   **Q.E.D.**

Thus all constructions for Bent functions are valid for H–Boolean and thus also for n-dimensional Hadamard matrix of order 2.

Theorem 5.3.9 ([10]) *For any Boolean function $g(y_1, \ldots, y_n)$, the function*

$$f(x_1, \ldots, x_n; y_1, \ldots, y_n) = \sum_{i=1}^{m} x_i y_i + g(y_1, \ldots, y_n)$$

is Bent.

Proof. Let $a = (a_1, \ldots, a_n)$ and $b = (b_1, \ldots, b_n)$ be two binary vector.

Case 1: If $b \neq 0$, then

$$f(x_1, \ldots, x_n; y_1, \ldots, y_n) + f(x_1 + a_1, \ldots, x_n + a_n; y_1 + b_1, \ldots, y_n + b_n)$$

$$= \sum_{i=1}^{n} b_i x_i + h(y_1, \ldots, y_n),$$

which is linear in the variables $x_1, x_2, \ldots, x_n$ and thus its Hamming weight is 2^{2n-1}.

Case 2: If $b = 0$, then $a \neq 0$ and

$$f(x_1, \ldots, x_n; y_1, \ldots, y_n) + f(x_1 + a_1, \ldots, x_n + a_n; y_1 + b_1, \ldots, y_n + b_n)$$

$$= \sum_{i=1}^{n} a_i y_i,$$

which is linear in the variables $y_1, y_2, \ldots, y_n$, and thus its Hamming weight is also 2^{2n-1}.

The proof is finished by the second statement of Theorem 5.3.7. **Q.E.D.**

A Bent function $f(x_1, x_2, \ldots, x_{2n})$ can also be equivalently treated as a (± 1)-valued vector of length 4^n. We call such a vector the Bent sequence ([12]) . In other word, a (± 1)-sequence $y = (y_1, y_2, \ldots, y_N)$, $N = 4^n$, is a Bent sequence if and only if its normalized Walsh transform $Y = (1/2^n) H_{2n} y$ is also a (± 1)-sequence, where H_m is the Walsh matrix of order 2^m defined in the first chapter of this book.

It is easy to check that there are eight Bent sequences of length 4, they are

$$B_4 = \{\ 111 - 1;\ 11 - 11;\ 1 - 111;\ -1111;$$
$$-1 - 1 - 11;\ -1 - 11 - 1;\ -11 - 1 - 1;\ 1 - 1 - 1 - 1\}.$$

Theorem 5.3.10 ([12], [13]) *Let m, n be positive even integers and let $y(1), y(2), \ldots, y(2^m)$ be Bent sequences of length 2^n. Furthermore, let z be the concatenation of the normalized Walsh transforms of these Bent sequences; that is*

$$z^T = (Y^T(1) Y^T(2) \ldots Y^T(2^m)).$$

The sequence z is Bent if and only if the sequence $(y(1)_i, y(2)_i, \ldots, y(2^m)_i)$ is Bent for all i, where $y(j)_i$ is the i-th bit of the sequence $y(j)$.

Proof. To show that z is Bent, we show that its normalized Walsh transform Z is a (± 1)-sequence. We write z as a $2^n \times 2^m$ matrix: $a = (Y(1)Y(2)\ldots Y(2^m))$. It is easy to check that

$$Z = \frac{1}{2^{(m+n)/2}} H_{m+n} z \quad \Longrightarrow \quad A = \frac{1}{2^{m/2}} \frac{1}{2^{n/2}} H_n a H_m$$

(i.e., that A is the 2-dimensional Walsh transform of a) . Thus,

$$\begin{aligned} A &= \frac{1}{2^{m/2}} \frac{1}{2^{n/2}} H_n (Y(1)Y(2)\ldots Y(2^m)) H_m \\ &= \frac{1}{2^{m/2}} (y(1)y(2)\ldots y(2^m)) H_m \\ &= \frac{1}{2^{m/2}} B H_m, \end{aligned}$$

where $B = (y(1)y(2)\ldots y(2^m))$ (note that the entries of B are all ± 1) . The rows of A are the normalized Walsh transforms of the rows of B, and are therefore (± 1)-sequences of length 2^m if and only if $(y(1)_i, y(2)_i, \ldots, y(2^m)_i)$ is Bent for all i.   **Q.E.D.**

By using Theorem 5.3.10 we can construct long Bent sequences from short ones. For example, if x is a Bent sequence of length 4^k, then the following four sequences are Bent of length 4^{k+1} ([12], [13]) :

$$(X^T X^T X^T - X^T), (X^T X^T - X^T X^T),$$
$$(X^T - X^T X^T X^T), (-X^T X^T X^T X^T).$$

If x and y are two Bent sequences of length 4^k, then the following six sequences are Bent of length 4^{k+1} ([12], [13]):

$$(X_1^T X_1^T X_2^T - X_2^T), (X_1^T X_2^T X_1^T - X_2^T), (X_1^T X_2^T - X_2^T X_1^T),$$
$$(X_2^T X_1^T X_1^T - X_2^T), (X_2^T X_1^T - X_2^T X_1^T), (X_2^T - X_2^T X_1^T X_1^T).$$

By the same proof as that for Theorem 5.3.10, we have the following theorem:

Theorem 5.3.11 ([14]) *If $x = (x_1, x_2, \ldots, x_{4^k})$ and $y = (y_1, y_2, \ldots, y_{4^k})$ are two Bent sequences of length 4^k, then the following two sequences, u and v, are Bent sequences of length 4^{k+1}, where*

$$\begin{aligned} u = (\ &x_1, x_2, x_3, x_4, y_1, y_2, y_3, y_4, x_5, x_6, x_7, x_8, y_5, y_6, y_7, y_8, \ldots, y_{4^k}, \\ &x_1, x_2, x_3, x_4, -y_1, -y_2, -y_3, -y_4, x_5, x_6, x_7, x_8, \\ &-y_5, -y_6, -y_7, -y_8, \ldots, -y_{4^k}) \end{aligned}$$

and

$$v = (\; x_1, x_2, x_3, x_4, y_1, y_2, y_3, y_4, x_5, x_6, x_7, x_8, y_5, y_6, y_7, y_8, \ldots, y_{4^k},$$
$$-x_1, -x_2, -x_3, -x_4, y_1, y_2, y_3, y_4, -x_5, -x_6, -x_7, -x_8,$$
$$y_5, y_6, y_7, y_8, \ldots, y_{4^k}\;).$$

Besides the above constructions, Bent sequences can also be produced by Kronecker product, dyadic shift, Boolean variable transform, threshold logic synthesis, and so on. (e.g., see [12], [15], [16], [17], [18].)

The definitions of SAC and Bent functions have been generalized as the following propagation criterion [9].

Definition 5.3.4 *A Boolean function $f(x_1, \ldots, x_n)$ is said to 'satisfy the propagation criterion of degree k' (PC of degree k) if $f(x_1, \ldots, x_n)$ changes with a probability of one half whenever i, $1 \leq i \leq k$, bits of $x = (x_1, \ldots, x_n)$ are complemented, i.e.,*

$$w(f(x) + f(x + a)) = 2^{n-1}, \quad \text{if} \;\; 1 \leq w(a) \leq k.$$

Thus the SAC (or equivalently H-Boolean functions) is equivalent to the PC of degree 1, and Bent functions satisfy PC of the maximum degree n. A PC of degree k, $k > 1$, is also a PC of degree $k - 1$, and hence a PC of degree 1, the H-Boolean function.

It seems plausible to study what happens if m bits are kept constant in functions that satisfy PC of degree k. This allows for the following more general classification of propagation characteristics of Boolean functions.

Definition 5.3.5 ([9]) *A Boolean function $f(x_1, \ldots, x_n)$ is said to satisfy the propagation criterion of degree k and order m (PC of degree k and order m) if any function obtained from $f(x_1, \ldots, x_n)$ by keeping m input bits constant satisfies the PC of degree k.*

A function $f(x_1, x_2, \ldots, x_n)$ satisfying PC of degree 1 and order m produces an n-dimensional Hadamard matrix $A = [(-1)^{f(x_1, x_2, \ldots, x_n)}]$, in which all of the $(n - m)$-dimensional sections are $(n - m)$-dimensional Hadamard matrices of order 2. Thus all k-dimensional sections of this matrix A are Hadamard matrices of order 2 if $k \geq (n - m)$. In particular, functions satisfying PC of degree 1 and order $n - 2$ are those H-Boolean functions that produce absolutely proper higher-dimensional Hadamard matrices of order 2. Thus, According to Theorem 5.1.13, they are of the form

$$f(x_1, x_2, \ldots, x_n) = \sum_{1 \leq i < j \leq n} x_i x_j + a + \sum_{i=1}^{n} b_i x_i.$$

Similarly, functions satisfying PC of degree 1 and order $(n - 3)$ are those functions of the form

$$f(x_1, x_2, \ldots, x_n) = \sum_{1 \leq i < j \leq n} x_i x_j + g(x_1, \ldots, x_n),$$

where

$$g(x_1, x_2, \ldots, x_n) = \sum_{1 \leq i < j \leq n} b_{ij} x_i x_j + a + \sum_{i=1}^{n} a_i x_i,$$

with the condition $\sum_i b_{ij} \leq 1$ and $\sum_j b_{ij} \leq 1$, i.e., $g(x_1, \ldots, x_n)$ satisfies $\deg(g(.)) \leq 2$ such that every variable x_i occurs at most once in the second order terms.

5.3.3 Reed–Muller Codes and H–Boolean Functions ([7])

Reed–Muller codes are one of the oldest and best understood families of error-correcting codes. This subsection explores connections between n-dimensional Hadamard matrices of order 2 and Reed–Muller codes[7].

Reed–Muller code can be defined very simply in terms of Boolean functions (see [10]):

Definition 5.3.6 *The r-th order binary Reed–Muller code $R(r, n)$ of length 2^n, for $0 \leq r \leq n$, is the set of all truth tables of Boolean functions of degree up to r.*

Let $H(2, r, n)$ denote an n-dimensional Hadamard matrix of order 2 in which all of the r-dimensional sections are r-dimensional Hadamard matrices of order 2. Thus a $H(2, 2, n)$ is an absolutely proper Hadamard matrix of order 2. An n-dimensional Hadamard matrix is clearly an $H(2, n, n)$. In general, an $H(2, r, n)$ is also an $H(2, r+i, n)$ for any $i = 1, 2, \ldots, n-r$. In fact, it is easy to prove that a (± 1)-valued higher-dimensional matrix is a Hadamard matrix if its lower-dimensional layers are Hadamard matrices.

Recall that every n-dimensional matrix of the form

$$X(f) = [(-1)^{f(x_1, x_2, \ldots, x_n)}]$$

corresponds to the Boolean function $f(x)$, where $x = (x_1, x_2, \ldots, x_n)$. Similarly, every 2^n-dimensional binary vector corresponds to a unique Boolean function $f(x)$: the binary vector $c(f) = (f(x))_x$ being the truth table of $f(x)$. Let $C(2, r, n)$ denote the set of vectors $c(f)$ where $X(f)$ is an $H(2, r, n)$.

Lemma 5.3.5 ([7]) *Let $n > r \geq 1$ be two integers, and let $f(x)$ be a Boolean function of n variables. Then $c(f) \in R(r-1, n)$ if and only if every r-dimensional section of $X(f)$ has even Hamming weight, where the Hamming weight of some section X, formulated by $w(X)$, refers to the number of 1s contained in that section.*

Proof. First consider the Boolean function $f(x) = x_1 x_2 \ldots x_m$ where $m \leq n$. We show that every r-dimensional section of $X(f)$ has even weight if and only if $m < r$. Suppose $m \geq r$, and let X be the m-dimensional section of $X(f)$ defined by setting

$$x_{r+1} = x_{r+2} = \ldots = x_m = 1 \quad \text{and} \quad x_{m+1} = x_{m+2} = \ldots = x_n = 0.$$

Then its Hamming weight is $w(X) = 1$. Conversely, suppose $m < r$, and consider any r-dimensional section X of $X(f)$. For some $t > m$, x_t is not fixed. Let Y_0 and Y_1 be the $(r-1)$-dimensional sections of X that are defined by setting $x_t = 0$ and $x_t = 1$, respectively. Then $w(X) = w(Y_0) + w(Y_1)$, and $Y_0 = Y_1$; so $w(X)$ is even.

Now we prove the general result. Without loss of generality we may assume $x_1 x_2 \ldots x_m$ is the highest degree term in $f(x)$. By the argument above, if $m < r$, $X(f)$ is the sum of n-dimensional matrices whose r-dimensional sections all have even Hamming weights. Conversely, if $m \geq r$, then the m-dimensional section defined by setting $x_{m+1} = x_{m+2} = \ldots = x_n = 0$ will have an odd Hamming weight, and, since this is the sum of Hamming weights of 2^{m-r} r-dimensional sections parallel to each other, there must be at least one r-dimensional section with odd Hamming weight. Since $c(f) \in R(r-1, n)$ if and only if the degree of $f(x)$ is less than r, the lemma is proved.   **Q.E.D.**

Theorem 5.3.12 ([7]) *Let $n \geq 2$. Then $C(2, 2, n)$ is a coset of $R(1, n)$, the first-order Reed-Muller code.*

Proof. Fix $c(f_1) \in C(2,2,n)$. Let

$$B = \{c(f_1) + c(f) : c(f) \in C(2,2,n)\}.$$

Observe that every 2-dimensional Hadamard matrix of order 2 has odd Hamming weight; so $c(f) \in C(2,2,n)$ if and only if every 2-dimensional section of $X(f)$ has odd Hamming weight, and, hence, $c(f) \in B$ if and only if every 2-dimensional section of $X(f)$ has even Hamming weight. Therefore, by Lemma 5.3.5, $B = R(1,n)$.   **Q.E.D.**

The above Theorem 5.3.12 states that the set of absolutely proper Hadamard matrices of order 2 is equivalent to a coset of the first-order Reed-Muller code. This has the following three implications ([7]) :

1. If the entries are taken in a prescribed order, each matrix (and it negation) corresponds to a row of an Hadamard matrix that is equivalent to the Sylvester matrix.

2. Absolutely proper Hadamard matrices of order 2 may be repaired by any of the methods used to decode first-order Reed-Muller codes (the Yates transform will do this in order $n2^n$ steps.)

3. The values taken by any entry and its n adjacent entries determine the entire matrix. In particular, since each of these adjacent entries lies in separate $(n-1)$-dimensional sections, the absolutely proper n-dimensional Hadamard matrix of order 2 is unique up to the complementation of $(n-1)$-dimensional sections.

By Theorem 5.1.3,

$$C(2,2,n) = \left\{ \sum_{1 \leq i < j \leq n} x_i x_j + a_0 + \sum_{i=1}^{n} a_i x_i \right\}.$$

In particular, $f(x) = \sum_{1 \leq i < j \leq n} x_i x_j$ is an $H(2,2,n)$. Put $m = \lfloor n/2 \rfloor$, and

$$g(x) = \begin{cases} \sum_{i=1}^{m} x_{2i-1} x_{2i}, & \text{for } n \text{ even} \\[2ex] \sum_{i=1}^{m} x_{2i-1} x_{2i} + x_n \sum_{i=1}^{2m} x_i & \text{for } n \text{ odd.} \end{cases}$$

Then the Walsh transform of $g'(x) = (-1)^{g(x)}$ is

$$G'(w) = \begin{cases} 2^m \prod_{i=1}^{m} (-1)^{w_{2i} w_{2i-1}} & \text{for } n \text{ even,} \\[2ex] 2^m (1 + (-1)^{m+|w|}) \prod_{i=1}^{m} (-1)^{w_{2i} w_{2i-1}} & \text{for } n \text{ odd.} \end{cases}$$

Now put $b_i = \lfloor (i-1)/2 \rfloor$, and let L be the linear map such that $y = Lx$, where $y_n = x_n$ and for $i = 1, 2, \ldots, m$,

$$y_{2i-1} = x_{2i-1} + (x_{2i+1} + x_{2i+2} + \ldots + x_{2m}),$$

$$y_{2i} = x_{2i} + (x_{2i+1} + x_{2i+2} + \ldots + x_{2m}).$$

Then

$$f(x) + \sum_{i=1}^{2m} b_i x_i = g(y).$$

Finally, put

$$b = \begin{cases} (b_1, b_2, \ldots, b_{2m}), & \text{for } n \text{ even,} \\[2mm] (b_1, b_2, \ldots, b_{2m}, 0), & \text{for } n \text{ odd.} \end{cases}$$

Then the Walsh transform of $f'(x) = (-1)^{f(x)}$ satisfies

$$F'(wL^T + b) = G'(w) = \begin{cases} 2^m \prod_{i=1}^m (-1)^{w_{2i} w_{2i-1}} & \text{for } n \text{ even,} \\[3mm] 2^m (1 + (-1)^{m+|w|}) \prod_{i=1}^m (-1)^{w_{2i} w_{2i-1}} & \text{for } n \text{ odd.} \end{cases}$$

$$(5.16)$$

It follows that when $n = 2m$, half of the $H(2, 2, n)$ has $2^{m-1}(2^m - 1)$ entries equal to -1, and the remainder has $2^{m-1}(2^m + 1)$ entries equal to -1. When $n = 2m + 1$, half of it has 2^{2m} entries equal to -1, one quarter of it has $2^{m-1}(2^m - 1)$ entries equal to -1, and the rest has $2^{m-1}(2^m + 1)$ entries equal to -1. The Equation 5.16 implies

Theorem 5.3.13 ([7]) *Every $H(2, 2, 2m)$ corresponds to a Bent Function, or equivalently, every absolutely proper $(2m)$-dimensional Hadamard matrix of order 2 corresponds to a Bent function.*

This theorem is also a simple implication of Theorem 5.1.3.

Theorem 5.3.14 ([7]) *If $n \geq r \geq 3$, then $C(2, r, n) \subset R(r - 1, n)$.*

Proof. Because of Theorem 5.1.3, the Hamming weight of every $H(2, n, n)$ is even. Now suppose that $X(f)$ is an $H(2, r, n)$, and $X(g)$ is a r-dimensional section of $X(f)$. Then $X(g)$ is Hadamard and the Hamming weight $w(g)$ is even. Indeed, by Lemma 5.3.5, $c(f) \in R(r - 1, n)$ as required. To show that $C(2, r, n) \neq R(r - 1, n)$, consider $f(x) = x_1 x_2 \ldots x_{r-1}$ and the

r-dimensional section of $X(f)$ defined by putting $x_{r+1} = \ldots = x_n = 0$. **Q.E.D.**

So, for $n \geq r \geq 3$, if no more than $2^{n-r} - 1$ entries of an $H(2, r, n)$ are corrupted, then the usual methods used for decoding a corrupted codeword of $R(r-1, n)$ can be used to repair the $H(2, r, n)$. Precisely, by the known result of the minimum Hamming distance of Reed-Muller code $R(r-1, n)$, we have the following corollary:

Corollary 5.3.1 ([7]) *If $n \geq r \geq 2$, then $C(2, r, n)$ has minimum Hamming distance of at least 2^{n-r+1}.*

Theorem 5.3.15 ([7]) *Let $n - 1 \geq t \geq r \geq 1$ be integers. Let $g(.)$ be any Boolean function of $(n-t)$ variables, and let $X(h)$ be an $H(2, r+1, n)$. If*

$$f(x_1, x_2, \ldots, x_n) = g(x_1, x_2, \ldots, x_{n-t}) + h(x_1, x_2, \ldots, x_n)$$

then $X(f)$ is an $H(2, n+r-t, n)$.

Proof. It is sufficient to show that any two opposed $(n - r - t - 1)$-dimensional sections are orthogonal. The proof can be made by two cases according to whether g depends on the opposed index i_m. If $m > n - t$, then g has no bearing on the inner product. If $m \leq n - t$, then the result is obtained by fixing the varied indexes among the indexes $i_1, i_2, \ldots, i_{n-t}$, and using the orthogonality properties of $X(h)$. We do this explicitly for $t = 1$ and leave the details of the general result to the interested reader. Without loss of generality, we need only show that the pair of $(n - 1)$-dimensional sections defined by setting $x_1 = 0$ and 1 and the pair obtained by setting $x_n = 0$ and 1 are orthogonal pairs. Equivalently, we must show that

$$w_1 = w(f(0, x_2, \ldots, x_n) + f(1, x_2, \ldots, x_n)) = 2^{n-2}$$

and

$$w_2 = w(f(x_1, \ldots, x_{n-1}, 0) + f(x_1, \ldots, x_{n-1}, 1)) = 2^{n-2}.$$

Now

$$w_2 = w(h(x_1, \ldots, x_{n-1}, 0) + h(x_1, \ldots, x_{n-1}, 1)) = 2^{n-2},$$

since $X(h)$ is Hadamard, and

$$w_1 = w(g(0, x_2, \ldots, x_{n-r}) + g(1, x_2, \ldots, x_{n-r}))$$

$$+h(0, x_2, \ldots, x_n) + h(1, x_2, \ldots, x_n))$$

$$= \sum_{i=2}^{n-r} (\sum_{a_i=0}^{1} w(g(0, a_2, \ldots, a_{n-r}) + g(1, a_2, \ldots, a_{n-r})$$

$$+h(0, a_2, \ldots, a_{n-r}, x_{n-r+1}, \ldots, x_n)$$

$$+h(1, a_2, \ldots, a_{n-r}, x_{n-r+1}, \ldots, x_n)))$$

$$= 2^{n-r-1} 2^{r-1},$$

since every $(r + 1)$-dimensional section of $X(h)$ is Hadamard.   **Q.E.D.**

As a corollary, we obtain an extension of Theorem 5.3.12.

Corollary 5.3.2 ([7]) *If $n \geq r \geq 2$, then $C(2, r, n)$ is a union of cosets of $R(1, n)$.*

Proof. Let $s = r - 1$. It is sufficient to prove that $X(f)$, where $f(x) = h(x) + \sum_{i=1}^{n} a_i x_i + a_0$, is an $H(2, s+1, n)$ whenever $X(h)$ is an $H(2, s+1, n)$. To do so, fix $h(x)$ and apply Theorem 5.3.15 (with $r = s$) successively to $g = g_i$, where $g_i = a_i x_i$, $i = 1, 2, \ldots, n$, and $g_0 = a_0$. Since $t = n - 1$, the result follows. **Q.E.D.**

Arguments similar to those used in the proof of Theorem 5.3.15 can be used to prove the following result.

Theorem 5.3.16 ([7]) *Let h and g be Boolean functions of, respectively, t and $n - t$ variables and let*

$$f(x_1, x_2, \ldots, x_n) = g(x_1, x_2, \ldots, x_{n-1})$$

$$+ \sum_{i \leq t,\, j > t} x_i x_j + h(x_{n-t+1}, \ldots, x_n).$$

If $X(h)$ is an $H(2, r, t)$, the $X(f)$ is an $H(2, n - t + r, n)$. Moreover, if $X(g)$ is an $H(2, s, n - t)$ and $u = \max(t + s, n - t + r)$, then $X(f)$ is an $H(2, u, n)$.

Corollary 5.3.3 ([7]) *Let $n \geq r \geq 2$. Then*

$$|C(2, r, n)| \geq \sum_{t=n-r+2}^{n} \binom{n}{t} \sum_{i=0}^{n-t} (-1)^i \binom{n-t}{i} 2^{t+i+2^{n-t-i}}.$$

Proof. For all $i = 1, 2, \ldots, n$, let $h_i(x) = x_i \sum_{j=1}^{n} x_j$. We will say f "has attribute i" if $f(x) = h_i(x) + g(x) + ax_i$, where $a = 0$ or 1, and $g(x)$ is independent of x_i. Now let $N = \{1, 2, \ldots, n\}$, and, for all $T \subset N$, let $S(T)$ denote the set of $H(2, n, n)$ of the form $X(f)$ where, for all $i \in T$, f has property i. The typical element of $S(T)$ is $X(f)$ where

$$f(x) = g(x) + \sum_{i \in T,\ j \in N-T} x_i x_j + \sum_{i \in T} a_i x_i + \sum_{i < j,\ i,j \in T} x_i x_j,$$

where $g(x)$ is independent of x_i for all $i \in T$. Hence, if $t = \mid T \mid$, $\mid S(T) \mid = 2^t 2^{2^{n-t}}$. By Theorem 5.3.16, every element of $S(T)$ is an $H(2, n-t+2, n)$. Now let $E(T)$ be the set of matrices of the form $X(f)$ where f has property i precisely when $i \in T$, then, by the Principle of Inclusion and Exclusion,

$$\mid E(T) \mid = \sum_{i=0}^{n-t} (-1)^i \binom{n-t}{i} 2^{t+i+2^{n-t-i}}.$$

Q.E.D.

For more details of higher-dimensional Hadamard matrices of order 2, the readers are recommended to the papers [20-24].

Bibliography

[1] P.J. Shlichta, *Higher-Dimensional Hadamard Matrices*, IEEE Trans. On Inform. Theory, Vol.IT-25, No.5, pp.566-572, 1979.

[2] Y. Xing and Yi Xian Yang, *On the H–Boolean Functions (II)*, J. of Electronics (China), Vol.19, No.2, pp214-216.

[3] Y.X. Yang, *On the H–Boolean Functions*, J. of Beijing Univ. of Posts and Telecomm, Vo.11, No.3, pp1-9, 1988.

[4] A.F. Webster and S.E. Tavares, On the Design of S-box, Advances in Cryptology, Proc. Crypto'85, Springer-Verlag, Berlin, 1986, pp523-534.

[5] S.A. Lloyd, *Characterising and Counting Functions Satisfying the Strict Avalanche Criterion of Order* $(n - 3)$, Proc. of the Second IMA Conf. on Cryptography and Coding, 1989, Clarendon Press, Oxford, 1992, pp165-172.

[6] S.A. Lloyd, *Counting Binary Functions with Certain Cryptographic Properties*, J. of Cryptology, (1992)5:107-131.

[7] R.Forre, *The Strict Avalanche Criterion: Spectral Properties of Boolean Functions and an Extended Definition*, Advances in Cryptology, Proc. Crypto'88, Springer-Verlag, Berlin, 1986, pp450-468.

[8] T.W.Cusick, *Boolean Functions Satisfying a Higher Order Strict Avalanche Criterion*, Proc. of Eurocrypt'93, pp86-95.

[9] B.Preneel, W.Vanleekwijk etal, *Propagation Characteristics of Boolean Functions*, Euro-crypt'90, 1991: 161-173.

[10] F.J.Macwilliams, N. Sloane, *The Theory of Error-Correcting Codes*, North-Holland, New York, 1977.

[11] Yi Xian Yang and X.D. Lin, *Coding and Cryptography* , PPT Press, Beijing, 1992.

[12] R. Yarlagadda and J.E. Hershey, *Analysis and Synthesis of Bent Sequences*, IEE Proc. Vol.136, Pt.E, No.2, pp112-123, 1989.

[13] C.M. Adams and S.E. Tavares, *Generating Bent Sequences*, IEEE Trans. Inform. Theory, Vol.36, No.5, pp1170-1173, 1990.

[14] B.A. Guo and C.N.Cai, *Generating and Counting a Class of Binary Bent functions Which is Neither Bent-Based Nor Linear-Based*, Chinese Science Bulletin, Vol.37, No.6, pp517-520, 1992.

[15] C.Carlet, *Two New Classes of Bent Functions*, Proc. of Eurocrypt'93, Lofthus, Norway, 1993, pp75-85.

[16] Yi Xian Yang and Z.M.Hu, *Dyadic Codes With Single Valued Correlations (I)*, Chinese J. of Electronics, Vol.16, No.6, pp50-55, 1988.

[17] Z.M. Hu and Yi Xian Yang, *Dyadic Codes With Single Valued Correlations (II)*, J. of China Institute of Communications, Vol.10, No.5, pp42-46, 1989.

[18] Z.M. Hu and Yi Xian Yang, *On the Dyadic Cross-Correlations of Dyadic Codes* J. of China Institute of Communications, Vol.14, No.1, pp15-21, 1993.

[19] W.Launey, *A Note on N-Dimensional Hadamard Matrices of Order 2^t and Reed-Muller Codes*, IEEE Trans. on Inform. Theory, Vol.27, No.3, pp664-667, 1991.

[20] Y.X. Yang, *N-Dimensional Hadamard Matrices of Order Two* , J. of Beijing Univ. of Posts and Telecomm, Vo.14, No.4, pp1-8, 1991.

[21] Yi Xian Yang and Zhen Ming Hu, *On the Classification of 4-dimensional Hadamard Matrices of Order 2*, J. of Systems Science and Mathematical Sciences, Vol.7, No.1, pp40-46, 1987.

[22] Xin An Pan and Yi Xian Yang, *On the Enumeration of 5-Dimensional Hadamard Matrices of Order 2*, J. of Beijing Univ. of Posts and Telecomm, Vo.10, No.4, pp11-19, 1987.

[23] S. Q. Li and Yi Xian Yang, *The Final Solution of Enumerating the 5-Dimensional Hadamard Matrices of Order 2*, J. of Beijing Univ. of Posts and Telecomm, Vo.11, No.2, pp17-21, 1988.

[24] J.Hammer and J.Seberry, *Higher-Dimensional Orthogonal Designs And Applications*, IEEE Trans. Inform. Theory, Vol.27, No.6, pp772-779, 1981.

Chapter 6
General Higher-Dimensional Hadamard Matrices

When m^n ($n \geq 2$ an integer) elements are given they can be arranged in the form of an n-dimensional cube of order m (in short, an n-cube). An n-cube can be mathematically described by the matrix form $A = [A(i_1, i_2, \ldots, i_n)]$, $0 \leq i_1, i_2, \ldots, i_n \leq m-1$. The elements which have all the same suffixes, with the exception of i_k and/or i_s, lie in the same two-dimensional layer parallel to a coordinate axis (a plane). Thus an n-cube can be treated as a set of m^{n-2} two-dimensional square matrices $B = [B(x, y)] = [A(a_1, a_2, \ldots, x, \ldots, y, \ldots, a_n)]$, $0 \leq x, y \leq m - 1$, where the a_is are prefixed integers. The elements which have all the same suffices, except i_k, lie in the same row (line). Thus an n-cube can be treated as a set of m^{n-1} lines of length m. The elements which have only one suffix in common lie in an $(n-1)$-dimensional layer. Thus an n-cube can also be treated as a set of m $(n-1)$-dimensional layers. With regard to practical applications, the most obvious advantage of higher-dimensional Hadamard matrices is the presence or absence of property. Some higher-dimensional Hadamard matrices, especially, those proper Hadamard matrices of dimension $n \geq 4$ and those n-dimensional Hadamard matrices of order two will prove advantageous in error correcting codes. Certain forms of higher-dimensional Hadamard matrices, e.g., H-Boolean functions, may be of value in security coding because of their resemblance to random matrices. Hadamard matrices are kinds of paradigm of random binary matrices in which the correlation values for any pair of parallel $(n-1)$-dimensional sections (i.e., zero) are the expected values for such correlations in a random matrix. It therefore seems plausible to regard any sufficiently large random binary matrix as potentially an Hadamard matrix with errors which can be located and corrected. Those Hadamard matrices derived by checking a random binary matrix might usefully combine error correction with immunity from unauthorized decoding, because of the absence of obvious pattern and their resemblance to random matrices.

This chapter concentrates on the theory of higher-dimensional Hadamard matrices and their generalizations. Definitions, properties, existences, constructions, and the other related topics will be systematically introduced in the coming sections.

6.1   Definitions, Existences and Constructions

The definition of n-dimensional Hadamard matrices has been informally stated in the last few chapters. Now we present, here, its mathematical definition.

Definition 6.1.1 ([1]) *A general n-dimensional Hadamard matrix of order m is an n-dimensional binary matrix*

$$H = [H(i_1, i_2, \ldots, i_n)], \quad 0 \leq i_k \leq m-1, \quad 1 \leq k \leq n,$$

of size $\overbrace{m \times \ldots \times m}^{n}$ in which all parallel $(n-1)$-dimensional sections (layers) are mutually orthogonal; that is, all $H(i_1, i_2, \ldots, i_n) = \pm 1$ and

$$\sum_{i_1=0}^{m-1} \cdots \sum_{i_{n-1}=0}^{m-1} H(i_1, i_2, \ldots, i_{n-1}, a) H(i_1, i_2, \ldots, i_{n-1}, b) = m^{(n-1)} \delta_{ab}$$

$$\sum_{i_1=0}^{m-1} \cdots \sum_{i_{n-2}=0}^{m-1} \sum_{i_n=0}^{m-1} H(i_1, i_2, \ldots, i_{n-2}, a, i_n) H(i_1, i_2, \ldots, i_{n-2}, b, i_n) = m^{(n-1)} \delta_{ab}$$

$$\vdots$$

$$\sum_{i_2=0}^{m-1} \cdots \sum_{i_n=0}^{m-1} H(a, i_2, i_3, \ldots, i_n) H(b, i_2, i_3, \ldots, i_n) = m^{(n-1)} \delta_{ab}. \quad (6.1)$$

6.1.1 n-Dimensional Hadamard Matrices of Order $2k$

Recall that the order m of a non-trivial 2-dimensional Hadamard matrix has to be $m = 2$ or $m = 4k$, whilst the order m of a non-trivial 3-dimensional Hadamard matrix can be $m = 2k$. In this subsection, it will be proved, at first, that the orders of the non-trivial general n-Dimensional Hadamard matrices are also necessary to be even integers. Then infinite families of $n(\geq 4)$-dimensional Hadamard matrices of order $2k$ will be constructed by algebraic approaches.

Theorem 6.1.1 *Let $H = [H(h_1,\ldots,h_n)]$, $0 \leq h_k \leq m-1$, $1 \leq k \leq n$, be an $n(\geq 3)$-dimensional Hadamard matrix of order m. If $m > 1$, then m is necessarily even.*

Proof. From Definition 6.1.1, we have the following

$$\sum_{0 \leq h_2,\ldots,h_n \leq m-1} H(a,h_2,\ldots,h_n)H(b,h_2,\ldots,h_n) = m^{n-1}\delta(a,b), \qquad (6.2)$$

which implies

$$\sum_{0 \leq h_2,\ldots,h_n \leq m-1} [H(0,h_2,\ldots,h_n) + H(1,h_2,\ldots,h_n)]H(0,h_2,\ldots,h_n)$$

$$= \sum_{0 \leq h_2,\ldots,h_n \leq m-1} [H(0,h_2,\ldots,h_n)H(0,h_2,\ldots,h_n)$$

$$+ H(0,h_2,\ldots,h_n)H(1,h_2,\ldots,h_n)]$$

$$= \sum_{0 \leq h_2,\ldots,h_n \leq m-1} [1 + H(0,h_2,\ldots,h_n)H(1,h_2,\ldots,h_n)]$$

$$= m^{n-1} \qquad \text{(By Equation (6.2))} . \qquad (6.3)$$

Because $H(h_1,\ldots,h_n) = \pm 1$, $H(0,h_2,\ldots,h_n) + H(1,h_2,\ldots,h_n)$ is even. In other words, the left hand side of Equation (6.3) is the summation of some even integers, and its right hand side, m^{n-1}, is also even. So the order m has to be even.   **Q.E.D.**

Theorem 6.1.2 *Let $1 \leq i,j,k,l \leq 6$ be integers with their binary expended forms*

$$i = \sum_{s=0}^{2} i_s 2^s, \quad j = \sum_{s=0}^{2} j_s 2^s, \quad k = \sum_{s=0}^{2} k_s 2^s, \quad \text{and} \quad l = \sum_{s=0}^{2} l_s 2^s.$$

Then the matrix $H = [H(i,j,k,l)]$, $1 \leq i,j,k,l \leq 6$, defined by

$$H(i,j,k,l) = (-1)^{i_0 j_0 + i_1 k_1 + i_2 l_2 + j_1 l_1 + j_2 k_0 + k_2 l_0}$$

is a 4-dimensional Hadamard matrix of order 6.

Proof. Let $a = \sum_{s=0}^{2} a_s 2^s$ and $b = \sum_{s=0}^{2} b_s 2^s$ be two different integers that $1 \leq a \neq b \leq 6$. Then

$$\sum_{1 \leq j,k,l \leq 6} H(a,j,k,l)H(b,j,k,l)$$

$$= \sum_{1 \leq j,k,l \leq 6} (-1)^{(a_0+b_0)j_0+(a_1+b_1)k_1+(a_2+b_2)l_2}$$

$$= [\sum_{j=1}^{6}(-1)^{(a_0+b_0)j_0}][\sum_{k=1}^{6}(-1)^{(a_1+b_1)k_1}][\sum_{l=1}^{6}(-1)^{(a_2+b_2)l_2}]. \qquad (6.4)$$

The non-equality $a \neq b$ implies $a_0 + b_0 = 1$, $a_1 + b_1 = 1$, or $a_2 + b_2 = 1$. These three possible cases are separately studied as follows:

Case 1: $a_0 + b_0 = 1$ implies $\sum_{j=1}^{6}(-1)^{(a_0+b_0)j_0} = 0$, and the first term in Equation (6.4) is vanished;

Case 2: $a_1 + b_1 = 1$ implies $\sum_{k=1}^{6}(-1)^{(a_1+b_1)k_1} = 0$, and the second term in Equation (6.4) is vanished;

Case 3: $a_2 + b_2 = 1$ implies $\sum_{l=1}^{6}(-1)^{(a_2+b_2)l_2} = 0$, and the third term in Equation (6.4) is vanished.

Thus the left hand side of Equation (6.4) always vanishes, i.e.,

$$\sum_{1 \leq j,k,l \leq 6} H(a,j,k,l)H(b,j,k,l) = 6^3\delta(a,b).$$

Similarly, it can be proved that the following equations are true:

$$\sum_{1 \leq i,k,l \leq 6} H(i,a,k,l)H(i,b,k,l) = \sum_{1 \leq i,j,l \leq 6} H(i,j,a,l)H(i,j,b,l)$$

$$= \sum_{1 \leq i,j,k \leq 6} H(i,j,k,a)H(i,j,k,b)$$

$$= 6^3\delta(a,b).$$

Hence the matrix $H = [H(i,j,k,l)]$ is a 4-dimensional Hadamard matrix of order 6.   **Q.E.D.**

In order to construct higher-dimensional Hadamard matrices of general order $2k$, we prove the following important lemma at first.

Lemma 6.1.1 *Let n and k be two integers such that $k \leq 2^{n-1}$. Then there exists a 2-dimensional $(0,1)$-valued matrix $A = [A(i,j)]$ of size $(2k) \times n$ which satisfy the following two conditions:*

1. *The rows of A are different from each other;*

2. *Each column of A contains k 0s and k 1s.*

A matrix satisfying both of these two conditions is called a column-balanced matrix, because each of its column is balanced by 1s and 0s.

Proof. We show a constructive proof for this lemma.

Let $A_0 = [A_0(i,j)]$, $0 \leq i \leq 2^n - 1$, $0 \leq j \leq n - 1$, be the $(0,1)$-valued matrix such that its i-th row $(A_0(i,0),\ldots,A_0(i,n-1))$ is the binary expended vector of the integer i, i.e., $i = \sum_{k=0}^{n-1} A_0(i,k)2^k$. It is easy to verify that this matrix A_0 is a column-balanced matrix of size $2^n \times n$.

Two binary vectors $(i_0,\ldots,i_{n-1})$ and $(j_0,\ldots,j_{n-1})$ are said to be a complementary-pair if

$$(i_0,\ldots,i_{n-1}) + (j_0,\ldots,j_{n-1}) = (1,\ldots,1).$$

The rows of A_0 clearly consist of 2^{n-1} complementary-pairs. For any integer r, $0 \leq r \leq 2^{n-1} - 1$, a column-balanced matrix of size $(2^n - 2r) \times n$ is obtained by deleting r complementary-pairs from the matrix A_0.   **Q.E.D.**

Theorem 6.1.3 *Let $n = 2s$ be even, $m \leq s - 1$, and $k \leq 2^{m-1}$. And let $B = [B(i,j)]$, $0 \leq i \leq 2k-1$, be a column-balanced matrix of size $(2k) \times m$. Then the following matrix $H = [H(h(0),\ldots,h(n-1))]$, $0 \leq h(i) \leq 2k-1$, $0 \leq i \leq n-1$, defined by*

$$H(h(0),\ldots,h(n-1)) = (-1)^{\sum_{i=0}^{s-1}\sum_{j=0}^{m-1} B(h(i),j)B(h((i+j)\bmod s+s),j)}$$

is an n-dimensional Hadamard matrix of order 2k.

Proof. For every prefixed integer l, $0 \leq l \leq n - 1$, let

$$\alpha(a,b) = \sum_{\substack{0 \leq h(0),\ldots,h(l-1),h(l+1),\ldots,h(n-1) \leq 2k-1}} H(h(0),\ldots,h(l-1),$$
$$a, h(l+1),\ldots,h(n-1))H(h(0),\ldots,h(l-1),$$
$$b, h(l+1),\ldots,h(n-1)), \tag{6.5}$$

where $0 \leq a,b \leq 2k - 1$.

It is sufficient to prove that $\alpha(a,b) = (2k)^{n-1}\delta(a,b)$. The following two cases are separately considered:

Case 1: If $0 \leq l \leq s - 1$, then by the definition of the matrix H, we have

$$\alpha(a, b) = \sum_{0 \leq h(0),\ldots,h(l-1),h(l+1),\ldots,h(n-1) \leq 2k-1} (-1)^{\sum_{j=0}^{m-1}[B(a,j)+B(b,j)]B(h((l+j) \bmod s+s),j)}$$

$$= \sum_{0 \leq h(r) \leq 2k-1,\ r \neq l,\ l+s,\ (l+1)\bmod s+s,\ \ldots,\ (l+m-1)\bmod s+s} \left\{ \sum_{0 \leq h(l+s),h((l+1)\bmod s+s),\ldots,h((l+m-1)\bmod s+s) \leq 2k-1} (-1)^{\sum_{j=0}^{m-1}[B(a,j)+B(b,j)]B(h((l+j)\bmod s+s),j)} \right\}$$

$$= \sum_{0 \leq h(r) \leq 2k-1,\ r \neq l,\ l+s,\ (l+1)\bmod s+s,\ \ldots,\ (l+m-1)\bmod s+s} \prod_{j=0}^{m-1} \left\{ \sum_{0 \leq h((l+j)\bmod s+s) \leq 2k-1} (-1)^{[B(a,j)+B(b,j)]B(h((l+j)\bmod s+s),j)} \right\}.$$

Because the rows of the matrix $B = [B(i, j)]$ are different from each other, $a \neq b$ implies the existence of some integer j_0, $0 \leq j_0 \leq m - 1$, such that $B(a, j_0) + B(b, j_0) = 1$, which implies, in further, the following

$$\sum_{0 \leq h((l+j_0)\bmod s+s) \leq 2k-1} (-1)^{[B(a,j_0)+B(b,j_0)]B(h((l+j_0)\bmod s+s),j_0)} = 0,$$

which is owed to the columns of $B = [B(i, j)]$ being balanced by 1s and 0s.

Hence, we have proved that if $0 \leq l \leq s-1$, then $\alpha(a, b) = (2k)^{n-1}\delta(a, b)$.

Case 2: If $s \leq l \leq n - 1$, then by the same way as that used in Case 1, it can be proved that the equation $\alpha(a, b) = (2k)^{n-1}\delta(a, b)$ is also true. The theorem follows.   **Q.E.D.**

Theorem 6.1.4 *Let $A = [A(i, j)]$, $0 \leq i, j \leq (2t)^s - 1$, be a 2-dimensional Hadamard matrix of order $(2t)^s$, where $s > 1$ and t are positive integers. Then the following matrix $H = [H(x_0, \ldots, x_{s-1}, y_0, \ldots, y_{s-1})]$,*

$0 \leq x_i, y_i \leq 2t - 1$, $0 \leq i \leq s - 1$, *defined by*

$$H(x_0, \ldots, x_{s-1}, y_0, \ldots, y_{s-1})$$
$$= A((2t)^{s-1}x_{s-1} + (2t)^{s-2}x_{s-2} + \ldots + 2tx_1$$
$$+ x_0, (2t)^{s-1}y_{s-1} + (2t)^{s-2}y_{s-2} + \ldots + 2ty_1 + y_0)$$

is a $(2s)$-dimensional Hadamard matrix of order $2t$.

Proof. By the definition of the matrix H, we have

$$\alpha(a, b) = \sum_{0 \leq x_0, \ldots, x_{s-1}, y_0, \ldots, y_{s-1} \leq 2t-1}$$

$$\times H(a, x_1, \ldots, x_{s-1}, y_0, \ldots, y_{s-1}) H(b, x_1, \ldots, x_{s-1}, y_0, \ldots, y_{s-1})$$

$$= \sum_{0 \leq x_0, \ldots, x_{s-1} \leq 2t-1} \left[\sum_{0 \leq y_0, \ldots, y_{s-1} \leq 2t-1} \right.$$

$$A((2t)^{s-1}x_{s-1} + \ldots + 2tx_1 + a, \ (2t)^{s-1}y_{s-1} + \ldots + 2ty_1 + y_0)$$

$$\left. A((2t)^{s-1}x_{s-1} + \ldots + 2tx_1 + b, \ (2t)^{s-1}y_{s-1} + \ldots + 2ty_1 + y_0) \right].$$

$$(6.6)$$

If $a \neq b$, then

$$(2t)^{s-1}x_{s-1} + (2t)^{s-2}x_{s-2} + \ldots + 2tx_1 + a$$
$$\neq (2t)^{s-1}x_{s-1} + (2t)^{s-2}x_{s-2} + \ldots + 2tx_1 + b.$$

Thus the inner summation of Equation (6.6) is

$$\sum_{0 \leq y_0, \ldots, y_{s-1} \leq 2t-1} [A((2t)^{s-1}x_{s-1} + \ldots + 2tx_1 + a, \ (2t)^{s-1}y_{s-1} + \ldots$$

$$+ 2ty_1 + y_0) A((2t)^{s-1}x_{s-1} + \ldots + 2tx_1 + b, \ (2t)^{s-1}y_{s-1} + \ldots$$

$$+ 2ty_1 + y_0)]$$

$$= \sum_{r=0}^{(2t)^s - 1} A((2t)^{s-1}x_{s-1} + \ldots + 2tx_1 + a, r) A((2t)^{s-1}x_{s-1} + \ldots$$

$$+ 2tx_1 + b, r) = 0,$$

where the last equation is due to the fact that the matrix $A = [A(i,j)]$ is a 2-dimensional Hadamard matrix. Thus we have

$$\sum_{r=0}^{(2t)^s-1} A(u,r)A(v,r) = 0, \text{ if } u \neq v.$$

By the same way it can be proved that for each l, $0 \leq l \leq 2s - 1$, we have the equation

$$\sum_z H(z_0, \ldots, z_{l-1}, a, z_{l+1}, \ldots, z_{2s-1})H(z_0, \ldots, z_{l-1}, b, z_{l+1}, \ldots, z_{2s-1})$$
$$= (2t)^{2s-1}\delta(a,b).$$

Hence the matrix H is a $(2s)$-dimensional Hadamard matrix of order $2t$. **Q.E.D.**

Lemma 6.1.2 If $0 \leq a, b, j \leq N - 1$ and $a \neq b$, then $(a + j) \bmod N \neq (b + j) \bmod N$.

The following theorem provides us a recursive construction of higher-dimensional Hadamard matrices from lower-dimensional ones.

Theorem 6.1.5 Let $H = [H(i_1, \ldots, i_n)]$, $0 \leq i_1, \ldots, i_n \leq N - 1$, be an n-dimensional Hadamard matrix of order N. Then the matrix $A = [A(i_1, \ldots, i_n, i_{n+1})]$, $0 \leq i_1, \ldots, i_n, i_{n+1} \leq N - 1$, defined by

$$A(i_1, \ldots, i_n, i_{n+1}) = H(i_1, \ldots, i_{n-1}, (i_n + i_{n+1}) \bmod N))$$

is an $(n + 1)$-dimensional Hadamard matrix of order N.

Proof. Let

$$\alpha(a,b) = \sum_{0 \leq i_2, \ldots, i_{n+1} \leq N-1} A(a, i_2, \ldots, i_{n+1})A(b, i_2, \ldots, i_{n+1})$$

$$= \sum_{0 \leq i_2, \ldots, i_{n+1} \leq N-1} H(a, i_2, \ldots, i_{n-1}, (i_n + i_{n+1}) \bmod N))$$

$$\times H(b, i_2, \ldots, i_{n-1}, (i_n + i_{n+1}) \bmod N))$$

$$= \sum_{i_{n+1}=0}^{N-1} \{ \sum_{0 \leq i_2, \ldots, i_n \leq N-1} H(a, i_2, \ldots, i_{n-1}, (i_n + i_{n+1}) \bmod N))$$

$$\times H(b, i_2, \ldots, i_{n-1}, (i_n + i_{n+1}) \bmod N)) \}. \tag{6.7}$$

For each i_{n+1}, $0 \leq i_{n+1} \leq N - 1$, and $a \neq b$, the inner summation of Equation (6.7) is

$$\sum_{0 \leq i_2,\ldots,i_n \leq N-1} H(a, i_2, \ldots, i_{n-1}, (i_n + i_{n+1}) \mathrm{mod} N)) H(b, i_2, \ldots, i_{n-1}, (i_n$$
$$+ i_{n+1}) \mathrm{mod} N))$$
$$= \sum_{0 \leq i_2,\ldots,i_n \leq N-1} H(a, i_2, \ldots, i_n) H(b, i_2, \ldots, i_n)$$

$$= 0,$$

where the last equation is due to the fact that $H = [H(i_1, \ldots, i_n)]$ is an n-dimensional Hadamard matrix of order N.

Similarly, it can be proved that for each l, $1 \leq l \leq n$,

$$\sum_i A(i_1, \ldots, i_{l-1}, a, i_{l+1}, \ldots, i_{n+1}) A(i_1, \ldots, i_{l-1}, b, i_{l+1}, \ldots, i_{n+1})$$
$$= N^n \delta(a, b).$$

Let

$$\beta(a, b) = \sum_{0 \leq i_1,\ldots,i_n \leq N-1} A(i_1, \ldots, i_n, a) A(i_1, \ldots, i_n, b)$$

$$= \sum_{0 \leq i_1,\ldots,i_n \leq N-1} H(i_1, \ldots, i_{n-1}, (i_n + a) \ \mathrm{mod} N)) H(i_1, \ldots, i_{n-1},$$
$$(i_n + b) \ \mathrm{mod} N))$$

$$= \sum_{i_n=0}^{N-1} \{ \sum_{0 \leq i_1,\ldots,i_{n-1} \leq N-1} H(i_1, \ldots, i_{n-1}, (i_n + a) \ \mathrm{mod} N))$$

$$\times H(i_1, \ldots, i_{n-1}, (i_n + b) \ \mathrm{mod} N)) \}. \tag{6.8}$$

By Lemma 6.1.2, $a \neq b$ implies $(i_n + b) \mathrm{mod} N \neq (i_n + a) \mathrm{mod} N$. Thus for each i_n, $0 \leq i_n \leq N - 1$, the inner summation of the last Equation (6.8) is

$$\sum_{0 \leq i_1,\ldots,i_{n-1} \leq N-1} H(i_1, \ldots, i_{n-1}, (i_n + a) \mathrm{mod} N)) H(i_1, \ldots,$$
$$i_{n-1}, (i_n + b) \mathrm{mod} N)) = 0.$$

Hence $\beta(a, b) = N^n \delta(a, b)$.   **Q.E.D.**

Recall that the famous Hadamard Conjecture states that 'For each positive integer k, there exists at least one 2-dimensional Hadamard of order $4k$.' Theorems 6.1.5 and 6.1.3 lead to the interesting statement: 'if the Hadamard Conjecture had been proved, then for each positive $n(\geq 4)$ and $2t$, there exists at least one n-dimensional Hadamard matrix of order $2t$, even though t is odd.'

Theorem 6.1.5 can be generalized further. For example, we have

Theorem 6.1.6 *Let* $H = [H(i_1, \ldots, i_n)]$, $0 \leq i_1, \ldots, i_n \leq N - 1$, *be an* n-*dimensional Hadamard matrix of order* N. *And let* r *be an integer satisfying* $\gcd(r, N) = 1$. *Then the matrix* $A = [A(i_1, \ldots, i_{n+1})]$, $0 \leq i_1, \ldots, i_{n+1} \leq N - 1$, *defined by*

$$A(i_1, \ldots, i_{n+1}) = H(i_1, \ldots, i_{n-1}, (i_n + r i_{n+1}) \bmod N)$$

is an $(n + 1)$-*dimensional Hadamard matrix of order* N.

Clearly, this theorem reduces to Theorem 6.1.5 if $r = 1$.

Proof. For $0 \leq a, b \leq N - 1$, let

$$\alpha(a, b) = \sum_{0 \leq i_2, \ldots, i_{n+1} \leq N-1} A(a, i_2, \ldots, i_{n+1}) A(b, i_2, \ldots, i_{n+1})$$

$$\beta(a, b) = \sum_{0 \leq i_1, \ldots, i_{n-1}, i_{n+1} \leq N-1} A(i_1, \ldots, i_{n-1}, a, i_{n+1}) A(i_1, \ldots, i_{n-1}, b, i_{n+1})$$

and

$$\gamma(a, b) = \sum_{0 \leq i_1, \ldots, i_n \leq N-1} A(i_1, \ldots, i_n, a) A(i_1, \ldots, i_n, b).$$

It is sufficient to prove that

$$\alpha(a, b) = \beta(a, b) = \gamma(a, b) = N^n \delta(a, b).$$

The equation $\alpha(a, b) = N^n \delta(a, b)$ can be proved by the same way as that used in the proof of Theorem 6.1.5.

If $a \neq b$, then

$$\beta(a, b) = \sum_{0 \leq i_1, \ldots, i_{n-1}, i_{n+1} \leq N-1}$$

$$\times H(i_1, \ldots, i_{n-1}, (a + r i_{n+1}) \bmod N)$$

$$\times H(i_1,\ldots,i_{n-1},(b+ri_{n+1})\bmod N)$$

$$= \sum_{i_{n+1}=0}^{N-1} \{ \sum_{0\le i_1,\ldots,i_{n-1}\le N-1}$$

$$\times H(i_1,\ldots,i_{n-1},(a+ri_{n+1})\bmod N)$$

$$\times H(i_1,\ldots,i_{n-1},(b+ri_{n+1})\bmod N)\}$$

$$= \sum_{i_{n+1}=0}^{N-1} 0$$

$$= 0,$$

where the last two equations are due to the facts (1) H is an Hadamard matrix, and (2) $(b+ri_{n+1})\bmod N \ne (a+ri_{n+1})\bmod N$, if $0 \le a \ne b \le N-1$.

Hence we have proved that $\beta(a,b) = N^n\delta(a,b)$.

If $0 \le a \ne b \le N-1$, and $\gcd(r,N)=1$, then $[(i_n+ra)-(i_n+rb)]\bmod N = [r(a-b)]\bmod N \ne 0$, or equivalently, $(i_n+ra)\bmod N \ne (i_n+rb)\bmod N$. Thus we have

$$\gamma(a,b) = \sum_{0\le i_1,\ldots,i_n\le N-1}$$

$$\times H(i_1,\ldots,i_{n-1},(i_n+ra)\bmod N)$$
$$\times H(i_1,\ldots,i_{n-1},(i_n+rb)\bmod N)$$

$$= \sum_{i_n=0}^{N-1} \{ \sum_{0\le i_1,\ldots,i_{n-1}\le N-1}$$

$$\times H(i_1,\ldots,i_{n-1},(i_n+ra)\bmod N)$$

$$\times H(i_1,\ldots,i_{n-1},(i_n+rb)\bmod N) \}$$

$$= \sum_{i_n=0}^{N-1} 0$$

$$= 0,$$

where the last two equations result from H being an Hadamard matrix.

Hence we have proved that $\gamma(a,b) = N^n \delta(a,b)$. The theorem follows.
Q.E.D.

Let p be a positive integer. Then every integer a, $0 \leq a \leq p^m - 1$, can be uniquely decomposed by $a = \sum_{i=0}^{m-1} a_i p^i$, $0 \leq a_i \leq p - 1$, $0 \leq i \leq m - 1$. Thus there is a one-to-one mapping from the integer a to the m-dimensional vector $(a_0, \ldots, a_{m-1})$. Similarly to the binary $(p = 2)$ case, we use the symbol a to represent both the integer a or its corresponding p-ary vector $(a_0, \ldots, a_{m-1})$. Let $a = (a_0, \ldots, a_{m-1})$ and $b = (b_0, \ldots, b_{m-1})$ be two integers, $0 \leq a, b \leq p^m - 1$. Then the p-dyadic summation between a and b, denoted by $a \oplus_p b$, is defined by the integer corresponding to $((a_0 + b_0) \bmod p, \ldots, (a_{m-1} + b_{m-1}) \bmod p)$. Thus the regular dyadic summation, bit-wise mod2 summation, is the special case of p-dyadic summation for $p = 2$.

Theorem 6.1.7 *Let* $A = [A(i_1, \ldots, i_n)]$, $0 \leq i_1, \ldots, i_n \leq p^m - 1$, *be an* n-*dimensional Hadamard matrix of order* p^m, *then the following matrix* $H = [H(i_1, \ldots, i_{n+1})]$, $0 \leq i_1, \ldots, i_{n+1} \leq p^m - 1$, *defined by*

$$H(i_1, \ldots, i_{n+1}) = A(i_1, \ldots, i_{n-1}, i_n \oplus_p i_{n+1}),$$

is an $(n+1)$-*dimensional Hadamard matrix of order* p^m.

The proof of Theorem 6.1.7 can be finished by the same way as that of Theorem 6.1.5.

Let $X = \{0, 1, \ldots, N - 1\}$ be an integer-set, and $f(x, y)$ a mapping from $X \times X$ to X such that the following conditions are satisfied:

1. *For any given* $a \in X$, *both* $f(a, y)$ *and* $f(x, a)$ *are one-to-one mapping from* X *to itself;*

2. *If* $y_1 \neq y_2$, *then* $f(x, y_1) \neq f(x, y_2)$ *for each* $x \in X$. *If* $x_1 \neq x_2$, *then* $f(x_1, y) \neq f(x_2, y)$ *for each* $y \in X$.

A mapping $f(x, y)$ satisfying the above two conditions are called an H-mapping.

Theorem 6.1.8 *Let* $A = [A(i_1, \ldots, i_n)]$, $0 \leq i_1, \ldots, i_n \leq N-1$ *be an* n-*dimensional Hadamard matrix of order* N, *and* $f(x,y)$ *an* H-*mapping. Then the following matrix* $H = [H(i_1, \ldots, i_{n+1})]$, $0 \leq i_1, \ldots, i_{n+1} \leq N-1$, *defined by*

$$H(i_1, \ldots, i_{n+1}) = A(i_1, \ldots, i_{n-1}, f(i_n, i_{n+1}))$$

is an $(n+1)$-*dimensional Hadamard matrix of order* N.

It is easy to see that Theorem 6.1.8 reduces to Theorems 6.1.5, 6.1.6, and 6.1.7, if we choose $f(x,y) = (x+y) \bmod N$, $f(x,y) = (x+ry) \bmod N$, $\gcd(r, N) = 1$, and $f(x,y) = x \oplus_p y$, respectively. The proof of Theorem 6.1.8 is almost the same as that of Theorem 6.1.5.

From the last chapter, we know that H-Boolean functions are equivalent to n-dimensional Hadamard matrices of order 2. The following theorem shows us another construction of the general n-dimensional Hadamard matrices by using the H-Boolean functions consisting of terms of degree 2.

Theorem 6.1.9 *Let* $f(x_1, \ldots, x_n)$ *be an* H-*Boolean function of* n *variables defined by*

$$f(x_1, \ldots, x_n) = \sum_{i<j, \ (i,j)\in B} x_i x_j,$$

where B *is a subset of* $\{(x,y) : 1 \leq x,y \leq n\}$. *And let* $a = [A(i,j)]$, $1 \leq i,j \leq N$, *be a 2-dimensional Hadamard matrix of order* N. *Then the following matrix* $C = [C(c(1), \ldots, c(n))]$, $1 \leq c(i) \leq N$, $1 \leq i \leq n$, *defined by*

$$C(c(1), \ldots, c(n)) = \prod_{1\leq i<j\leq n, \ (i,j)\in B} A(c(i), c(j))$$

is an n-*dimensional Hadamard matrix of order* N.

Proof. Let k, $1 \leq k \leq n$. It is sufficient to prove that

$$\alpha(a,b) = \sum_{1\leq c(1),\ldots,c(k-1),c(k+1),\ldots,c(n)\leq N} C(c(1), \ldots, c(k-1), a,$$

$$c(k+1), \ldots, c(n))C(c(1), \ldots, c(k-1), b, c(k+1), \ldots, c(n))$$

$$= N^{n-1}\delta(a,b).$$

Rewrite the Boolean function $f(x_1, \ldots, x_n)$ as

$$f(x_1, \ldots, x_n) = x_k \sum_{j>k,(k,j)\in B} x_j + x_k \sum_{i<k,(i,k)\in B} x_i$$

$$+ \sum_{i<j,(i,j)\in B, i\neq k, j\neq k} x_i x_j.$$

Because every H-Boolean function depends on each of their variables, the set

$$\{j > k : (k,j) \in B\} \cup \{i < k : (i,k) \in B\}$$

is non-empty. Thus by employing the identity $A(i,j)^2 = 1$, and $a \neq b$, we have

$$\alpha(a,b) = \sum_{1\leq c(1),\ldots,c(k-1),c(k+1),\ldots,c(n)\leq N}$$

$$\{\prod_{j<k,(k,j)\in B} A(a,c(j))A(b,c(j))\} \times \{\prod_{i<k,(i,k)\in B} A(c(i),a)A(c(i),b)\}$$

$$= \rho[\prod_{j>k,(k,j)\in B} (\sum_{c(j)=1}^{N} A(a,c(j))A(b,c(j)))]$$

$$\times[\prod_{i<k,(i,k)\in B} (\sum_{c(i)=1}^{N} A(c(i),a)A(c(i),b))]$$

$$= 0 \quad \text{(for } A = [A(i,j)] \text{ is an Hadamard matrix) },$$

where ρ is a constant defined by $\rho = N^{r_1+r_2}$, with $r_1 = |\{j : j > k, (k,j) \notin B\}|$, and $r_2 = |\{i : i < k, (i,k) \notin B\}|$.

Thus we have proved that $\alpha(a,b) = N^{r_1+r_2}\delta(a,b)$. The theorem follows.   **Q.E.D.**

The m-sequences, or equivalently the longest linear shift register sequences, are very popular (± 1)-valued sequences because they are the best pseudo-random sequences and have good correlation functions. One of the interested properties about the m-sequences is[2]:

Lemma 6.1.3 *If $a_0, \ldots, a_{N-1}$, $N = 2^k - 1$, is an m-sequence of length*

N, then the matrix

$$A = \begin{bmatrix} 1 & 1 & 1 \cdots & \cdots & 1 \\ 1 & a_0 & a_1 \cdots a_{N-2} & a_{N-1} \\ 1 & a_1 & a_2 \cdots a_{N-1} & a_0 \\ 1 & a_2 & a_3 \cdots & a_0 & a_1 \\ \vdots & \vdots & \vdots & \vdots & \vdots \\ 1 & a_{N-1} & a_0 \cdots a_{N-3} & a_{N-2} \end{bmatrix}$$

is a 2-dimensional Hadamard matrix of order $N + 1$.

This lemma provides us another construction of higher-dimensional Hadamard matrices:

Theorem 6.1.10 *Let $a_0, \ldots, a_{N-1}$, $N = 2^k - 1$, be an m-sequence of length N. And let r, $1 \le r \le n$, be an integer. Then the matrix $A = [A(i_1, \ldots, i_n)]$, $0 \le i_1, \ldots, i_n \le N$, defined by*

$$A(i_1, \ldots, i_n) = (a_{(i_1 + \ldots + i_n - 2) \bmod N})^{\delta((i_1 + \ldots + i_r)(i_{r+1} + \ldots + i_n))}$$

is an n-dimensional Hadamard matrix of order $N + 1$, where $\delta(x) = 1$, iff $x = 0$.

Proof. From Lemma 6.1.3, we know that the matrix

$$B = [B(i,j)] = [(a_{(i+j-2) \bmod N})^{\delta(ij)}]$$

is a 2-dimensional Hadamard matrix of order $N + 1$. The proof is finished by recursively applying Theorem 6.1.5 to the matrix B.   **Q.E.D.**
Remark. The m-sequences used in Theorem 6.1.10 can be replaced by every (± 1)-valued sequence provided that its periodic out-of-phase auto-correlation is the constant -1.

Theorem 6.1.11 *Let A and B be two n-dimensional Hadamard matrices. Then their direct multiplication $A \otimes B$ is also an n-dimensional Hadamard matrix.*

A more general result will be proved later, so we omitted here the proof of Theorem 6.1.11.

Theorem 6.1.12 *Let $B = [B(i_1, \ldots, i_n)]$ be an n-dimensional Hadamard matrix of order N. Then the matrix $A = [A(i_1, \ldots, i_n)]$, $0 \leq i_1, \ldots, i_n \leq N - 1$, defined by*

$$A(i_1, \ldots, i_n) = B(i_1, \ldots, i_n)(-1)^{\sum_{u=0}^{M-1} \sum_{v=0}^{n-1} a_{uv} i(u,v)}$$

is an n-dimensional Hadamard matrix of order N, where $a_{uv} = 0$ or 1, $N \leq 2^M$, and $i_k = (i(M-1,k), i(M-2,k), \ldots, i(0,k))$, the binary expended vector of the integer i_k.

In the case of $n = 2$, Theorem 6.1.12 implies that every Hadamard matrix is transformed to another Hadamard matrix if some rows and/or columns are minused.

Theorem 6.1.13 *Let $\tau(.)$ be a permutation of the set $\{1, 2, \ldots, n\}$, and let $A = [A(a(1), \ldots, a(n))]$ be an n-dimensional Hadamard matrix. Then the matrix $B = [B(a(1), \ldots, a(n))]$ defined by*

$$B(a(1), \ldots, a(n)) = A(a(\tau(1)), \ldots, a(\tau(n)))$$

is also an n-dimensional Hadamard matrix.

In the case of $n = 2$, Theorem 6.1.13 implies that the transpose of an Hadamard matrix is also an Hadamard matrix.

Theorem 6.1.14 *Let $f_1(.), \ldots, f_n(.)$ be one-to-one mappings of the set $\{0, 1, \ldots, m-1\}$. And let $A = [A(i_1, \ldots, i_n)]$, $0 \leq i_k \leq m-1$, $1 \leq k \leq n$, be an n-dimensional Hadamard matrix of order m. Then the matrix $B = [B(i_1, \ldots, i_n)]$ defined by*

$$B(i_1, \ldots, i_n) = A(f_1(i_1), \ldots, f_n(i_n))$$

is also an n-dimensional Hadamard matrix of order m.

In the case of $n = 2$, Theorem 6.1.14 implies that every Hadamard matrix is transformed to another Hadamard matrix if the rows and/or columns are permutated.

Theorem 6.1.15 *Let $A = [A(i_1, \ldots, i_n)]$ and $B = [B(j_1, \ldots, j_m)]$ be two Hadamard matrices of the same order N and dimensions n and m, respectively. Then the matrix $C = [C(i_1, \ldots, i_n, j_1, \ldots, j_m)]$ defined by*

$$C(i_1, \ldots, i_n, j_1, \ldots, j_m) = A(i_1, \ldots, i_n) \times B(j_1, \ldots, j_m)$$

is an (m + n)-dimensional Hadamard matrix of order N.

The proofs of Theorems 6.1.12 to 6.1.15 are trivial, so we omitted the details here.

Theorem 6.1.16 Let $B = [B(b(1), \ldots, b(n))]$, $0 \le b(i) \le m - 1$, $1 \le i \le n$, be an n-dimensional matrix of order m. And let $A = [A(a(1), \ldots, a(n))]$, $0 \le a(i) \le m^n - 1$, $1 \le i \le n$, be an n-dimensional matrix of order m^n which is defined by

$$A(a(1), \ldots, a(n)) = \prod_{i=1}^{n} B(a(1, i), a(2, i), \ldots, a(n, i))$$

where $(a(1, i), a(2, i), \ldots, a(n, i))$ is the m-ary expended vector of the integer $a(i)$, i.e., $a(i) = \sum_{j=1}^{n} a(j, i) m^{j-1}$. Then A is an Hadamard matrix if and only if B is an Hadamard matrix.

Proof. $\Longleftarrow$: Let $0 \le a \ne b \le m^n - 1$. And let $(a_1, \ldots, a_n)$ and $(b_1, \ldots, b_n)$ be the m-ary expended vectors of the integers a and b, respectively. Then there is at least one k, $1 \le k \le n$, such that $a_k \ne b_k$. Hence

$$\sum_{a(2),\ldots,a(n)} A(a, a(2), \ldots, a(n)) A(b, a(2), \ldots, a(n))$$

$$= \sum_{0 \le a(i,j) \le m-1,\ 2 \le i \le n,\ 1 \le j \le n} \prod_{p=1}^{n} B(a_p, a(2, p), \ldots, a(n, p))$$

$$\prod_{p=1}^{n} B(b_p, a(2, p), \ldots, a(n, p))$$

$$= \sum_{0 \le a(i,j) \le m-1,\ 2 \le i \le n,\ 1 \le j \le n,\ j \ne k} \rho[\sum_{0 \le a(i,k) \le m-1,\ 2 \le i \le n}$$

$$B(a_k, a(2, k), \ldots, a(n, k)) B(b_k, a(2, k), \ldots, a(n, k))]$$

$$= \sum \rho \times 0, \quad \text{(for } B \text{ is an Hadamard matrix)}$$

$$= 0,$$

where ρ is a constant that independent of $a(i, k)$.

Similarly, it can be proved that for each r, $1 \le r \le n$,

$$\sum_{a(.)} A(a(1), \ldots, a(r-1), a, a(r+1), \ldots, a(n))$$

$$\times A(a(1),\ldots,a(r-1),b,a(r+1),\ldots,a(n)) = (m^n)^{n-1}\delta(a,b).$$

Thus B is an n-dimensional Hadamard matrix of order m^n.

$\Longrightarrow$: If B is not an Hadamard matrix, then there exists a pair a and b, $0 \le a \ne b \le m-1$, such that

$$\sum_{b(2),\ldots,b(n)} B(a,b(2),\ldots,b(n))B(b,b(2),\ldots,b(n)) = r \ne 0.$$

Let u and v be two integers satisfying $u = \sum_{i=0}^{n-1} am^i$ and $v = \sum_{i=0}^{n-1} bm^i$, respectively. Thus $0 \le u \ne v \le m^n - 1$, and

$$\sum_{a(2),\ldots,a(n)} A(u,a(2),\ldots,a(n))A(v,a(2),\ldots,a(n))$$

$$= \sum_{0\le a(i,j)\le m-1,\ 2\le i\le n,\ 1\le j\le n} \prod_{k=1}^{n} \{B(a,a(2,k),\ldots,a(n,k))$$

$$\times B(b,a(2,k),\ldots,a(n,k))\}$$

$$= \prod_{k=1}^{n} \{ \sum_{0\le a(2,k),\ldots,a(n,k)\le m-1} B(a,a(2,k),\ldots,a(n,k))$$

$$\times B(b,a(2,k),\ldots,a(n,k))\}$$

$$= \prod_{k=1}^{n} r = r^n \ne 0.$$

Therefore A is not an Hadamard matrix. The theorem follows.   **Q.E.D.**

Theorem 6.1.17 *The following are true:*

1. *Let* $A = [A(a(1),\ldots,a(2n))]$, $1 \le a(i) \le m$, $1 \le i \le 2n$, *be a* $(2n)$-*dimensional* (± 1)-*valued matrix of order* m. *If* $AA' = m^n I$. *Then* A *is a* $(2n)$-*dimensional Hadamard matrix, where* A' *refers to the transpose of the matrix* A, *and* I *the* $(2n)$-*dimensional unit matrix of order* m. *(Remark: It has been known that* $AA' = m^2 I$ *is both necessary and sufficient conditions of* A *being a 2-dimensional Hadamard matrix. However, it should be pointed out that if* $n > 2$, *there do exist* $(2n)$-*dimensional Hadamard matrix* A *satisfying* $AA' \ne m^n I$*).*

2. Let $B = [B(b(1), \ldots, b(2n+1))]$, $1 \le b(i) \le m$, $1 \le i \le 2n+1$, be a $(2n+1)$-dimensional (± 1)-valued matrix of order m. If B simultaneously satisfies the following two equations:

$$BB' = m^n I$$

and, for $1 \le a, b \le m$,

$$\sum_{1 \le b(1), \ldots, b(2n) \le m} B(b(1), \ldots, b(2n), a) B(b(1), \ldots, b(2n), b) = m^{2n} \delta(a, b).$$

Then the B is a $(2n+1)$-dimensional Hadamard matrix, where B' refers to the transpose of the matrix B, and I the $(2n+1)$-dimensional unit matrix of order m.

Proof. The proofs for these two statements are almost the same, so we prove the first one only.

$$\alpha(a, b) = \sum_{1 \le a(2), \ldots, a(2n) \le m} A(a, a(2), \ldots, a(2n)) A(b, a(2), \ldots, a(2n))$$

$$= \sum_{1 \le a(2), \ldots, a(2n) \le m} A(a, a(2), \ldots, a(n), a(n+1), \ldots, a(2n))$$

$$\times A'(a(n+1), \ldots, a(2n), b, a(2), \ldots, a(n))$$

$$\text{(by the definition of } A')$$

$$= \sum_{1 \le a(2), \ldots, a(n) \le m} \left\{ \sum_{1 \le a(n+1), \ldots, a(2n) \le m} A(a, a(2), \ldots, a(n), \right.$$

$$a(n+1), \ldots, a(2n))$$

$$\left. \times A'(a(n+1), \ldots, a(2n), b, a(2), \ldots, a(n)) \right\}$$

$$= \sum_{1 \le a(2), \ldots, a(n) \le m} m^n \delta(a, b) \text{ (because of } AA' = m^n I)$$

$$= m^{2n-1} \delta(a, b).$$

Now let

$$\beta(a, b) = \sum_{1 \le a(1), \ldots, a(n), a(n+2), \ldots, a(2n) \le m} A(a(1), \ldots, a(n), a, a(n+2), \ldots,$$

$$a(2n))A(a(1), \ldots, a(n), b, a(n+2), \ldots, a(2n)).$$

$AA' = m^n I$ implies $A' = m^n A^{-1}$. Thus $A'A = m^n A^{-1}A = m^n I$, where A^{-1} refers to the inverse matrix of A. Hence the equation $\beta(a, b) = m^{2n-1}\delta(a, b)$ can be proved by the same way as the $\alpha(a, b) = m^{2n-1}\delta(a, b)$.

Similarly, we can prove that, for each $1 \leq r \leq 2n$,

$$\sum_{a(.)} A(a(1), \ldots, a(r-1), a, a(r+1), \ldots, a(2n))$$

$$\times A(a(1), \ldots, a(r-1), b, a(r+1), \ldots, a(2n)) = m^{2n-1}\delta(a, b).$$

Thus A is a $(2n)$-dimensional Hadamard matrix of order m.   **Q.E.D.**

6.1.2   Proper and Improper n-Dimensional Hadamard Matrices

Absolutely proper and improper higher-dimensional Hadamard matrices are two extreme special cases of the general Hadamard matrices that are defined by:

Definition 6.1.2 ([1])   *An absolutely proper n-dimensional Hadamard matrix is a (± 1)-valued matrix in which all two-dimensional sections in all possible axis-normal orientations are Hadamard matrices.*

Similar to the consequences of Definition 6.1.2, it is easy to see that ([1])

1. *All intermediate-dimensional sections of an absolutely proper n-dimensional Hadamard matrix are also absolutely proper Hadamard matrices;*

2. *absolutely proper Hadamard matrices are themselves Hadamard matrices. In fact, an n-dimensional Hadamard matrix is specified if either all $(n-1)$-dimensional sections in one direction are Hadamard matrices and also are mutually orthogonal or if all $(n-1)$-dimensional sections in two directions are Hadamard matrices.*

It has been proved, in the last chapter, that

Theorem 6.1.18 *An n-dimensional (± 1)-valued matrix of order 2 is an absolutely proper Hadamard matrix if and only if this matrix is defined by $A = [A(x_1, \ldots, x_n)]$, $0 \le x_1, \ldots, x_n \le 1$,*

$$A(x_1, \ldots, x_n) = (-1)^{f(x_1, \ldots, x_n)}$$

for some Boolean function of the form

$$f(x_1, \ldots, x_n) = \sum_{1 \le i, j \le n} x_i x_j + a + \sum_{i=1}^{n} b_i x_i$$

where $a, b_1, \ldots, b_n \in \{0, 1\}$.

Besides this theorem, another important result about the absolutely proper Hadamard matrix is:

Theorem 6.1.19 *Let $A = [A(i, j)]$, $0 \le i, j \le m - 1$, be a 2-dimensional Hadamard matrix of order m. Then the following matrix $C = [C(c(1), \ldots, c(n))]$, $0 \le c(k) \le m - 1$, $1 \le k \le n$, defined by*

$$C(c(1), \ldots, c(n)) = \prod_{1 \le i < j \le n} A(c(i), c(j)), \qquad (6.9)$$

is an n-dimensional absolutely proper Hadamard matrix of order m.

Proof. It is sufficient to prove that all 2-dimensional sections, in all possible axis-normal orientations, are Hadamard matrices. In other words, for any prefixed $1 \le i, j \le n$, and $c(k)$, $k \ne i$, $k \ne j$, we have to prove that the 2-dimensional section

$$D = [D(c(i), c(j))] \qquad (6.10)$$
$$= [C(c(1), \ldots, c(i), \ldots, c(j), \ldots, c(n))], \quad 0 \le c(i), c(j) \le m - 1,$$

is an Hadamard matrix of order m.

From Equations (6.9) and (6.10) , we have

$$D(c(i), c(j)) = A(c(i), c(j)) \times \{ \prod_{i < k \le n, \ k \ne j} A(c(i), c(k)) \}$$

$$\times \{ \prod_{1 \le k < i} A(c(k), c(i)) \} \times \{ \prod_{j < k \le n} A(c(j), c(k)) \}$$

$$\times \{ \prod_{1 \le k < j, \ k \ne i} A(c(k), c(j)) \}$$

$$\times \{ \prod_{1 \le r < s \le n, \ r \ne i, j \ s \ne i, j} A(c(r), c(s)) \}.$$

Thus because $A(i,j)^2 = 1$, we know that, for $0 \leq a \neq b \leq m-1$,

$$\sum_{c(j)=0}^{m-1} D(a, c(j))D(b, c(j))$$

$$= \sum_{c(j)=0}^{m-1} \{ \prod_{i<k\leq n} A(a, c(k))A(b, c(k))\} \times \{ \prod_{1\leq k<i} A(c(k), a)A(c(k), b)\}$$

$$= \{ \prod_{i<k\leq n,\ k\neq j} A(a, c(k))A(b, c(k))\} \times \{ \prod_{1\leq k<i} A(c(k), a)A(c(k), b)\} \times$$

$$\sum_{c(j)=0}^{m-1} A(a, c(j))A(b, c(j))$$

$$= \{ \prod_{i<k\leq n,\ k\neq j} A(a, c(k))A(b, c(k))\} \times \{ \prod_{1\leq k<i} A(c(k), a)A(c(k), b)\} \times 0$$

(for A is an Hadamard matrix)

$$= 0.$$

Hence $\sum_{c(j)=0}^{m-1} D(a, c(j))D(b, c(j)) = m\delta(a, b)$.

Similarly, it can be proved that

$$\sum_{c(i)=0}^{m-1} D(c(i), a)D(c(i), b) = m\delta(a, b).$$

Thus D is indeed an Hadamard matrix. The theorem follows.　　**Q.E.D.**

Let A be an n-dimensional absolutely proper Hadamard matrix of order $m > 2$. The requirement of 'A's 2-dimensional sections are Hadamard matrices' implies that $m = 4k$, for some integer k. On the other hand, Theorem 6.1.19 confirms that every 2-dimensional Hadamard matrix produces an n-dimensional absolutely proper Hadamard matrix of the same order. Hence if the Hadamard conjecture was proved, there would exist at least one n-dimensional absolutely proper Hadamard matrix of order $4k$, for each $n \geq 2$ and $4k$.

The other concept opposite to the absolutely proper Hadamard matrix is the following absolutely improper one.

Definition 6.1.3 ([1]) *An absolutely improper n-dimensional Hadamard matrix is an Hadamard matrix in which no lower-dimensional section is an Hadamard matrix.*

An absolutely improper n-dimensional Hadamard matrix of order m^2 may be formed by $n-1$ successive direct multiplications of n two-dimensional Hadamard matrices of order m in appropriately different orientations, e.g., the $i_1 i_2$, $i_2 i_3$, ..., $i_{n-1} i_n$, and $i_n i_1$ planes. To be precise, we have the following theorem.

Theorem 6.1.20 *Let $A = [A(i,j)]$, $0 \le i, j \le m-1$, be a 2-dimensional Hadamard matrix. And let A_1, A_2, ..., and A_n be the n-dimensional matrices produced by treating A as the n-dimensional ones of orders $m \times m \times 1 \times \ldots \times 1$, $1 \times m \times m \times 1 \times \ldots \times 1$, ..., $1 \times 1 \times \ldots \times m \times m$, and $m \times 1 \times \ldots \times 1 \times m$, respectively. Then the following direct multiplication matrix B, defined by*

$$B = [B(b(1),\ldots,b(n))] := A_1 \otimes A_2 \otimes \ldots \otimes A_n,$$

is an n-dimensional absolutely improper Hadamard matrix of order m^2, where $0 \le b(1),\ldots,b(n) \le m^2 - 1$.

Proof. At first we prove that B is an Hadamard matrix. In fact, from the definition of direct multiplication, we know that

$$B(b(1),\ldots,b(n)) = A([b(1)]_m, [b(2)]_m) \times A\left(\left\lfloor \frac{b(2)}{m} \right\rfloor, [b(3)]_m\right)$$

$$\times A\left(\left\lfloor \frac{b(3)}{m} \right\rfloor, [b(4)]_m\right) \times \ldots \times A\left(\left\lfloor \frac{b(n-1)}{m} \right\rfloor, [b(n)]_m\right)$$

$$\times A\left(\left\lfloor \frac{b(1)}{m} \right\rfloor, \left\lfloor \frac{b(n)}{m} \right\rfloor\right). \tag{6.11}$$

Because of the identity $A(i,j)^2 = 1$,

$$\sum_{0 \le b(2),\ldots,b(n) \le m^2-1} B(a, b(2),\ldots,b(n)) B(b, b(2),\ldots,b(n))$$

$$= \sum_{0 \le b(2),\ldots,b(n) \le m^2-1} A([a]_m, [b(2)]_m)$$

$$\times A\left(\left\lfloor \frac{a}{m} \right\rfloor, \left\lfloor \frac{b(n)}{m} \right\rfloor\right) \times A([b]_m, [b(2)]_m) \times A\left(\left\lfloor \frac{b}{m} \right\rfloor, \left\lfloor \frac{b(n)}{m} \right\rfloor\right)$$

$$= \sum_{0 \le b(2),\ldots,b(n-1) \le m^2-1} \left[\sum_{b(2)=0}^{m^2-1} A([a]_m, [b(2)]_m) \times A([b]_m, [b(2)]_m) \right]$$

$$\times \Big[\sum_{b(2)=0}^{m^2-1} A\Big(\Big\lfloor \frac{a}{m}\Big\rfloor, \Big\lfloor \frac{b(n)}{m}\Big\rfloor\Big) A\Big(\Big\lfloor \frac{b}{m}\Big\rfloor, \Big\lfloor \frac{b(n)}{m}\Big\rfloor\Big) \Big]. \tag{6.12}$$

If $a \neq b$, then either $[a]_m \neq [b]_m$ or $\lfloor \frac{a}{m}\rfloor \neq \lfloor \frac{b}{m}\rfloor$. In the first case, the summation in the first bracket of Equation (6.12) vanishes, otherwise the summation in the second bracket of Equation (6.12) vanishes.

By the same way it can be proved that, for each $1 \leq r \leq n$,

$$\sum_{B(.)} B(b(1), \ldots, b(r-1), a, b(r+1), \ldots, b(n))$$

$$\times B(b(1), \ldots, b(r-1), b, b(r+1), \ldots, b(n)) = m^{2(n-1)} \delta(a, b).$$

Thus B is an n-dimensional Hadamard matrix of order m^2.

Then we prove that no lower section of B is Hadamard matrix.

Let $E = [E(e(1), \ldots, e(k))]$, $k < n$, be a k-dimensional section of B. Without loss of the generality, we assume that E is obtained by fixing the coordinate $b(i_0)$, $1 \leq i_0 \leq n$, but not $b(i_0 + 1)$. Thus, we have

$$E(e(1), \ldots, e(k)) = A(\{b(i_0)\}, \{e(j)\}) \times A(\{e(j)\}, (.)) \times (:),$$

where $\{x\}$ stands for either $[x]_m$ or $\lfloor x/m\rfloor$; $(.)$ is a constant if $b(i_0 + 2)$ has been fixed during the construction of E, otherwise $(.) = \lfloor e(j+1)\rfloor$; and $(:)$ is a term independent of $e(j)$.

Then

$$\rho(a, b) = \sum_{0 \leq e(1), \ldots, e(j-1), e(j+1), \ldots, e(k) \leq m^2-1} E(e(1), \ldots, e(j-1), a,$$

$$e(j+1), \ldots, e(k)) E(e(1), \ldots, e(j-1), b, e(j+1), \ldots, e(k))$$

$$= \sum_{0 \leq e(1), \ldots, e(j-1), e(j+1), \ldots, e(k) \leq m^2-1} A(\{b(i_0)\}, \{a\})$$

$$\times A(\{b(i_0)\}, \{b\}) \times A(\{a\}, \{.\}) A(\{b\}, \{.\}).$$

If $(.)$ is a constant, then

$$|\rho(a, b)| = m^{2(k-1)} \neq m^{2(k-1)} \delta(a, b).$$

If $(.) = \{e(j+1)\}$, then

$$\rho(a,b) = A(\{b(i_0)\}, \{a\}) \times A(\{b(i_0)\}, \{b\})$$

$$\times \left[\sum_{0 \leq e(1),\ldots,e(j-1),e(j+2),\ldots,e(k) \leq m^2-1} \right.$$

$$\left. \left(\sum_{e(j+1)=0}^{m^2-1} A(\{a\}, \{e(j+1)\}) A(\{b\}, \{e(j+1)\}) \right) \right]. \quad (6.13)$$

Because we can always find two integers a_0 and b_0 such that $a_0 \neq b_0$ and $\{a_0\} = \{b_0\}$, from Equation (6.13), we have

$$\rho(a_0, b_0) = m^{2(k-1)} \neq 0 = m^{2(k-1)} \delta(a_0, b_0).$$

Hence we have proved that the k-dimensional section E is not Hadamard matrix. The theorem follows.   **Q.E.D.**

A more general version of Theorem 6.1.20 is the following

Theorem 6.1.21 *Let $A = [A(a(1), \ldots, a(k))]$ be a k-dimensional Hadamard matrix of order m. And let A_1, A_2, ..., and A_n, $n > k$, be the n-dimensional matrices produced by treating A as the n-dimensional ones of orders $m \times \ldots \times m \times 1 \times \ldots \times 1$, $1 \times m \times \ldots \times m \times 1 \times \ldots \times 1$, ..., $1 \times 1 \ldots \times 1 \times m \times \ldots \times m$, $m \times 1 \times 1 \ldots \times 1 \times m \times \ldots \times m$, ..., $m \times \ldots \times m \times 1 \times \ldots \times 1 \times m$, respectively. Then the following direct multiplication matrix B, defined by*

$$B = [B(b(1), \ldots, b(n))] := A_n \otimes A_{n-1} \otimes \ldots \otimes A_1,$$

is an n-dimensional absolutely improper Hadamard matrix of order m^k.

If $k = 2$, this theorem clearly reduces to Theorem 6.1.20.

Between the two extreme cases of absolutely proper and absolutely improper Hadamard matrices, there exist a wide variety of intermediate degrees of property. For example, an n-dimensional Hadamard matrix A is called 'proper in some direction', if all of its $(n-1)$-dimensional sections in that direction are Hadamard matrices, otherwise A is called improper in that direction. These partially proper n-dimensional Hadamard matrix can be constructed from a lower-dimensional one, by some 'bootstrap' sequence of cyclic section-permutations.

6.1.3 Generalized Higher-Dimensional Hadamard Matrices

Let $A = [A(i_1, \ldots, i_n)]$, $0 \le i_k \le a_k - 1$, $1 \le k \le n$, be an n-dimensional matrix of size $a_1 \times \ldots \times a_n$. Recall that a necessary condition for this A to be Hadamard matrix is that its size satisfying $a_1 = \ldots = a_n$, i.e., the length of every side is the same constant. While in this subsection, we will generalize the concept of Hadamard matrices as those which have different side lengths.

Definition 6.1.4 *An n-dimensional (± 1)-valued matrix $A = [A(i_1, \ldots, i_n)]$, $0 \le i_k \le a_k - 1$, $1 \le k \le n$, of size $a_1 \times \ldots \times a_n$ is called a generalized Hadamard matrix if the following two conditions are simultaneously satisfied:*

1. *At least two of $a_1, a_2, \ldots, a_n$ are larger than 1;*

2. *If $a_i > 1$, then*

$$\sum_{a(1),\ldots,a(i-1),a(i+1),\ldots,a(n)}$$

$$\times A(a(1), \ldots, a(i-1), a, a(i+1), \ldots, a(n))$$

$$\times A(a(1), \ldots, a(i-1), b, a(i+1), \ldots, a(n))$$

$$= (\prod_{j=1, j \neq i}^{n})a_j)\delta(a, b).$$

Clearly, a generalized Hadamard matrix of size $m \times \ldots \times m$ is in fact a regular higher-dimensional Hadamard matrix studied in last subsections. In addition, an n-dimensional Hadamard matrix of order m can be treated as an $(n + k)$-dimensional Hadamard matrix of size $1 \times \ldots \times m \ldots \times 1 \ldots m \ldots \times 1$. Particularly, the matrices $A_1, \ldots, A_n$ used in Theorems 6.1.21 and 6.1.20 are all generalized Hadamard matrices. In other words, these two theorems assert that the direct multiplication of some generalized Hadamard matrices may produce another Hadamard matrices. The following theorem proves the rightness of this assertion in general.

Theorem 6.1.22 *Let* $A = [A(i_1, \ldots, i_n)]$, $0 \le i_k \le a_k - 1$, $1 \le k \le n$, *and* $B = [B(j_1, \ldots, j_n)]$, $0 \le j_k \le b_k - 1$, $1 \le k \le n$, *be n-dimensional generalized Hadamard matrices of size* $a_1 \times \ldots \times a_n$, *and* $b_1 \times \ldots \times b_n$, *respectively. Then their direct multiplication* $C = B \otimes A$ *is an n-dimensional generalized Hadamard matrices of size* $(a_1 b_1) \times \ldots \times (a_n b_n)$.

Proof. Let $C = [C(c(1), \ldots, c(n))]$, $0 \le c(i) \le a_i b_i - 1$, $1 \le i \le n$. By the definition of direct multiplication, we have

$$C(c(1), \ldots, c(n)) = A([c(1)]_{a_1}, \ldots, [c(n)]_{a_n}) \times B\left(\left\lfloor \frac{c(1)}{a_1} \right\rfloor, \ldots, \left\lfloor \frac{c(n)}{a_n} \right\rfloor\right).$$

If $a_i b_i > 1$, and $0 \le a \ne b \le a_i b_i - 1$, then

$$\rho(a, b) = \sum_{c(1), \ldots c(i-1), c(i+1), \ldots, c(n)} C(c(1), \ldots c(i-1), a, c(i+1), \ldots, c(n))$$

$$\times C(c(1), \ldots c(i-1), b, c(i+1), \ldots, c(n))$$

$$= \sum_{c(1), \ldots c(i-1), c(i+1), \ldots, c(n)} B\left(\left\lfloor \frac{c(1)}{a_1} \right\rfloor, \ldots, \left\lfloor \frac{c(i-1)}{a_{i-1}} \right\rfloor, \left\lfloor \frac{a}{a_i} \right\rfloor,\right.$$

$$\left\lfloor \frac{c(i+1)}{a_{i+1}} \right\rfloor, \ldots, \left\lfloor \frac{c(n)}{a_n} \right\rfloor \right)$$

$$\times B\left(\left\lfloor \frac{c(1)}{a_1} \right\rfloor, \ldots, \left\lfloor \frac{c(i-1)}{a_{i-1}} \right\rfloor, \left\lfloor \frac{b}{a_i} \right\rfloor, \left\lfloor \frac{c(i+1)}{a_{i+1}} \right\rfloor, \ldots, \left\lfloor \frac{c(n)}{a_n} \right\rfloor \right)$$

$$\times A([c(1)]_{a_1}, \ldots, [c(i-1)]_{a_{i-1}}, [a]_{a_i}, [c(i+1)]_{a_{i+1}}, \ldots, [c(n)]_{a_n})$$

$$\times A([c(1)]_{a_1}, \ldots, [c(i-1)]_{a_{i-1}}, [b]_{a_i}, [c(i+1)]_{a_{i+1}}, \ldots, [c(n)]_{a_n}).$$

Now we prove $\rho(a, b) = 0$ in the following two separate cases:

Case 1. $a_i = 1$. Thus $b_i > 1$.

Because of $[a]_{a_i} = [b]_{a_i} = 0$, $\lfloor \frac{a}{a_i} \rfloor = a$, $\lfloor \frac{b}{a_i} \rfloor = b$, and $A(a(1), \ldots, a(n))^2 = 1$, we have

$$\rho(a, b) = \sum_{c(1), \ldots c(i-1), c(i+1), \ldots, c(n)} B\left(\left\lfloor \frac{c(1)}{a_1} \right\rfloor, \ldots, \left\lfloor \frac{c(i-1)}{a_{i-1}} \right\rfloor, \left\lfloor \frac{a}{a_i} \right\rfloor,\right.$$

$$\left\lfloor \frac{c(i+1)}{a_{i+1}} \right\rfloor, \ldots, \left\lfloor \frac{c(n)}{a_n} \right\rfloor \right)$$

$$\times B\left(\left\lfloor\frac{c(1)}{a_1}\right\rfloor,\ldots,\left\lfloor\frac{c(i-1)}{a_{i-1}}\right\rfloor,\left\lfloor\frac{b}{a_i}\right\rfloor,\left\lfloor\frac{c(i+1)}{a_{i+1}}\right\rfloor,\ldots,\left\lfloor\frac{c(n)}{a_n}\right\rfloor\right)$$

$$= b_1\times\ldots\times b_{i-1}\times b_{i+1}\times\ldots\times b_n$$

$$\times\sum_{b(1),\ldots b(i-1),b(i+1),\ldots,b(n)} B(b(1),\ldots b(i-1),a,b(i+1),\ldots,b(n))$$

$$\times B(b(1),\ldots b(i-1),b,b(i+1),\ldots,b(n))$$

$$= 0,$$

where the last equation is due to the fact that B is an n-dimensional generalized Hadamard matrix.

Case 2. $a_i > 1$. Thus

$$\rho(a,b) = [\sum_{0\leq k_1\leq b_1-1,\ldots,0\leq k_{i-1}\leq b_{i-1}-1,0\leq k_{i+1}\leq b_{i+1}-1,\ldots,0\leq k_n\leq b_n-1}$$

$$B(k_1,\ldots,k_{i-1},\left\lfloor\frac{a}{a_i}\right\rfloor,k_{i+1},\ldots,k_n)\times B(k_1,\ldots,k_{i-1},\left\lfloor\frac{b}{a_i}\right\rfloor,$$

$$k_{i+1},\ldots,k_n)]$$

$$\times[\sum_{0\leq l_1\leq a_1-1,\ldots,0\leq l_{i-1}\leq a_{i-1}-1,0\leq l_{i+1}\leq a_{i+1}-1,\ldots,0\leq l_n\leq a_n-1}$$

$$A(l_1,\ldots,l_{i-1},[a]_{a_i},l_{i+1},\ldots,l_n)\times A(l_1,\ldots,l_{i-1},[b]_{a_i},l_{i+1},\ldots,l_n)]$$

$$(6.14)$$

$a\neq b$ implies either $[a]_{a_i}\neq[b]_{a_i}$ or $\lfloor a/a_i\rfloor\neq\lfloor b/a_i\rfloor$.

If $[a]_{a_i}\neq[b]_{a_i}$, then the summation in the second bracket of Equation 6.14 is zero. If $\lfloor a/a_i\rfloor\neq\lfloor b/a_i\rfloor$, then the summation in the first bracket of Equation (6.14) is zero.

Thus we have proved that $\rho(a,b) = 0$, if $a\neq b$. The proof is finished. **Q.E.D.**

A trivial corollary of Theorem 6.1.22 is

Corollary 6.1.1 *The direct multiplication of two n-dimensional Hadamard matrices is also an Hadamard matrix, which is proper in some direction if its parent matrices are proper in that direction.*

The other application of Theorem 6.1.22 is that it provides us a recursive construction of higher-dimensional Hadamard matrices from lower-dimensional ones. For example, we have the following two theorems.

Theorem 6.1.23 *Let A and B be two n-dimensional generalized Hadamard matrices of sizes $a_1 \times \ldots \times a_n$ and $b_1 \times \ldots \times b_n$, respectively. If $a_1 b_1 = \ldots = a_n b_n = m$, then $A \otimes B$ is an n-dimensional regular Hadamard matrix of order m.*

Lemma 6.1.4 *Let $A = [A(a(1), \ldots, a(n))]$ be an n-dimensional generalized Hadamard matrix of size $a_1 \times \ldots \times a_n$. Then,*

1. *A is also an $(n+k)$-dimensional generalized Hadamard matrix of size $1 \times \ldots \times a_1 \times \ldots \times 1 \ldots \times a_n \times \ldots \times 1$;*

2. *If $\tau(.)$ is a permutation of the set $\{1, 2, \ldots, n\}$, then the matrix $B = [A(a(\tau(1))), \ldots, a(\tau(n)))]$ is an n-dimensional generalized Hadamard matrix of size $a_{\tau(1)} \times \ldots \times a_{\tau(n)}$.*

Theorem 6.1.24 *Every n-dimensional generalized Hadamard matrix of size $a_1 \times \ldots \times a_n$ produces an n-dimensional regular Hadamard matrix of order $\prod_{i=1}^{n} a_i$.*

Proof. Let A be an n-dimensional generalized Hadamard matrix of size $a_1 \times \ldots \times a_n$. And let $\tau(x) = x + 1$, if $1 \le x \le n - 1$; $\tau(n) = 1$. This $\tau(.)$ is clearly a permutation of the set $\{1, 2, \ldots, n\}$. Thus, by Theorem 6.1.22 and Lemma 6.1.4, the direct multiplication of the matrices $A \otimes A(\tau) \otimes \ldots \otimes A(\tau^k) \otimes \ldots \otimes A(\tau^{n-1})$ is the required n-dimensional regular Hadamard matrix of order $\prod_{i=1}^{n} a_i$, where $A(\tau^k) = [A(a(\tau^k(1)), \ldots, a(\tau^k(n)))]$. **Q.E.D.**

6.2   Higher-Dimensional Hadamard Matrices Based on Perfect Binary Arrays

6.2.1 n-dimensional Hadamard Matrices Based On PBAs

The definition of 2-dimensional perfect binary arrays(PBA) can be generalized for higher-dimensional cases which will be used to construct the general higher-dimensional Hadamard matrices.

Definition 6.2.1 ([3]) *Let $A = [A(a(1), \ldots, a(n))]$, $0 \leq a(j) \leq a_j - 1$, $1 \leq j \leq n$, be a (± 1)-valued n-dimensional matrix of size $a_1 \times a_2 \times \ldots \times a_n$. A is called a perfect binary array, abbreviated as a $\mathrm{PBA}(a_1, a_2, \ldots, a_n)$ if its n-dimensional cyclic auto-correlation is a δ-function, i.e., for all $(u(1), \ldots, u(n)) \neq (0, \ldots, 0)$, with $0 \leq u(i) \leq a_i - 1$, we have*

$$R_A(u(1), \ldots, u(n)) = \sum_{j_1=0}^{a_1-1} \cdots \sum_{j_n=0}^{a_n-1} A(j_1, \ldots, j_n)$$
$$\cdot A(j_1 + u(1), \ldots, j_n + u(n)) = 0,$$

where $j_i + u(i) \equiv (j_i + u(i)) \bmod a_i$, $1 \leq i \leq n$.

This definition is clearly a natural generalization of the PBAs studied in the chapter of three-dimensional Hadamard matrices. In this book we consider only the non-trivial cases, i.e., at least one of a_i is larger than 1. Note that if τ is a permutation of $\{1, 2, \ldots, n\}$, then $A = [A(a(1), \ldots, a(n))]$ is a $\mathrm{PBA}(a_1, a_2, \ldots, a_n)$ if and only if $A(\tau) := [A(a(\tau(1)), \ldots, a(\tau(r)))]$ is a $\mathrm{PBA}(a_{\tau(1)}, \ldots, a_{\tau(n)})]$.

Theorem 6.2.1 *Let $A = [A(a(1), \ldots, a(n))]$, $0 \leq a(i) \leq a_i - 1$, $1 \leq i \leq n$, be a $\mathrm{PBA}(a_1, a_2, \ldots, a_n)$. And let a_{n+1} be an integer satisfying $a_{n+1} \leq \mathrm{LCM}(a_1, a_2, \ldots, a_n)$, the least common multiple of $a_1, a_2, \ldots, a_n$. Then the following $(n+1)$-dimensional matrix $H = [H(h(1), \ldots, h(n+1))]$, $0 \leq h(i) \leq a_i - 1$, $1 \leq i \leq n + 1$, defined by*

$$H(h(1), \ldots, h(n+1)) = A((h(1) + h(n+1)) \bmod a_1, \ldots,$$
$$(h(n) + h(n+1)) \bmod a_n)$$

is an $(n+1)$-dimensional generalized Hadamard matrix of size $a_1 \times \ldots \times a_{n+1}$.

Particularly, if $a_1 = a_2 = \ldots = a_n = m$, then a_{n+1} can be chosen to be the integer m. Hence the corresponding H is an $(n + 1)$-dimensional Hadamard matrix of order m.

Proof. On one hand, if $0 \leq a \neq b \leq a_{n+1} - 1$, then there exists an integer i, $1 \leq i \leq n$, such that $a \not\equiv b \bmod(a_i)$, which is due to $a_{n+1} \leq \text{LCM}(a_1, a_2, \ldots, a_n)$. Thus

$$\sum_{h(1)=0}^{a_1-1} \cdots \sum_{h(n)=0}^{a_n-1} H(h(1), \ldots, h(n), a) H(h(1), \ldots, h(n), b)$$

$$= \sum_{h(1)=0}^{a_1-1} \cdots \sum_{h(n)=0}^{a_n-1} A((h(1) + a) \bmod a_1, \ldots, (h(i) + a) \bmod a_i, \ldots,$$

$$(h(n) + a) \bmod a_n)$$

$$\times A((h(1) + b) \bmod a_1, \ldots, (h(i)$$

$$+ b) \bmod a_i, \ldots, (h(n) + b) \bmod a_n) = 0.$$

The last equation is due to the facts that (1) A is a PBA$(a_1, a_2, \ldots, a_n)$, and (2) $a \not\equiv b \bmod a_i$.

On the other hand, if $1 \leq r \leq n$, and $0 \leq a \neq b \leq a_r - 1$, then

$$\sum_{h(1)=0}^{a_1-1} \cdots \sum_{h(r-1)=0}^{a_{r-1}-1} \sum_{h(r+1)=0}^{a_{r+1}-1} \cdots \sum_{h(n+1)=0}^{a_{n+1}-1} H(h(1), \ldots, h(r-1), a,$$

$$h(r + 1), \ldots, h(n + 1))$$

$$\times H(h(1), \ldots, h(r-1), b, h(r + 1), \ldots, h(n + 1))$$

$$= \sum_{h(1)=0}^{a_1-1} \cdots \sum_{h(r-1)=0}^{a_{r-1}-1} \sum_{h(r+1)=0}^{a_{r+1}-1} \cdots \sum_{h(n+1)=0}^{a_{n+1}-1} A((h(1) + h(n + 1)) \bmod a_1,$$

$$\ldots, (h(r-1) + h(n + 1)) \bmod a_{r-1}, (h(r) + a) \bmod a_r,$$

$$(h(r + 1) + h(n + 1)) \bmod a_{r+1},$$

$$\ldots, (h(n) + h(n+1))\mathrm{mod}a_n)$$

$$\times A((h(1) + h(n+1))\mathrm{mod}a_1, \ldots, (h(r-1) + h(n+1))\mathrm{mod}a_{r-1},$$

$$(h(r) + b)\mathrm{mod}a_r, (h(r+1) + h(n+1))\mathrm{mod}a_{r+1},$$

$$\ldots, (h(n) + h(n+1))\mathrm{mod}a_n)$$

$$= \sum_{h(n+1)=0}^{a_{n+1}-1} [\sum_{h(1)=0}^{a_1-1} \cdots \sum_{h(r-1)=0}^{a_{r-1}-1} \sum_{h(r+1)=0}^{a_{r+1}-1} \cdots \sum_{h(n)=0}^{a_n-1}$$

$$A(h(1), \ldots, h(r-1), h(r) + a, h(r+1), \ldots, h(n))$$

$$\times A(h(1), \ldots, h(r-1), h(r) + b, h(r+1), \ldots, h(n))]$$

$$= 0.$$

Thus H is indeed a generalized Hadamard matrix.   **Q.E.D.**

Up to now, Theorem 6.1.22, Lemma 6.1.4, Theorem 6.1.24 together with Theorem 6.2.1 imply the following important assertion, which motivates us to construct as many higher-dimensional perfect binary arrays as possible.

Corollary 6.2.1 *Each non-trivial higher-dimensional perfect binary array of any size produces infinite families of higher-dimensional Hadamard matrices.*

The following theorem is also true which can be proved by the same way as that of Theorem 6.1.22.

Theorem 6.2.2 *Let $A = [A(a(1), \ldots, a(n))]$, $0 \le a(i) \le m-1$, $1 \le i \le n$, be a $PBA(m, m, \ldots, m)$. Then the following matrix $H = [H(h(1), \ldots, h(n+1))]$, $0 \le h(i) \le m-1$, $1 \le i \le n+1$, defined by*

$$H(h(1), \ldots, h(n+1)) = A(h(1) + h(2), h(2) + h(3), \ldots, h(n) + h(n+1)),$$

is an $(n+1)$-dimensional Hadamard matrix of order m.

6.2.2   Construction and Existence of Higher-Dimensional PBAs

Let $A = [A(a(1), \ldots, a(n))]$, $0 \le a(i) \le a_i - 1$, $1 \le i \le n$, be a PBA$(a_1, \ldots, a_n)$. Its n-dimensional discrete Fourier transform (DFT) is defined by ([4])

$$F(f(1), \ldots, f(n)) = \sum_{a(1), \ldots, a(n)} A(a(1), \ldots, a(n)) \exp\left[-j2\pi\left(\sum_{k=1}^{n} \frac{f(k)a(k)}{a_k}\right)\right],$$

$$\text{(6.15)}$$

where $0 \le f(i) \le a_i - 1$, $1 \le i \le n$.

Thus the mean value, M, of the array A is calculated by

$$M = \sum_{a(1), \ldots, a(n)} A(a(1), \ldots, a(n)) = F(0, \ldots, 0). \tag{6.16}$$

Furthermore, the volume, E, of the array A results from the DFT of the periodic autocorrelation correlation function $R_A(u(1), \ldots, u(n))$, which is a $\delta(u(1), \ldots, u(n))$, since

$$DFT(R_A(u(1), \ldots, u(n))) = |F(f(1), \ldots, f(n))|^2 = E. \tag{6.17}$$

Using Equations (6.16) and (6.17) as well as the definition of volume $E = \prod_{k=1}^{n} a_k$, the mean value M is specified by

$$M = \sqrt{E} = \sqrt{\prod_{k=1}^{n} a_k}. \tag{6.18}$$

Equation (6.18) makes the following results to be true ([4], [5]) :

1. PBAs *are not* (±1)-*balanced, i.e, its mean value* M *is an non-zero integer;*

2. *The volume of each* PBA *is a square number, i.e.,* $E = m^2$, *for some integer* m;

3. *The number of* 1 *and* -1 *in a* PBA *is given by* $(E \pm \sqrt{E})/2$;

4. *In order to vanish the out-of-phase correlations of a* PBA, *the number of elements contained in each* PBA *must be even. Thus* $E = 4m^2 \in \{4, 16, 36, 64, 100, 144, \ldots\}$. *The following Theorem 6.2.3 will show a more comprehensive prove for this result.*

Theorem 6.2.3 ([3], [4]) *If the (± 1)-valued matrix $A=[A(i_1, i_2, \ldots, i_r)]$, $0 \le i_j \le s_j - 1$, $1 \le j \le r$, is a $PBA(s_1, s_2, \ldots, s_r)$, then its volume $\prod_{i=1}^{r} s_i = 4N^2$ for an integer N.*

Proof. At first we prove that the volume must be a square of an integer. In fact, on one hand, because the out-of-phase autocorrelation of A is zero,

$$\sum_{u_1=0}^{s_1-1} \cdots \sum_{u_r=0}^{s_r-1} R_A(u_1, \ldots, u_r) = R_A(0, \ldots, 0) = \prod_{i=1}^{r} s_i.$$

On the other hand,

$$\sum_{u_1=0}^{s_1-1} \cdots \sum_{u_r=0}^{s_r-1} R_A(u_1, \ldots, u_r) = \sum_{u_1=0}^{s_1-1} \cdots \sum_{u_r=0}^{s_r-1} \sum_{j_1=0}^{s_1-1} \cdots \sum_{j_r=0}^{s_r-1} A(j_1, \ldots, j_r)$$

$$\times A(j_1 + u_1, \ldots, j_r + u_r)$$

$$= [\sum_{u_1=0}^{s_1-1} \cdots \sum_{u_r=0}^{s_r-1} A(u_1, \ldots, u_r)]^2.$$

Therefore the volume is a perfect square. Now it is sufficient to prove that the volume is even. In fact, by the definition of PBA, we have $R_A(1, 0, \ldots, 0) = 0$. However, by the definition of auto-correlation, we know that $R_A(1, 0, \ldots, 0)$ is the sum of $\prod_{i=1}^{r} s_i$ terms, each of which is ± 1. Thus this number of terms must be even.   **Q.E.D.**

Examples of $PBA(2, 2)$ and $PBA(6, 6)$ have been presented in the chapter of three-dimensional Hadamard matrices. In addition, $(+ + + -)$ is a $PBA(4)$, and

$$\begin{bmatrix} - + + - + + + + + - + - \\ + + + + - + + - + - + - \\ + + - - + + - - - - - + \end{bmatrix}$$

is a $PBA(3, 12)$. Starting from these four smaller PBAs, we can construct infinite families of PBAs by the following constructions.

Theorem 6.2.4 ([4], [5]) *Let $A = [A(a(1), \ldots, a(n+k))]$ be a $PBA(a_1, \ldots, a_n, d_1, \ldots, d_k)$, and $B = [B(b(1), \ldots, b(k+m))]$ a $PBA(e_1, \ldots, e_k, b_1, \ldots, b_m)$. If $\gcd(e_i, d_i) = 1$, $1 \le i \le k$, then the matrix $C = [C(c(1), \ldots, c(n+m+k))]$, $0 \le c(i) \le a_i - 1$, if $1 \le i \le n$; $0 \le$*

$c(i) \leq d_{i-n}e_{i-n} - 1$, if $1 + n \leq i \leq n + k$; $0 \leq c(i) \leq b_{i-n-k} - 1$, if $1 + n + k \leq i \leq n + m + k$, defined by

$$C(c(1), \ldots, c(n + m + k)) = A(c(1), \ldots, c(n), (c(n + 1))\mathrm{mod}d_1,$$
$$\ldots, (c(n + k))\mathrm{mod}d_k)B((c(n + 1))\mathrm{mode}_1, \ldots, (c(n + k))\mathrm{mode}_k,$$
$$c(n + k + 1), \ldots, c(n + k + m))$$

is a $\mathrm{PBA}(a_1, \ldots, a_n, (d_1e_1), \ldots, (d_ke_k), b_1, \ldots, b_m)$.

Proof. The periodic correlation of C is

$$R(u) = \sum_{c(.)} C(c(1), \ldots, c(n + m + k))$$
$$\times C(c(1) + u_1, \ldots, c(n + m + k) + u_{n+m+k}). \qquad (6.19)$$

Case 1. If $(u_1, \ldots, u_n) \neq (0, \ldots, 0)$, then

$$R(u) = \sum_{c(n+1), \ldots, c(n+k+m)}$$

$$B((c(n + 1))\mathrm{mode}_1, \ldots, (c(n + k))\mathrm{mode}_k, c(n + k + 1),$$

$$\ldots, c(n + k + m))B((c(n + 1) + u_{n+1})\mathrm{mode}_1,$$

$$\ldots, (c(n + k) + u_{n+k})\mathrm{mode}_k,$$

$$c(n + k + 1) + u_{n+k+1}, \ldots, c(n + k + m) + u_{n+k+m})$$

$$\left[\sum_{c(1), \ldots, c(n)} A(c(1), \ldots, c(n), (c(n + 1)\mathrm{mod}d_1, \ldots, (c(n + k))\mathrm{mod}d_k) \right.$$

$$A(c(1) + u_1, \ldots, c(n) + u_n, (c(n + 1) + u_{n+1})\mathrm{mod}d_1,$$

$$\left. \ldots, (c(n + k) + u_{n+k})\mathrm{mod}d_k) \right]$$

$$= 0 \quad \text{(for } A \text{ is a PBA)}.$$

Case 2. If $(u_{n+k+1}, \ldots, u_{n+k+m}) \neq (0, \ldots, 0)$, then by the same way as that of Case 1, it can be proved that $R(u) = 0$.

Case 3. Otherwise $(u_{n+1}, \ldots, u_{n+k}) \neq (0, \ldots, 0)$. Because $0 \leq u_{n+i} \leq e_i d_i - 1$, $1 \leq i \leq k$, and $\gcd(e_i, d_i) = 1$, we know that either $(u_{n+1} \bmod e_1, \ldots, u_{n+k} \bmod e_k) \neq (0, \ldots, 0)$ or $(u_{n+1} \bmod d_1, \ldots, u_{n+k} \bmod d_k) \neq (0, \ldots, 0)$. Without loss of the generality, we assume that $(u_{n+1} \bmod d_1, \ldots, u_{n+k} \bmod d_k) \neq (0, \ldots, 0)$. Then

$$
R(u) = \rho \Big[\sum_{b(.)} B(b(1), \ldots, b(k), b(k+1), \ldots, b(k+m))
$$

$$
\times B(b(1) + u_{n+1}, \ldots, b(k) + u_{n+k}, b(k+1), \ldots, b(k+m)) \Big]
$$

$$
\Big[\sum_{a(.)} A(a(1), \ldots, a(n), a(n+1), \ldots, a(n+k))
$$

$$
\times A(a(1), \ldots, a(n), a(n+1) + u_{n+1}, \ldots, a(n+k) + u_{n+k}) \Big]
$$

$$
= 0 \quad \text{(the summation in second bracket is zero)} .
$$

Therefore the identity $R(u) = 0$ has been proved in each case.   **Q.E.D**

The construction stated in Theorem 6.2.4 is called 'the periodic production'. Particularly, if $k = 0$, then this theorem results in the following corollary.

Corollary 6.2.2 ([4], [5]) *Let* $A = [A(a(1), \ldots, a(n))]$ *be a* PBA$(a_1, \ldots, a_n)$, *and* $B = [B(b(1), \ldots, b(m))]$ *a* PBA$(b_1, \ldots, b_m)$. *Then the following matrix* $C = [C(c(1), \ldots, c(n+m))]$, $0 \leq c(i) \leq a_i - 1$, *if* $1 \leq i \leq n$; $0 \leq c(i) \leq b_{i-n} - 1$, *if* $1 + n \leq i \leq n + m$, *defined by*

$$
C(c(1), \ldots, c(n+m)) = A(c(1), \ldots, c(n)) B(c(n+1), \ldots, c(n+m))
$$

is a PBA$(a_1, \ldots, a_n, b_1, \ldots, b_m)$.

Applying this corollary to the known PBA(4), we find PBA$(4 \times \ldots \times 4)$; applying it to PBA(6, 6) we find PBA$(6 \times \ldots \times 6)$ of even dimension; applying it to PBA(12, 3) and PBA(4) we find PBA(12, 12); applying it to PBA(2, 2) we find PBA$(2, \ldots, 2)$ of even dimension, and so on.

Theorem 6.2.5 *Let* $A = [A(a(1), \ldots, a(n))]$, $0 \leq a(i) \leq a_i - 1$, $1 \leq i \leq n$, *be a* PBA$(a_1, \ldots, a_n)$. *And let* $b_1, \ldots, b_n$ *and* $c_1, \ldots, c_n$ *be integers* such that $\gcd(b_i, a_i) = 1$, $1 \le i \le n$. Then the following matrix $H = [H(h(1), \ldots, h(n))]$, $0 \le h(i) \le a_i - 1$, $1 \le i \le n$, defined by

$$H(h(1), \ldots, h(n)) = A(b_1 h(1) + c_1, \ldots, b_n h(n) + c_n)$$

is also a $PBA(a_1, \ldots, a_n)$.

The construction described in this theorem is called 'sampling' or 'decimation'. Its proof is trivial.

Besides the sampling, there are many other invariance operations, which produce further perfect binary arrays of the same volume as the mother array. The example invariance operations include cyclic shifts, inverting all signs, reflection, rotation, permutation of the coordinates and so on.

Let $A = [A(a(1), \ldots, a(n))]$, $0 \le a(i) \le a_i - 1$, $1 \le i \le n$, be an n-dimensional (± 1)-valued array of size $a_1 \times \ldots \times a_n$. If $a_1 = b_1 \times b_2$, where $\gcd(b_1, b_2) = 1$, $b_1 > 1$ and $b_2 > 1$, then, by the Chinese remainder theorem, for each $a(1) \in Z_{a_1}$, there exists one and only one pair of integers $(b(1), b(2)) \in Z_{b_1} \times Z_{b_2}$ such that $a(1) \equiv b(1) \bmod b_1$ and simultaneously $a(1) \equiv b(2) \bmod b_2$, and vice versa. Thus the n-dimensional matrix A corresponds to an $(n+1)$-dimensional matrix $B = [B(b(1), \ldots, b(n+1))]$ of size $b_1 \times b_2 \times a_2 \times \ldots \times a_n$ since

$$B(b(1), \ldots, b(n+1)) = A(a(1), b(3), \ldots, b(n+1))$$

where $a(1) \equiv b(1) \bmod b_1$ and $a(1) \equiv b(2) \bmod b_2$. This matrix B is called 'the folding of A', or equivalently the matrix A is called 'the refolding of B'.

Theorem 6.2.6 *An n-dimensional matrix A is a perfect binary array if and only if its folding (resp. refolding) is a perfect binary array. Hence, if $\gcd(s, t) = 1$, then there exists a $PBA(s, t, s_1, \ldots, s_n)$ if and only if there exists a $PBA((st), s_1, \ldots, s_n)$.*

Proof. We prove the forward assertion only, i.e., the folding of a PBA is also a PBA. In fact, if $(t(1), \ldots, t(n+1)) \ne (0, \ldots, 0)$, $0 \le t(1) \le b_1 - 1$, $0 \le t(2) \le b_2 - 1$, $0 \le t(i) \le a_i - 1$, $3 \le i \le n + 1$, then the periodic correlation is

$$\sum_{b(1), \ldots, b(n+1)} B(b(1), \ldots, b(n+1)) . B(b(1) + t(1), \ldots, b(n+1) + t(n+1))$$

$$= \sum_{a(1)=0}^{a_1-1} \sum_{b(3),\dots,b(n+1)} A(a(1), b(3), \dots, b(n+1))$$

$$\times A(a(1) + t'(1), b(3) + t(3), \dots, b(n+1) + t(3))$$

$$= 0,$$

where in the last two equations, $t'(1)$ is determined from $t(1)$ and $t(2)$ by $t'(1) \equiv t(1) \mathrm{mod} b_1$ and $t'(2) \equiv t(2) \mathrm{mod} b_2$. Thus $t'(1) \mathrm{mod} a_1 = 0$ if and only if $t(1) \mathrm{mod} b_1 = t(2) \mathrm{mod} b_2 = 0$. The last equation results from A being a PBA.   **Q.E.D.**

With the fact that the array's volume remains unchanged, Theorem 6.2.6 transforms a PBA of larger-dimension and smaller size to a PBA of smaller-dimension and larger size, and vice versa. For example, applying the forward part of Theorem 6.2.6, a PBA$(12, 3)$ leads to a PBA$(3, 4, 3)$, and a PBA$(6, 6)$ leads to a PBA$(2, 3, 2, 3)$, etc.

Recall that a (v, k, λ)-difference set, D, is a subset of an additive group, G, of order v such that D contains k elements and the number of solutions to the equation

$$d' - d = g, \text{ for } d, d' \in D, d \neq d',$$

is λ for each nonzero $g \in G$. In particular, a $(4N^2, 2N^2 - N, N^2 - N)$-difference set is called 'a Menon difference set ([6])'. An integer t is called 'a (numerical) multiplier of D' if

$$\{td : d \in D\} = \{d + h : d \in D\} \text{ for some } h \in G,$$

and if $h = 0$, then D is fixed by the multiplier t. A difference set fixed by the multiplier -1 is called 'symmetric'. A well-known identity for a (v, k, λ)-difference set states that

$$k(k-1) = \lambda(v-1). \tag{6.20}$$

Definition 6.2.2 ([3], [6]) *Let* $A = [A(j_1, \dots, j_n)]$ *be a binary array of size* $s_1 \times \dots \times s_n$. *The set equivalent of* A *is the subset* $\nu(A)$ *of the abelian group* $Z_{s_1} \times \dots \times Z_{s_n}$ *given by*

$$\nu(A) = \{(j_1, \dots, j_n) : A(j_1, \dots, j_n) = -1\}.$$

If $(u_1, \ldots, u_n) \in Z_{s_1} \times \ldots \times Z_{s_n}$, denote by $\lambda_A(u_1, \ldots, u_n)$ the number of solutions to the equation

$$(j'_1, \ldots, j'_n) - (j_1, \ldots, j_n) = (u_1, \ldots, u_n), \ for \ (j'_1, \ldots, j'_n), (j_1, \ldots, j_n) \in \nu(A).$$

The mapping from A to $\nu(A)$ is invertible. The following lemma shows that a binary array with two-valued auto-correlation is equivalent to a difference set in $Z_{s_1} \times \ldots \times Z_{s_n}$.

Lemma 6.2.1 ([3], [6]) *Let* $A = [A(j_1, \ldots, j_n)]$ *be a binary array of size* $s_1 \times \ldots \times s_n$. *And let* $\mid \nu(A) \mid = k$. *Then, for all* $0 \le u_i \le s_i - 1$, *the periodic autocorrelation of* A *is*

$$R_A(u_1, \ldots, u_n) = \prod_{i=1}^{n} s_i - 4(k - \lambda_A(u_1, \ldots, u_n)).$$

Proof. By Definition 6.2.2, $\lambda_A(u_1, \ldots, u_n)$ is equal to the number of occurrences of the product of -1 with -1 on the right hand side of

$$R_A(u_1, \ldots, u_n) = \sum_{j_1=0}^{s_1-1} \cdots \sum_{j_n=0}^{s_n-1} A(j_1, \ldots, j_n) A(j_1 + u_1, \ldots, j_n + u_n) = 0.$$

The number of occurrences of the product of -1 with 1, of 1 with -1, and of 1 with 1 is then respectively $k - \lambda_A(u_1, \ldots, u_n)$, $k - \lambda_A(u_1, \ldots, u_n)$, and $\prod_{i=1}^{n} s_i - 2k + \lambda_A(u_1, \ldots, u_n)$.   **Q.E.D.**

The following theorem shows that a PBA is equivalent to a Menon difference set in an abelian group.

Theorem 6.2.7 ([3], [6]) *A is a* PBA$(s_1, \ldots, s_n)$ *if and only if* $\nu(A)$ *is a* $(4N^2, 2N^2 - N, N^2 - N)$-*difference set in the abelian group* $Z_{s_1} \times \ldots \times Z_{s_n}$, *where* $4N^2 = \prod_{i=1}^{n} s_i$.

Proof. Suppose A is a PBA$(s_1, \ldots, s_n)$, and let $\mid \nu(A) \mid = k$. Then by Lemma 6.2.1 and Definition 6.2.1, for all $(u_1, \ldots, u_n) \ne (0, \ldots, 0)$, $0 \le u_i \le s_i - 1$, we have

$$\lambda_A(u_1, \ldots, u_n) = k - \prod_{i=1}^{n} s_i/4.$$

Therefore by the definitions of difference set and $\nu(A)$, we know that $\nu(A)$ is a $(\prod_{i=1}^{n} s_i, k, k - \prod_{i=1}^{n} s_i/4)$-difference set in $Z_{s_1} \times \ldots \times Z_{s_n}$. By

Theorem 6.2.3, $\prod_{i=1}^{n} s_i = 4N^2$ for an integer N. Then by Equation (6.20), $k = 2N^2 - N$. Hence $\nu(A)$ is a Menon difference set.

Conversely, suppose that $\nu(A)$ is a $(4N^2, 2N^2 - N, N^2 - N)$-difference set in $Z_{s_1} \times \ldots \times Z_{s_n}$, where $4N^2 = \prod_{i=1}^{n} s_i$. Then by the definitions of difference set and $\nu(A)$, we know that for all $(u_1, \ldots, u_n) \neq (0, \ldots, 0)$, $0 \leq u_i \leq s_i - 1$, we have

$$\lambda_A(u_1, \ldots, u_n) = N^2 - N.$$

The result follows from Lemma 6.2.1 and Definition 6.2.1.   **Q.E.D.**

The equivalence between PBAs and Menon difference sets are useful for the construction of PBAs and especially for the proof of the existence and non-existence results. Now we turn to introduce some Menon difference sets in the abelian group $Z_{s_1} \times \ldots \times Z_{s_n}$.

Definition 6.2.3 ([6], [7], [15]) *Let A, B, C, D be (± 1)-valued arrays of size $s_1 \times \ldots \times s_n$. $\{A, B, C, D\}$ is called 'an $s_1 \times \ldots \times s_n$ binary supplementary quadruple (BSQ)' if the following are satisfied for all $0 \leq u_i \leq s_i - 1$, $1 \leq i \leq n$,*

1. $(R_A + R_B + R_C + R_D)(u_1, \ldots, u_n) = 0$ *if* $(u_1, \ldots, u_n) \neq (0, \ldots, 0)$;

2. $(R_{WX} + R_{YZ})(u_1, \ldots, u_n) = 0$ *for all* $\{W, X, Y, Z\} = \{A, B, C, D\}$,

where $R_X(.)$ and $R_{XY}(.)$ refer to the periodic auto- and cross-correlations, respectively.

In the next subsection, the following useful theorem will be proved (See Corollary 6.2.5.

Theorem 6.2.8 ([3], [8], [6]) *If there exists an $s_1 \times \ldots \times s_n$ BSQ, then there exists a Menon difference in the group*

$$Z_{2^{a_1}} \times Z_{2^{a_2}} \times \ldots \times Z_{2^{a_k}} \times Z_{s_1} \times Z_{s_2} \times \ldots \times Z_{s_n}$$

where $\sum_i a_i = 2a + 2 \geq 2$ and $a_i \leq a + 2$ for all i. Thus we have a PBA$(2^{a_1}, \ldots, 2^{a_k}, s_1, \ldots, s_n)$, by Theorem 6.2.7.

Let G be the group of the form

$$Z_{3^b}^2 \cong \langle y, z \rangle, \quad y^{3^b} = z^{3^b} = 1.$$

And let

$$D_{1,i} = \langle yz^i \rangle, \quad i = 0, 1, \ldots, \frac{3^{b+1} - 1}{3 - 1} - 2,$$

and

$$D_{3j,1} = \langle y^{3j} z \rangle, \quad j = 0, 1, \ldots, \frac{3^b - 1}{3 - 1}.$$

Thus

$$D_{1,m} = D_{1,m+3^b} \quad and \quad D_{3r,1} = D_{3(r+3^{b-1}),1}.$$

In the following, we will use D_0, D_1, D_2, and D_3 to respectively represent the

$$D_k = \bigcup_{i=0}^{(3^b-1)/2-1} z^i D_{1,3i+k}, \quad \text{if} \ \ k = 0, 1, 2$$

and

$$D_3 = \bigcup_{j=0}^{(3^b-1)/2} y^j D_{3j,1}.$$

Lemma 6.2.2 ([6], [9], [10]) *The above D_k has no repeated elements.*

Proof. We will prove the case of $k = 0$ only, because the other cases can be proved by the same way. Suppose there was a repeated element. Then there exist i, i', m, m' such that $z^i(yz^{3i})^m = z^{i'}(yz^{3i'})^{m'}$. In order for this to occur, $m = m'$, since the same power of y must be present. Considering the powers of z, we have

$$z^{i+3mi} = z^{i'+3mi'} \quad or \quad i(1 + 3m) \equiv i'(1 + 3m) \ (\text{mod} 3^b).$$

Since $1 + 3m$ is invertible mod 3^b, we can conclude that $i \equiv i' \ (\text{mod} 3^b)$; the restrictions on the i and i' imply that they are the same. Thus, these two elements are not really distinct, so no elements are repeated.   **Q.E.D.**

Lemma 6.2.3 ([8], [6]) *If χ is a character of order 3^b on $\langle y, z \rangle$, then $|\chi(D_k)| = 3^b$ for one value of k, and 0 for the others.*

Proof. Note that $G/\text{Ker}(\chi)$ is a cyclic group. Let $x\text{Ker}(\chi)$ be a generator of $G/\text{Ker}(\chi)$; the order of $x\text{Ker}(\chi)$ must be 3^b since it is the size of the factor group. Thus the order of x is also 3^b since it is the maximum order in G. The subgroups $\langle x \rangle$ and $\text{Ker}(\chi)$ intersect only in the identity (no power of x smaller than 3^b can be in $\text{Ker}(\chi)$), and their product is all of

G; thus G must be a direct product of $< x >$ and $\text{Ker}(\chi)$. The fact that G has rank 2 implies that $\text{Ker}(\chi)$ must be a cyclic group of order 3^b.

Since we have established the fact that $\text{Ker}(\chi)$ is cyclic, we need to observe that all of the cyclic subgroups of G of order 3^b are of the form $D_{i,j}$ for some i, j. Thus χ is principal on one $D_{i',j'}$ and non-principal on all of the others. If that $D_{i',j'}$ appears only once, then the character sum (in mod ulus) is the size of the set which is equal to 3^b. If the $D_{i',j'}$ appears twice, then we suppose that χ has order 3^b on the element z (it must have order 3^b on either y or z, and the y argument is the same) . The character sum is 3^b times $\chi(z^i) + \chi(z^{i+3^{b-1}})$. Since χ is a homomorphism, this can be rewritten as $\chi(z^i)(1 + \chi(z^{3^{b-1}}))$; the $\chi(z^{3^{b-1}})$ is a primitive third root of unity, so $\left| \chi(z^i)(1 + \chi(z^{3^{b-1}})) \right| = 1$. Thus the character sum is (in mod ulus) 3^b.   **Q.E.D.**

Lemma 6.2.4 ([8], [6], [10]) *If χ is a character of $< y, z >$ that is nonprincipal but of order less than 3^b, then $|\chi(D_k)| = 3^b$ for one k, and 0 for the others.*

Proof. Let ξ be a primitive 3^bth root of unity, and suppose that χ is a character of order less than 3^b. Then

$$\chi(y) = \xi^{e3^v}, \quad and \quad \chi(z) = \xi^{f3^t}, \quad 1 \le t, v \le b, \quad (t,v) \ne (b,b),$$

where e and f are nonzero integers which are not divisible by 3.

Consider the case of $v < t$ (the other cases of $v > t$ and $v = t$ are similar). Nothing of the form yz^x is in the kernel of the character because there is no way to satisfy the equation $e3^v + xf3^t \equiv 0 \pmod{3^b}$ when $v < t$. Thus χ is nonprincipal on every subgroup $D_{i,j}$ contained in the sets D_0, D_1, and D_2, so the character sum is 0 over these parts. The kernel does contain elements of the form $y^{3x}z$ whenever $3xe3^v + f3^t \equiv 0 \pmod{3^b}$, or whenever $x \equiv -fe^{-1}3^{t-v-1} \pmod{3^{b-v-1}}$. The character χ is principal on the subgroups $D_{3j,1}$ associated with those solutions, and non-principal on all the other subgroups $D_{3j,1}$ in D_3. There are q solutions x to this equation with $0 \le x \le (3^b-1)/(3-1)$, where q is either $3^v+3^{v-1}+\ldots+3+1$ or $3^v + 3^{v-1} + \ldots + 3 + 1 + 1$. (In general, the number of solutions to the congruence $x \equiv a \pmod{b}$ with $0 \le x \le c$ is $\lfloor c/b \rfloor$ or $\lfloor c/b \rfloor +1$) . Hence the character sum over D_3 is

$$3^b(\chi(y^x) + \chi(y^{x+3^{b-v-1}}) + \chi(y^{x+2.3^{b-v-1}}) + \ldots + \chi(y^{x+(q-1).3^{b-v-1}}))$$

$$= 3^b \chi(y^x)(1 + \chi(y^{3^{b-v-1}}) + \ldots + \chi(y^{(q-1)3^{b-v-1}})).$$

Since $\chi(y) = \xi^{e3^v}$, we get

$$\chi(D_3) = \begin{cases} 3^b \chi(y^x)(1 + \xi^{e3^{b-1}} + \xi^{2e3^{b-1}} + 1 + \ldots + \xi^{2e3^{b-1}} + 1) \\ \qquad\qquad\qquad\qquad\qquad\qquad q \equiv 1 \pmod 3; \\ \\ 3^b \chi(y^x)(1 + \xi^{e3^{b-1}} + \xi^{2e3^{b-1}} + 1 + \ldots + \xi^{2e3^{b-1}} + 1 + \xi^{e3^{b-1}}) \\ \qquad\qquad\qquad\qquad\qquad\qquad q \equiv 2 \pmod 3 \end{cases}$$

$$= \begin{cases} 3^b \chi(y^x) & q \equiv 1 \pmod 3; \\ \\ 3^b \chi(y^x)(1 + \xi^{e3^{b-1}}) & q \equiv 2 \pmod 3. \end{cases}$$

In either case, the mod plus of this sum is 3^b since $(1 + \xi^{e3^{b-1}}) = -\xi^{2 \cdot e3^{b-1}}$ has mod plus 1. Thus $|\chi(D_3)| = 3^b$ and all the others are 0.   **Q.E.D.**

Lemma 6.2.5 ([8], [6]) *Let $A = v^{-1}(D_0)$, $B = v^{-1}(D_1)$, $C = v^{-1}(D_2)$, and $D = v^{-1}(D_3)$ be four subsets of G Then the sets $\{A, B, C, D\}$ form a $3^b \times 3^b$ BSQ.*

Proof. We need to translate the definition of BSQ into character theoretic terms. The first condition $(R_A + R_B + R_C + R_D)(u_1, u_2) = 0$, can be shown by considering the group ring expression as

$$D_0 D_0^{(-1)} + D_1 D_1^{(-1)} + D_2 D_2^{(-1)} + D_3 D_3^{(-1)} - 3^{2b}.$$

By Lemmas 6.2.3 and 6.2.4, any nonprincipal character on G has a sum of 3^b (in mod plus) on one of the D_k, and it will be 0 on the others. Thus, the character sum on the expression is 0, which implies that the group ring expression is cG for some c. The fact that

$$|D_0| = |D_1| = |D_2| = 3^b(3^b - 1)/2 \text{ and } |D_3| = 3^b(3^b + 1)/2$$

implies that

$$c = \{\sum |D_i|^2 - 3^{2b}\}/|G|$$

$$= \{3 \cdot 3^{2b}(3^b - 1)^2/4 + 3^{2b}(3^b + 1)^2/4 - 3^{2b}\}/3^{2b}$$

$$= 3^{2b} - 3^b.$$

The number of times that the nonidentity element (u_1, u_2) appears in the expression is $3^{2b} - 3^b$, and this corresponds to the number of times that $X[j_1, j_2] = X[j_1 + u + 1, j_2 + u_2] = -1$ in the autocorrelation equation for $X = A, B, C,$ or D (we call these $(-1, -1)$ pairs) . There are $2.3^{2b} - 3^b$ times when $X[j_1, j_2] = -1$, so there are $2.3^{2b} - 3^b - (3^{2b} - 3^b) = 3^{2b}$ pairs of the form $(-1, 1)$ and $(1, -1)$. Finally, there are a total of 4.3^{2b} pairs and so there are $4.3^{2b} - 3^{2b} - 3^{2b} - (3^{2b} - 3^b) = 3^{2b} + 3^b$ $(1, 1)$ pairs. Thus, the autocorrelation equation becomes

$$(R_A + R_B + R_C + R_D)(u_1, u_2) = (3^{2b} - 3^b)(-1)(-1) + 3^{2b}(-1)(1)$$
$$+3^{2b}(1)(-1) + (3^{2b} + 3^b)(1)(1) = 0$$

for all $(u_1, u_2) \neq (0, 0)$.

The proof of the second condition involves studying the group ring expression $D_0 D_1^{(-1)} + D_2 D_3^{(-1)}$. Using the same arguments as above, we find that the character sum over this equation is 0 for any nonprincipal character, which implies that $D_0 D_1^{(-1)} + D_2 D_3^{(-1)} = cG$ for some c. A counting argument yields that $c = (3^{2b} - 3^b)/2$, and this number is also the number of $(-1, -1)$ pairs (by this we mean the number of times that $X[j_1, j_2] = Y[j_1 + u_1, j_2 + u_2] = -1$ for $(X, Y) = (A, B)$ or (C, D)) in the sum $R_{AB} + R_{CD}$. Since $|D_0| = (3^{2b} - 3^b)/2$, and $|D_2| = (3^{2b} - 3^b)/2$, there are

$$((3^{2b} - 3^b)/2 + (3^{2b} - 3^b)/2) - (3^{2b} - 3^b)/2 = (3^{2b} - 3^b)/2$$

$(-1, 1)$ pairs. Similar counts yield $(3^{2b} + 3^b)/2$ pairs of both $(1, -1)$ and $(1, 1)$. Thus

$$(R_{AB} + R_{CD})(u_1, u_2) = (3^{2b} - 3^b)/2(-1)(-1) + (3^{2b} - 3^b)/2(-1)(1)$$
$$+(3^{2b} - 3^b)/2(1)(-1) + (3^{2b} - 3^b)/2(1)(1)$$
$$= 0 \text{ (for every } (u_1, u_2)) .$$

We can shuffle the four sets any way we want, and we will get the same result for the autocorrelation equation. Thus, these four sets satisfy the definition of BSQ.   **Q.E.D.**

Now we are ready to state one of the main results about the existence of PBAs.

Theorem 6.2.9 ([3], [5]) *There exists*

$$\text{PBA}(2^{a_1}, \ldots, 2^{a_k}, 3^{b_1}, 3^{b_1}, \ldots, 3^{b_n}, 3^{b_n}),$$

where $\sum_i a_i = 2a + 2$, $a \geq 0$, $a_i \leq a + 2$. *Particularly, there exists* $\text{PBA}(2^{1+a}3^b, 2^{1+a}3^b)$ *for all* $a, b \geq 0$. *(This result has been used to construct the 3-dimensional Hadamard matrices of order* 2.3^b *in the chapter of 3-Dimensional Hadamard Matrices.)*

Proof. Lemma 6.2.5 together with Theorems 6.2.8 and 6.2.7 implies the existence of $\text{PBA}(2^{a_1}, \ldots, 2^{a_k}, 3^{b_1}, 3^{b_1})$. Then the proof is finished by Theorem 6.2.4. Particularly, the $\text{PBA}(2^{1+a}3^b, 2^{1+a}3^b)$ is constructed by using Theorem 6.2.6 to form a $\text{PBA}(2^a, 2^a, 3^b, 3^b)$.   **Q.E.D.**

Another important existence result about PBAs is based on the following lemma, which has been proved in [11] or reviewed in [6].

Lemma 6.2.6 ([11]) *There exists BSQ of size* $p \times p \times p \times p$ *for each positive integer* p *satisfying* $p \equiv 3 \mod 4$.

This lemma together with Lemma 6.2.5, Theorems 6.2.8, 6.2.7, and 6.2.4 results in the following larger class of PBAs.

Theorem 6.2.10 *There exists*

$$\text{PBA}(2^{a_1}, \ldots, 2^{a_k}, 3^{b_1}, 3^{b_1}, \ldots, 3^{b_n}, 3^{b_n}, p_1^{c_1}, p_1^{c_1}, p_1^{c_1}, p_1^{c_1}, \ldots, p_m^{c_m}, p_m^{c_m}, p_m^{c_m}, p_m^{c_m})$$

where $\sum_i a_i = 2a + 2$, $a \geq 0$, $a_i \leq a + 2$, *and* $p_i \equiv 3 \mod 4$.

Besides the above existence results, there are also many results about the nonexistence of PBAs. For example, it has been proved that ([3], [4], [11], [12], [13], [14], [15], [16])

1. *There exists no PBAs of volume* $4p^2$ *for any prime* $p > 3$;

2. *Suppose there exists a Menon difference in* $Z_{p^a} \times Z_{p^{2y-a}} \times H$, *where* $0 \leq a \leq 2y$, H *an abelian group of even order* h, $p > 2$, *a prime self-conjugate* mod $\exp(H)$, *and* $p \nmid h$. *Then* $a = y$, *and if* $p > 3$, *then*

   (a) *if* $a = 1$, $p^2 < h$ *and* $(p + 1)|h$;

(b) *if* $a > 1$, $p < h$.

3. *If there exists a* PBA$(4N^2)$ *with* $N > 1$, *then* N *is odd, not a prime power and* $N \geq 55$;

4. *Suppose there exists a* $(4N^2, 2N^2 - N, N^2 - N)$-*difference set in an abelian group* G. *If* N *is even, then the Sylow 2-subgroup of* G *is not cyclic;*

5. *There does not exist* PBA(s, t), *if* $(s, t)=$ $(2, 32)$, $(4, 64)$, $(2, 18)$, $(6, 54)$, $(12, 27)$, $(36, 81)$, $(2, 72)$, $(4, 36)$, $(8, 18)$, $(4, 100)$, $(8, 50)$, $(8, 98)$, $(24, 54)$, $(32, 98)$, $(6, 96)$, $(18, 32)$, $(32, 50)$, $(36, 64)$, $(54, 96)$, $(64, 100)$, $(10, 90)$, $(12, 75)$, $(18, 50)$, $(20, 45)$, $(18, 98)$, $(28, 63)$, $(50, 98)$, $(36, 100)$, $(50, 72)$, $(72, 98)$, $(28, 28)$, $(44, 44)$, $(76, 76)$, $(92, 92)$, $(30, 30)$, $(42, 42)$, $(70, 70)$, $(90, 90)$, $(21, 84)$, $(84, 84)$, $(10, 10)$, $(14, 14)$, $(7, 28)$, $(14, 56)$, $(66, 66)$, $(50, 50)$, $(98, 98)$, *and so on;*

6. *There does not exist* PBA$(2, 2, 5, 5)$, PBA$(2, 2, 3, 3, 9)$, PBA$(4, 3, 3, 9)$, *and so on.*

6.2.3 Generalized Perfect Arrays

For details of the generalized perfect arrays the readers are recommended to the papers [3] and [10].

Generalized perfect array (GPA) is a generalization of the perfect, quasiperfect and doubly quasiperfect arrays. This subsection generalize the elementary recursive constructions for the 2-dimensional cases. Another main aim of this subsection is to show Theorem 6.2.8, one of the most important result on the construction of PBAs. The main contents of this subsection are from [3] and [10]. In order to save space, many tedious proofs have to be omitted.

An n-dimensional m-ary array of size $a_1 \times \ldots \times a_n$ is a matrix $A = [A(a(1), \ldots, a(n))]$, $0 \leq a(i) \leq a_i - 1$, $1 \leq i \leq n$, such that

$$A(a(1), \ldots, a(n)) \in \begin{cases} \{\pm 1, \ldots, \pm m/2\} & \text{if } m \text{ is even} \\ \{0, \pm 1, \ldots, \pm(m-1)/2\} & \text{if } m \text{ is odd} \end{cases}$$

(When $m = 2$ the matrix A becomes the binary array studied in last subsections) . The energy and sum of the m-ary array A are defined by

([3], [10])

$$E_A = \sum_{a(1)=0}^{a_1-1} \cdots \sum_{a(n)=0}^{a_n-1} (A(a(1), \ldots, a(n)))^2$$

and

$$S_A = \sum_{a(1)=0}^{a_1-1} \cdots \sum_{a(n)=0}^{a_n-1} A(a(1), \ldots, a(n))$$

respectively. Thus in binary cases, $E_A = \prod_{i=1}^n a_i$, which is also the volume of this matrix.

Definition 6.2.4 ([3], [10]) *Let $A = [A(a(1), \ldots, a(n))]$ and $B = [B(b(1), \ldots, b(n))]$,
$0 \le a(i), b(i) \le s_i - 1$, $1 \le i \le n$, be two m-ary arrays of the same size $s_1 \times \ldots \times s_n$.*

1. *The periodic cross-correlation function of A and B is defined by*

$$R_{AB}(u_1, \ldots, u_n) = \sum_{j_1=0}^{s_1-1} \cdots \sum_{j_n=0}^{s_n-1} A(j_1, \ldots, j_n)B(j_1 + u_1, \ldots, j_n + u_n),$$

   *where $0 \le u_i \le s_i - 1$, $j_i + u_i = (j_i + u_i) \bmod s_i$, $1 \le i \le n$. If $A = B$,
   then $R_{AB}(.) := R_A(.)$ is the periodic auto-correlation function of A.*

2. *The adjoining of A with B in dimension r, $1 \le r \le n$, is an m-ary
   array, $\alpha^{(r)}(A, B) = [C(j_1, \ldots, j_n)]$, of size $s_1 \times \ldots \times s_{r-1} \times (2s_r) \times s_{r+1} \times \ldots \times s_n$ defined by*

$$C(j_1, \ldots, j_{r-1}, j_r + ys_r, j_{r+1}, \ldots, j_n) = \begin{cases} A(j_1, \ldots, j_n) & \text{if } y = 0 \\ B(j_1, \ldots, j_n) & \text{if } y = 1. \end{cases}$$

Clearly the energy of $\alpha^{(r)}(A, B)$ is $E_A + E_B$, the summation of its parent's energy.

Definition 6.2.5 ([3], [10]) *Let $A = [A(a(1), \ldots, a(n))]$, $0 \le a(i) \le a_i - 1$,
$1 \le i \le n$, be an m-ary array of size $a_1 \times \ldots \times a_n$, and let $z = (z_1, \ldots, z_n)$
be a $(0, 1)$-valued vector, which is called "the type vector".*

1. *The expansion of A with respect to the type vector z is the following
   m-ary array, $\epsilon(A, z) := E = [E(e(1), \ldots, e(n))]$, of size $(z_1 + 1)a_1 \times \ldots \times (z_n + 1)a_n$, defined by*

$$E(j_1 + y_1 a_1, \ldots, j_n + y_n a_n) = (-1)^{\sum_i y_i} A(j_1, \ldots, j_n),$$

where $0 \leq j_i \leq a_i - 1$, $0 \leq y_i \leq z_i$, $1 \leq i \leq n$.

2. *The m-ary array A is called "a generalized perfect array (GPA) of type z", abbreviated as a $\mathrm{GPA}(m; a_1, \ldots, a_n)$ of type z, if the periodic auto-correlation of $\epsilon(A, z) = E$ satisfies*

$$R_E(u_1, \ldots, u_n) \neq 0, \ \ only \ if \ u_i \equiv 0 (\bmod a_i) \ for \ all \ i \ .$$

where $0 \leq u_i \leq (z_i + 1)a_i - 1$. Particularly, if $m = 2$, the GPA is called GPBA.

For example, if A is a 2-dimensional m-ary array of size $a_1 \times a_2$, then

$$B_0 := A = \epsilon(A; 0, 0); \ \ B_1 := \alpha^{(1)}(A, -A) = \begin{bmatrix} A \\ -A \end{bmatrix} = \epsilon(A; 1, 0);$$

$$B_2 := \alpha^{(2)}(A, -A) = [A \ - A] = \epsilon(A; 0, 1)$$

and

$$B_3 := \alpha^{(1)}(\alpha^{(2)}(A, -A), \alpha^{(2)}(-A, A)) = \begin{bmatrix} A & -A \\ -A & A \end{bmatrix} = \epsilon(A; 1, 1).$$

Thus a PBA is a GPA of type $(0, \ldots, 0)$; a rowwise quasiperfect binary array (RQPBA) is a 2-dimensional GPBA of type $(1, 0)$; a columnwise quasiperfect binary array (CQPBA) is a 2-dimensional GPBA of type $(0, 1)$; a doubly quasiperfect binary array (DQPBA) is a 2-dimensional GPBA of type $(1, 1)$.

By the above definitions, it is not difficult to prove that ([3], [10])

1. *If $A = [A(a(1), \ldots, a(n))]$ is a $\mathrm{PBA}(a_1, \ldots, a_n)$, then the matrix*

$$B = [B(b(1), \ldots, b(n-1))] = \left[\sum_{a(n)=0}^{a_n - 1} A(a(1), \ldots, a(n)) \right]$$

is a GPA;

2. *If π is a permutation of $\{1, \ldots, n\}$, then $A = [A(a(1), \ldots, a(n))]$ is a $\mathrm{GPA}(m; a_1, \ldots, a_n)$ of type $(z_1, \ldots, z_n)$ if and only if $[A(a(\pi(1)), \ldots, a(\pi(n)))]$ is a $\mathrm{GPA}(m; a_{\pi(1)}, \ldots, a_{\pi(n)})$ of type $(z_{\pi(1)}, \ldots, z_{\pi(n)})$, i.e., the dimensions may be reordered.*

Lemma 6.2.7 ([3], [10]) *Let A be an m-ary array of size $a_1 \times \ldots \times a_n$, and $z = (z_1, \ldots, z_n)$ be a type vector. Then*

$$\epsilon(A; 1, z) = \alpha^{(1)}(\epsilon(A; 0, z), -\epsilon(A; 0, z)).$$

From this lemma, it is known that the array $\epsilon(A; z_1, \ldots, z_n)$ may be formed by recursively adjoining A with its negative copy $-A$ for each dimension i satisfying $z_i = 1$.

Lemma 6.2.8 ([3], [10]) *Let A be an m-ary array of size $a_1 \times \ldots \times a_n$, $z = (z_1, \ldots, z_n)$ be a type vector, and $E = \epsilon(A; z)$. Then*

$$R_E(y_1 a_1, \ldots, y_n a_n) = E_E (-1)^{\sum_i y_i} = E_A (-1)^{\sum_i y_i} \prod_i (z_i + 1),$$

where $0 \le y_i \le z_i$, $1 \le i \le n$, E_X the energy function of X, and R_E the periodic autocorrelation of E.

Lemma 6.2.9 ([3], [10]) *Let $A = [A(j, j_1, \ldots, j_n)]$ be an m-ary array of size $(st) \times a_1 \times \ldots \times a_n$, $z = (z_1, \ldots, z_n)$ be a type vector. And let $B = [B(j, j_1, \ldots, j_n)]$ be the $(1 + 2s\lfloor m/2 \rfloor)$-ary array defined by*

$$B(j, j_1, \ldots, j_n) = \sum_{r=0}^{s-1} A(j + rt, j_1, \ldots, j_n),$$

where $0 \le j \le t - 1$, $0 \le j_i \le a_i - 1$, $1 \le i \le n$. If A is a GPA$(m; (st), a_1, \ldots, a_n)$ of type $(0, z)$, then B is a GPA$(1 + 2s\lfloor m/2 \rfloor; t, a_1, \ldots, a_n)$ of type $(0, z)$ and $E_B = E_A$, $S_A = S_B$, i.e., the energy and sum functions remain unchanged.

Definition 6.2.6 ([3], [10]) *Let G be an additive group of order mn, and H a subgroup of G of order n. And let D be a subset of G containing k elements. D is called an $(n, m, k, \lambda', \lambda)$-relative difference set in G relative to H if the number of solutions of the equation*

$$d' - d = g, \quad \text{for} \quad d, d' \in D, \; d \ne d',$$

1. *is λ' for each nonzero $g \in H$, and*

2. *is λ for each $g \in G - H$.*

When H is a normal subgroup and $\lambda' = 0$, denote D by $D(m, n, k, \lambda)$.

Theorem 6.2.11 ([3], [10]) *Let A be a binary array of size $a_1 \times \ldots \times a_n$, $z = (z_1, \ldots, z_n) \neq (0, \ldots, 0)$ a type vector, and $E = \epsilon(A; z)$. Define the following groups G, H, and K, where H is a subgroup of G and K is a subgroup of H:*

$$G = Z_{(z_1+1)a_1} \times \ldots \times Z_{(z_n+1)a_n}$$
$$H = \{(h_1, \ldots, h_n) : h_i = y_i a_i \text{ and } 0 \leq y_i \leq z_i\}$$
$$K = \{(k_1, \ldots, k_n) : k_i = y_i a_i, \ 0 \leq y_i \leq z_i, \text{ and } \sum_i y_i \text{ is even}\}.$$

Let D be the subset of the factor group G/K given by

$$D = \{K + (j_1, \ldots, j_n) : (j_1, \ldots, j_n) \in \nu(E)\},$$

where $\nu(E)$ is the set equivalence of E defined in the last subsection.

Then A is a non-trivial GPBA$(a_1, \ldots, a_n)$ of type z if and only if D is a $D(E_A, 2, E_A, E_A/2)$ in G/K relative to H/K, where E_A is the energy of A.

Proof. It is easy to check that H is a subgroup of G, and K is a subgroup of H. Since G is abelian, the subgroups H and K are normal and the factor groups G/K and H/K are well defined. The arrays A and E are binary and so by Lemma 6.2.8 and the definition of energy,

$$E_E = E_A \prod_i (z_i + 1) = \prod_i (z_i + 1)a_i. \qquad (6.21)$$

Suppose that A is a non-trivial GPBA$(a_1, \ldots, a_n)$ of type z. Then Lemma 6.2.8 together with the definition of GPA results in

$$R_E(u_1, \ldots, u_n) = \begin{cases} E_E & \text{if } (u_1, \ldots, u_n) \in K \\ -E_E & \text{if } (u_1, \ldots, u_n) \in H - K \\ 0 & \text{if } (u_1, \ldots, u_n) \in G - H. \end{cases}$$

By Definitions 6.2.5, and 6.2.2, $\nu(E)$ is a subset of G. Since $z \neq (0, \ldots, 0)$, by Lemma 6.2.7 we may write $E = \alpha^{(r)}(B, -B)$ for some binary array B and $1 \leq r \leq n$. Then by Definition 6.2.4, exactly half of the elements of E are -1 and so by Equation (6.21), $|\nu(E)| = E_E/2$. Therefore by Lemma 6.2.1,

$$\lambda_E(u_1, \ldots, u_n) = \begin{cases} E_E/2 & \text{if } (u_1, \ldots, u_n) \in K \\ 0 & \text{if } (u_1, \ldots, u_n) \in H - K \\ E_E/4 & \text{if } (u_1, \ldots, u_n) \in G - H. \end{cases}$$

Recall from Definition 6.2.2 that for $(u_1, \ldots, u_n) \in G$, $\lambda_E(u_1, \ldots, u_n)$ is the number of solutions of the equation

$$(i_1, \ldots, i_n) - (j_1, \ldots, j_n) = (u_1, \ldots, u_n)$$

for $(j_1, \ldots, j_n), (i_1, \ldots, i_n) \in \nu(E)$.

Therefore for $K + (u_1, \ldots, u_n) \in G/K$, the number of solutions to the equation

$$(K + (i_1, \ldots, i_n)) - (K + (j_1, \ldots, j_n)) = K + (u_1, \ldots, u_n)),$$

for $K + (j_1, \ldots, j_n)$, $K + (i_1, \ldots, i_n) \in D$, is equal to

$$\begin{cases} 0 & \text{if } K + (u_1, \ldots, u_n)) \in H/K, \ (u_1, \ldots, u_n) \neq (0, \ldots, 0) \\ E_E/(4|K|) & \text{if } K + (u_1, \ldots, u_n)) \in (G/K) - (H/K). \end{cases}$$

Therefore by Definition 6.2.6, D is a $D(|G/K|/|H/K|, |H/K|, |D|, E_E/(4|K|))$ in G/K relative to H/K.

Now $|H| = 2|K| = \prod_i (z_i + 1)$ and so $|H/K| = |H|/|K| = 2$. By Equation (6.21) , $E_E/(4|K|) = E_A/2$ and $|G/K|/|H/K| = |G|/|H| = E_E/|H| = E_A$. Also, if $(j_1, \ldots, j_n) \in \nu(E)$, then by Definition 6.2.5,

$$(k_1, \ldots, k_n) + (j_1, \ldots, j_n) \in \nu(E) \text{ for all } (k_1, \ldots, k_n) \in K,$$

so $|D| = |\nu(E)|/|K| = E_A$.

Conversely, suppose that D is a $D(E_A, 2, E_A, E_A/2)$ in G/K relative to H/K. Then by Definitions 6.2.2 and 6.2.6, and Equation (6.21) ,

$$\lambda_E(u_1, \ldots, u_n) = |K|E_A/2 = E_E/4, \quad \text{if } (u_1, \ldots, u_n) \in G - H.$$

The result follows from Lemma 6.2.1 and Definition 6.2.5.   **Q.E.D.**

Definition 6.2.7 ([3], [10]) *Let $A = [A(a(1), \ldots, a(n))]$ and $B = [B(b(1), \ldots, b(n))]$ be two m-ary arrays of size $a_1 \times \ldots \times a_n$, and let $1 \leq r \leq n$. The interleaving of A with B in dimension r is the m-ary array , $\iota^{(r)}(A, B) := C = [C(c(1), \ldots, c(n))]$, of size $a_1 \times \ldots \times a_{r-1} \times (2a_r) \times a_{r+1} \times \ldots \times a_n$ defined by*

$$C(j_1, \ldots, j_{r-1}, 2j_r + y_r, j_{r+1}, \ldots, j_n) = \begin{cases} A(j_1, \ldots, j_r) & \text{if } y_r = 0 \\ B(j_1, \ldots, j_r) & \text{if } y_r = 1, \end{cases}$$

where $0 \leq j_i \leq a_i - 1$, $1 \leq i \leq n$. Clearly the energy of $\iota^{(r)}(A, B)$ is equal to $E_A + E_B$.

Definition 6.2.8 ([3], [10]) *Let $A = [A(a(1), \ldots, a(n))]$ be an m-ary array of size $a_1 \times \ldots \times a_n$, and let $1 \leq r \leq n$. The alternate sign-change of A in dimension r is the m-ary array, $\phi^{(r)}(A) := B = [B(b(1), \ldots, b(n))]$, of size $a_1 \times \ldots \times a_n$ defined by*

$$B(j_1, \ldots, j_n) = (-1)^{j_r} A(j_1, \ldots, j_n),$$

where $0 \leq j_i \leq a_i - 1$, $1 \leq i \leq n$. Clearly the energy of $\phi^{(r)}(A)$ is also E_A.

Definition 6.2.9 ([3], [10]) *Let $A = [A(i_1, i_2, a(1), \ldots, a(n))]$ be an m-ary array of size $s \times t \times a_1 \times \ldots \times a_n$, where $\gcd(s, t) = 1$, so $xt + sy = 1$ for some (x, y). The folding of A is the m-ary array, $\mu(A) := B = [B(j, b(1), \ldots, b(n))]$, of size $(st) \times a_1 \times \ldots \times a_n$ defined by*

$$B(j, j_1, \ldots, j_n) = A(jx, jy, j_1, \ldots, j_n),$$

where $0 \leq j \leq st - 1$, $0 \leq j_i \leq s_i - 1$, $1 \leq i \leq n$, jx and jy are regarded as reduced modulo s and t, respectively.

Note that if $x't + y's = 1$, then $x \equiv x' \pmod{s}$ and $y \equiv y' \pmod{t}$, so $\mu(A)$ is independent of the particular pair (x, y) used. Note also that the mapping from A to $\mu(A)$ is invertible. The energy of $\mu(A)$ is E_A.

Definition 6.2.10 ([3], [10]) *Let $A = [A(a(1), \ldots, a(n))]$ be an m-ary array of size $a_1 \times \ldots \times a_n$. The enlargement of A is the m-ary array, $I(A) := B = [B(0, j_1, \ldots, j_n)]$, of size $1 \times a_1 \times \ldots \times a_n$ defined by*

$$B(0, j_1, \ldots, j_n) = A(j_1, \ldots, j_n), \text{ for } 0 \leq j_i \leq a_i - 1.$$

The energy of $I(A)$ is also E_A. It is easy to see that A is a GPA$(m; a_1, \ldots, a_n)$ of type z if and only if $I(A)$ is a GPA$(m; 1, a_1, \ldots, a_n)$ of type (z_0, z)

Definition 6.2.11 ([3], [10]) *Let $A = [A(j_1, \ldots, j_n)]$, and $B = [B(j_{n+1}, \ldots, j_{n+m})]$ be respectively an $a_1 \times \ldots \times a_n$ m_1-ary and an $a_{n+1} \times \ldots \times a_{n+m}$ m_2-ary array. The tensor product of A with B is the $a_1 \times \ldots \times a_{n+m}$ $f(m_1, m_2)$-ary array $\prod(A, B) := C = [C(j_1, \ldots, j_{n+m})]$ defined by*

$$C(j_1, \ldots, j_{n+m}) = A(j_1, \ldots, j_n) B(j_{n+1}, \ldots, j_{n+m})$$

for all $0 \leq j_i \leq a_i - 1$, $1 \leq i \leq m + n$. From now on, we use $f(x, y)$ to represent

$$f(x, y) = \begin{cases} xy/2 & \text{if } x \text{ and } y \text{ are even} \\ 1 + 2\lfloor x/2 \rfloor \cdot \lfloor y/2 \rfloor & \text{otherwise.} \end{cases}$$

Note that $I(A) = \prod([+], A)$. The energy of $\prod(A, B)$ is $E_A E_B$.

Definition 6.2.12 ([3], [10]) *Let* $A = [A(r, j_1, \ldots, j_n)]$, *and* $B = [B(j, j_{n+1}, \ldots, j_{n+m})]$ *be respectively an* $s \times a_1 \times \ldots \times a_n$ m_1-*ary and an* $t \times a_{n+1} \times \ldots \times a_{n+m}$ m_2-*ary array. The* K-*product of* A *with* B *is the* $f(m_1, m_2)$-*ary array,* $\kappa(A, B) := C = [C(j, c(1), \ldots, c(n + m))]$, *of size* $(st) \times a_1 \times \ldots \times a_{n+m}$ *defined by*

$$C(rt + j, j_1, \ldots, j_{n+m}) = A(r, j_1, \ldots, j_n) B(j, j_{n+1}, \ldots, j_{n+m})$$

where $0 \leq r \leq s - 1$, $0 \leq j \leq t - 1$, $0 \leq j_i \leq a_i - 1$, $1 \leq i \leq n + m$.

Note that if $t = 1$, then $\kappa(A, B) = \prod(A, B')$, where $B = I(B')$. The energy of $\kappa(A, B)$ is clearly $E_A E_B$.

Definition 6.2.13 ([3], [10]) *Let* $A = [A(r, j, j_1, \ldots, j_n)]$ *be an* m-*ary array of size* $s \times t \times a_1 \times \ldots \times a_n$. *And let* c *be an integer such that* $ct \equiv 0 \pmod{s}$. *The shear of* A *and the transform of* A *with respect to* c *are respectively the* m-*ary arrays,* $\sigma(A; c) := [B(r, j, j_1, \ldots, j_n)]$ *and* $\tau(A; c) = D = [D(r, j, j_1, \ldots, j_n)]$ *of size* $s \times t \times a_1 \times \ldots \times a_n$ *defined by*

$$B(r, j, j_1, \ldots, j_n) = A(r - cj, j, j_1, \ldots, j_n)$$

and

$$D(r, j, j_1, \ldots, j_n) = (-1)^y A(r - cj, j, j_1, \ldots, j_n)$$

where

$$y = \begin{cases} 0 & \text{if } (r - cj) \bmod (2s) < s \\ 1 & \text{if } (r - cj) \bmod (2s) \geq s \end{cases}$$

$0 \leq r \leq s - 1$, $0 \leq j \leq t - 1$, $0 \leq j_i \leq a_i - 1$, $1 \leq i \leq n$, *and* $r - cj$ *is regarded as* $(r - cj) \bmod s$.

Note that the energy of both $\sigma(A; c)$ and $\tau(A; c)$ is the same as that of A. The condition $ct \equiv 0 \pmod{s}$ is necessary for the mappings τ and σ to be well-defined.

In the next two lemmas, we examine the effect of composing certain pairs of mappings, including the expansion mapping ϵ and the adjoining mapping $\alpha^{(r)}$.

Lemma 6.2.10 ([3], [10]) *Let A, B be m-ary arrays of size $a_1 \times \ldots \times a_n$, and $z = (z_1, \ldots, z_n)$ a type vector. Let $A' = \epsilon(A; z)$ and $B' = \epsilon(B; z)$. Then*

1. $\epsilon(\alpha^{(1)}(A, B); z) = \alpha^{(1)}(A', B')$, *if $z_1 = 0$;*

2. $\epsilon(\iota^{(1)}(A, B); z) = \iota^{(1)}(A', B')$;

3. $\epsilon(\phi^{(1)}(A); z) = \phi^{(1)}(A')$, *if $z_1 = 0$;*

4. $\phi^{(1)}(\alpha^{(1)}(A, B)) = \alpha^{(1)}(\phi^{(1)}(A), (-1)^{a_1}\phi^{(1)}(B))$;

5. $I(\iota^{(1)}(A, B))\iota^{(2)}(I(A), I(B))$.

Lemma 6.2.11 ([3], [10]) *Let $z = (z_1, \ldots, z_n)$ and $z' = (z_{n+1}, \ldots, z_{n+m})$ be two type vectors.*

1. *Let A be an m-ary array of size $s \times a_1 \times \ldots \times a_n$. Then*

$$\epsilon(\phi^{(1)}(A); 1, z) = \phi^{(1)}(\alpha^{(1)}(\epsilon(A; 0, z), (-1)^{s+1}\epsilon(A; 0, z)));$$

2. *Let A be an m-ary array of size $s \times t \times a_1 \times \ldots \times a_n$ and suppose $\gcd(s, t) = 1$. Then*

$$\epsilon(\mu(A); 0, z) = \mu(\epsilon(A; 0, 0, z));$$

3. *Let A and B be respectively an $s \times a_1 \times \ldots \times a_n$ m_1-ary and an $a_{n+1} \times \ldots \times a_{n+m}$ m_2-ary array. Then*

$$\epsilon(\prod(A, B); z, z') = \prod(\epsilon(A; z), \epsilon(B, z'));$$

4. *Let A and B be respectively an $s \times a_1 \times \ldots \times a_n$ m_1-ary and a $t \times a_{n+1} \times \ldots \times a_{n+m}$ m_2-ary array. Let $z_0 = 0$ or 1. Then*

$$\epsilon(\kappa(A, B); z_0, z, z') = \kappa(\epsilon(A; z_0, z), \epsilon(B; 0, z'));$$

5. *Let A be an m-ary array of size $s \times t \times a_1 \times \ldots \times a_n$ and c an integer such that $ct \equiv s \pmod{2s}$. Then*

$$\epsilon(\tau(A; c); 1, 1, z) = \sigma(\alpha^{(2)}(\epsilon(A; 1, 0, z), \epsilon(A; 1, 0, z)); c).$$

Now we begin to establish expressions for the correlation functions of various array mappings. These expressions will be used soon to prove the construction theorems for the generalized perfect arrays.

Lemma 6.2.12 ([3], [10]) *Let A, B, C, and D be four m-ary arrays of size $a_1 \times \ldots \times a_n$. Then for all $0 \le u_i \le a_i - 1$, $y = 0$ or 1,*

1. *The periodic cross-correlation between $E = \alpha^{(1)}(A, (-1)^w A)$ and $F = \alpha^{(1)}(B, B)$ satisfies*

$$R_{EF}(u_1 + ya_1, u_2, \ldots, u_n) = \begin{cases} 2R_{AB}(u_1, \ldots, u_n) & \text{if } w = 0 \\ 0 & \text{if } w = 1; \end{cases}$$

2. *The periodic cross-correlation between $E = \iota^{(1)}(A, B)$ and $F = \iota^{(1)}(C, D)$ satisfies*

$$R_{EF}(2u_1 + y, u_2, \ldots, u_n)$$
$$= \begin{cases} R_{AC}(u_1, \ldots, u_n) + R_{BD}(u_1, \ldots, u_n) & \text{if } y = 0 \\ R_{AD}(u_1, \ldots, u_n) + R_{BC}(u_1 + 1, u_2 \ldots, u_n) & \text{if } y = 1; \end{cases}$$

3. *The periodic cross-correlation between $E = \phi^{(1)}(A)$ and $F = \phi^{(1)}(B)$ satisfies*

$$R_{EF}(u_1, \ldots, u_n) = (-1)^{u_1} R_{AB}(u_1, \ldots, u_n);$$

4. *The periodic cross-correlation between $E = I(A)$ and $F = I(B)$ satisfies*

$$R_{EF}(0, u_1, \ldots, u_n) = R_{AB}(u_1, \ldots, u_n);$$

5. *The periodic auto-correlation of $E = \alpha^{(1)}(A, A)$ satisfies*

$$R_E(u_1 + ya_1, u_2, \ldots, u_n) = 2R_A(u_1, \ldots, u_n);$$

6. *The periodic auto-correlation of $E = \iota^{(1)}(A, B)$ satisfies*

$$R_E(2u_1 + y, u_2, \ldots, u_n)$$
$$= \begin{cases} R_A(u_1, \ldots, u_n) + R_B(u_1, \ldots, u_n) & \text{if } y = 0 \\ R_{AB}(u_1, \ldots, u_n) + R_{BA}(u_1 + 1, u_2 \ldots, u_n) & \text{if } y = 1; \end{cases}$$

7. *The periodic auto-correlation of $E = \phi^{(1)}(A)$ satisfies*

$$R_E(u_1,\ldots,u_n) = (-1)^{u_1} R_A(u_1,\ldots,u_n);$$

8. *The periodic auto-correlation of $E = I(A)$ satisfies*

$$R_E(0, u_1,\ldots,u_n) = R_A(u_1,\ldots,u_n);$$

9. *The periodic auto-correlation of $E = \alpha^{(1)}(A, -A)$ satisfies*

$$R_E(u_1 + ya_1, u_2, \ldots, u_n)$$

$$= \begin{cases} 2(-1)^y [P_A^{(1)}(u_1,\ldots,u_n) \\ \qquad -P_A^{(1)}(a_1 - u_1,\ldots,a_n - u_n)] & \text{if } u_1 \neq 0 \\ 2(-1)^y P_A^{(1)}(u_1,\ldots,u_n) & \text{if } u_1 = 0, \end{cases}$$

where $P_A^{(1)}(u_1,\ldots,u_n)$ is the semi-periodic autocorrelation of A in dimension 1 which is defined by

$$P_A^{(1)}(u_1,\ldots,u_n) = \sum_{j_1=0}^{a_1-u_1-1} \sum_{j_2=0}^{a_2-1} \cdots \sum_{j_n=0}^{a_n-1} A(u_1,\ldots,u_n)$$
$$\times A(u_1 + j_1,\ldots,u_n + j_n);$$

10. *The periodic crosscorrelation between $E = \alpha^{(1)}(I(A), I(B))$ and $F = \alpha^{(1)}(I(C), I(D))$ satisfies*

$$R_{EF}(u, u_1,\ldots,u_n)$$

$$= \begin{cases} (R_{AC} + R_{BD})(u_1,\ldots,u_n) & \text{if } u = 0 \\ (R_{AD} + R_{BC})(u_1,\ldots,u_n) & \text{if } u = 1; \end{cases}$$

11. *The periodic autocorrelation of $E = \alpha^{(1)}(I(A), I(B))$ satisfies*

$$R_E(u, u_1,\ldots,u_n) = \begin{cases} (R_A + R_B)(u_1,\ldots,u_n) & \text{if } u = 0 \\ (R_{AB} + R_{BA})(u_1,\ldots,u_n) & \text{if } u = 1; \end{cases}$$

12. *The semi-periodic autocorrelation of $E = \alpha^{(1)}(I(A), I(B))$ in dimension 1 satisfies*

$$P_E^{(1)}(u, u_1,\ldots,u_n) = \begin{cases} (R_A + R_B)(u_1,\ldots,u_n) & \text{if } u = 0 \\ R_{AB}(u_1,\ldots,u_n) & \text{if } u = 1. \end{cases}$$

The next lemma gives expressions for the periodic correlation function of further array mappings.

Lemma 6.2.13 ([3], [10]) *The following equations are true:*

1. *Let A be an m-ary array of size $s \times t \times a_1 \times \ldots \times a_n$ and suppose that $xt + ys = 1$. Then for all $0 \leq u \leq st - 1$, $0 \leq u_i \leq a_i - 1$,*

$$R_{\mu(A)}(u, u_1, \ldots, u_n) = R_A(ux, uy, u_1, \ldots, u_n).$$

2. *Let A and C be m_1-ary arrays of size $a_1 \times \ldots \times a_n$ and B and D be m_2-ary arrays of size $a_{n+1} \times \ldots \times a_{n+m}$. Then for all $0 \leq u_i \leq a_i - 1$, $1 \leq i \leq n + m$,*

$$R_{\prod(A,B)\prod(C,D)}(u_1, \ldots, u_{n+m}) = R_{AC}(u_1, \ldots, u_n)R_{BD}(u_{n+1}, \ldots, u_{n+m})$$
$$R_{\prod(A,B)}(u_1, \ldots, u_{n+m}) = R_A(u_1, \ldots, u_n)R_B(u_{n+1}, \ldots, u_{n+m}).$$

3. *Let A be an m_1-ary array of size $s \times a_1 \times \ldots \times a_n$ and B an m_2-ary array of size $t \times a_{n+1} \times \ldots \times a_{n+m}$. Then for all $0 \leq u \leq s - 1$, $0 \leq w \leq t - 1$, $0 \leq u_i \leq a_i - 1$, $1 \leq i \leq n + m$,*

$$R_{\kappa(A,B)}(ut + w, u_1, \ldots, u_{n+m})$$

$$= \begin{cases} R_A(u, u_1, \ldots, u_n)P_B^{(1)}(w, u_{n+1}, \ldots, u_{n+m}) \\ \quad + R_A(u + 1, u_1, \ldots, u_n) \\ \quad \times P_B^{(1)}(t - w, a_{n+1} - u_{n+1}, \ldots, a_{n+m} - u_{n+m}) \text{ if } w \neq 0 \\ \\ R_A(u, u_1, \ldots, u_n)P_B^{(1)}(w, u_{n+1}, \ldots, u_{n+m}) \qquad \text{if } w = 0. \end{cases}$$

4. *Let A be an m-ary array of size $s \times t \times a_1 \times \ldots \times a_n$ and c an integer such that $ct \equiv 0 (\mathrm{mod} s)$. Then for all $0 \leq u \leq s - 1$, $0 \leq w \leq t - 1$, $0 \leq u_i \leq a_i - 1$, $1 \leq i \leq n$,*

$$R_{\sigma(A;c)}(u, w, u_1, \ldots, u_n) = R_A(u - cw, w, u_1, \ldots, u_n).$$

Up to now, we are ready to present construction theorems for the generalized perfect arrays.

Theorem 6.2.12 ([3], [10]) *Let $z = (z_0, \ldots, z_n)$ be a type vector. And let s and t be integers coprimed with each other, i.e., $\gcd(s, t) = 1$. Then A is a $GPA(m; s, t, a_1, \ldots, a_n)$ of type $(0, 0, z)$ if and only if $\mu(A)$ is a $GPA(m; (st), a_1, \ldots, a_n)$ of type $(0, z)$.*

Proof. Use the definition of GPAs, the second statement of Lemma 6.2.11 and the first statement of Lemma 6.2.13. Note that $u \equiv 0 \pmod{st}$ if and only if $ux \equiv 0 \pmod{s}$ and $uy \equiv 0 \pmod{t}$, where x and y are determined by $tx + ys = 1$.   **Q.E.D.**

In the binary cases, Theorem 6.2.12 states that there exists a $PBA(s, t, a_1, \ldots, a_n)$ if and only if there exists a $PBA((st), a_1, \ldots, a_n)$, whenever $\gcd(s, t) = 1$. Thus Theorem 6.2.12 generalizes the known folding construction of PBAs.

The following theorem is a generalized version of the periodic production construction of PBAs.

Theorem 6.2.13 ([3], [10]) *Let $z = (z_1, \ldots, z_n)$ and $z' = (z_{n+1}, \ldots, z_{n+m})$ be type vectors. Then the following first two statements hold if and only if the third one hold:*

1. *A is a $GPA(m_1; a_1, \ldots, a_n)$ of type z;*

2. *B is a $GPA(m_2; a_{n+1}, \ldots, a_{n+m})$ of type z';*

3. *$\prod(A, B)$ is a $GPA(f(m_1, m_2); a_1, \ldots, a_{n+m})$ of type (z, z').*

Proof. Let $A' = \epsilon(A; z)$, $B' = \epsilon(B, z')$, $C = \prod(A, B)$, and $C' = \epsilon(C; z, z')$. Then by the third statement of Lemma 6.2.11, $C' = \prod(A', B')$, and so from the second statement of Lemma 6.2.13, we have, for all $0 \le u_i \le (z_i + 1)a_i - 1$, $1 \le i \le n + m$,

$$R_{C'}(u_1, \ldots, u_{n+m}) = R_{A'}(u_1, \ldots, u_n) R_{B'}(u_{n+m}, \ldots, u_{n+m}). \qquad (6.22)$$

Therefore by the definition of GPAs, the first two statements imply the third one.

Conversely, suppose that the third statement is true. Then Equation (6.22) implies

$$R_{C'}(u_1, \ldots, u_n, 0, \ldots, 0) = x R_{A'}(u_1, \ldots, u_n),$$

where

$$x = R_{B'}(0, \ldots, 0) = E_B \prod_{i=n+1}^{n+m} (z_i + 1) \neq 0.$$

Therefore the first statement follows. The second statement follows similarly.   **Q.E.D.**

We shall find it useful to express, in terms of the semi-periodic auto-correlation function, conditions equivalent to an array being a GPA of two related types simultaneously.

Lemma 6.2.14 ([3], [10]) *Let $A = [A(j, a(1), \ldots, a(n))]$ be an m-ary array of size $s \times a_1 \times \ldots \times a_n$, and $z = (z_1, \ldots, z_n)$ a type vector. Let $A' = \epsilon(A; 0, z)$. Then the following are equivalent:*

1. *A is a GPA$(m; s, a_1, \ldots, a_n)$ of type $(0, z)$ and type $(1, z)$;*

2. *For $0 \leq u \leq s - 1$, $0 \leq u_i \leq (z_i + 1)a_i - 1$, $1 \leq i \leq n$,*

   $P_{A'}^{(1)}(u, u_1, \ldots, u_n) \neq 0$ only if $u = 0$ and $u_i \equiv 0 (\mathrm{mod}\, a_i)$ for all i.

Theorem 6.2.14 ([3], [10]) *Let $z = (z_1, \ldots, z_n)$, $z' = (z_{n+1}, \ldots, z_{n+m})$ and (z_0) be three type vectors and let $s > 1$. Then the following (1) and (2) hold if and only if (3) holds:*

1. *A is a GPA$(m_1; s, a_1, \ldots, a_n)$ of type (z_0, z);*

2. *B is a GPA$(m_2; t, a_{n+1}, \ldots, a_{n+m})$ of type $(0, z')$ and type $(1, z')$;*

3. *$\kappa(A, B)$ is a GPA$(f(m_1, m_2); (st), a_1, \ldots, a_{n+m})$ of type (z_0, z, z').*

Proof. Let $A' = \epsilon(A; z_0, z)$, $B' = \epsilon(B; 0, z')$, $C = \kappa(A, B)$, and $C' = \epsilon(C; z_0, z, z')$. By the fourth statement of Lemma 6.2.11, $C' = \kappa(A', B')$. By the third statement of Lemma 6.2.13, for all $0 \leq u \leq (z_0 + 1)s - 1$, $0 \leq w \leq t - 1$, $0 \leq u_i \leq (z_i + 1)a_i - 1$, $1 \leq i \leq n + m$,

$R_{C'}(ut + w, u_1, \ldots, u_{n+m})$

$$= \begin{cases} P_{B'}^{(1)}(t - w, (z_{n+1} + 1)a_{n+1} - u_{n+1}, \ldots, \\ \quad (z_{n+m} + 1)a_{n+m} - u_{n+m}) \times R_{A'}(u + 1, u_1, \ldots, u_n) + \\ \quad R_{A'}(u, u_1, \ldots, u_n) P_{B'}^{(1)}(w, u_{n+1}, \ldots, u_{n+m}) \quad \text{if } w \neq 0 \quad (6.23) \\ \\ R_{A'}(u, u_1, \ldots, u_n) P_{B'}^{(1)}(w, u_{n+1}, \ldots, u_{n+m}) \qquad \text{if } w = 0. \end{cases}$$

By the definition of GPAs, the first statement is equivalent to

$$R_{A'}(u, u_1, \ldots, u_n) \neq 0 \; only \; if \; u \equiv 0 (\mathrm{mod}s), u_i \equiv 0 (\mathrm{mod}a_i). \quad (6.24)$$

By Lemma 6.2.14, the second statement is equivalent to

$$P_{B'}^{(1)}(w, u_{n+1}, \ldots, u_{n+m}) \neq 0 \; only \; if \; w \equiv 0 (\mathrm{mod}t), u_i \equiv 0 (\mathrm{mod}a_i).$$
$$(6.25)$$

Suppose (1) and (2) hold. Then by Equation (6.23)

$$R_{C'}(ut + w, u_1, \ldots, u_{n+m}) \neq 0$$

only if

$$R_{A'}(u, u_1, \ldots, u_n) \neq 0 \; and \; P_{B'}^{(1)}(w, u_{n+1}, \ldots, u_{n+m}) \neq 0,$$

which hold only if

$$u \equiv 0 (\mathrm{mod}s), \; w = 0, \; and \; u_i \equiv 0 (\mathrm{mod}a_i).$$

Therefore the third statement holds.

Conversely, suppose the third statement holds. If let $w = u_i = 0$ for all
$n + 1 \leq i \leq n + m$, then Equation (6.23) becomes

$$R_{C'}(ut, u_1, \ldots, u_n, 0, \ldots, 0) = x R_{A'}(u, u_1, \ldots, u_n),$$

where $x = P_{B'}^{(1)}(0, \ldots, 0) = R_{B'}(0, \ldots, 0) \neq 0$. Thus the first statement is true. Similarly, the second statement is also true.   **Q.E.D.**

Because if $t = 1$, then $\kappa(A, B) = \prod(A, B')$, where $B = I(B')$, so that Theorem 6.2.13 is a special case of Theorem 6.2.14.

Theorem 6.2.15 ([3], [10]) *Let s be an odd integer, $z = (z_1, \ldots, z_n)$ be a type vector. Then A is a GPA$(m; s, a_1, \ldots, a_n)$ of type $(0, z)$ if and only if $\phi^{(1)}(A)$ is a GPA$(m; s, a_1, \ldots, a_n)$ of type $(1, z)$.*

Proof. Let $A' = \epsilon(A; 0, z)$, $B' = \epsilon(\phi^{(1)}(A); 1, z)$, and $C = \alpha^{(1)}(A', A')$. By the first statement of Lemma 6.2.11, $B' = \phi^{(1)}(C)$. By the fifth and seventh statements of Lemma 6.2.12, for all $0 \leq u \leq s - 1$, $0 \leq u_i \leq (z_i + 1)a_i - 1$, $1 \leq i \leq n$, $y = 0$ or 1,

$$R_{B'}(u + ys, u_1, \ldots, u_n) = (-1)^{u+ys} R_C(u + ys, u_1, \ldots, u_n)$$

$$= 2(-1)^{u+y} R_{A'}(u, u_1, \ldots, u_n).$$

Q.E.D.

We next show under what conditions a GPA may be transformed into another GPA which is of the same type but of different size by altering the interleaving of its component arrays.

Lemma 6.2.15 ([3], [10]) *Let A and B be m-ary arrays of size $a_1 \times \ldots \times a_n$, $n \geq 2$, and $z = (z_1, \ldots, z_n)$ be a type vector. Let $A' = \epsilon(A; z)$ and $B' = \epsilon(B; z)$. Then any two of the following equations imply the other*

1. $C = \iota^{(1)}(A, B)$ *is a* GPA$(m; 2a_1, \ldots, a_n)$ *of type z;*

2. $D = \iota^{(2)}(A, B)$ *is a* GPA$(m; a_1, 2a_2, a_3, \ldots, a_n)$ *of type z;*

3. $R_{A'B'}(u_1, \ldots, u_n) = R_{A'B'}(u_1 + 1, u_2 - 1, u_3, \ldots, u_n)$ *for all $0 \leq u_i \leq (z_i + 1)a_i - 1$, $1 \leq i \leq n$.*

The following two theorems form the heart of the recursive constructions of GPAs.

Theorem 6.2.16 ([3], [10]) *Let A and B be m-ary arrays of size $a_1 \times \ldots \times a_n$, $n \geq 2$, and $z = (z_1, \ldots, z_n)$ be a type vector. Let $C = \alpha^{(1)}(A, A)$, $D = \alpha^{(1)}(B, -B)$. Then the following equations are equivalent:*

1. A *is a* GPA$(m; a_1, \ldots, a_n)$ *of type $(0, z)$, B is a* GPA$(m; a_1, \ldots, a_n)$ *of type $(1, z)$ and $E_A = E_B$;*

2. $E = \iota^{(1)}(C, D)$ *is a* GPA$(m; (4a_1), \ldots, a_n)$ *of type $(0, z)$;*

3. $F = \iota^{(2)}(C, D)$ *is a* GPA$(m; (2a_1), (2a_2), a_3, \ldots, a_n)$ *of type $(0, z)$*

Proof. Let $A' = \epsilon(A; 0, z)$, $B' = \epsilon(B; 1, z)$, $C' = \epsilon(C; 0, z)$, $D' = \epsilon(D; 0, z)$, $\overline{B} = \epsilon(B; 1, z)$, and $E' = \epsilon(E; 0, z)$.

At first we show that (2) is equivalent to (3) . By the first statement of Lemma 6.2.10, $C' = \alpha^{(1)}(A', A')$ and $D' = \alpha^{(1)}(\overline{B}, -\overline{B})$. Therefore by the sixth statement of Lemma 6.2.12 and the known identity $R_{XY}(u_1, \ldots, u_n) = R_{YX}(a_1 - u_1, \ldots, a_n - u_n)$, we have

$$0 = R_{D'C'}(u_1, \ldots, u_n) = R_{C'D'}(u_1, \ldots, u_n) \tag{6.26}$$

for all $0 \leq u_1 \leq 2a_1 - 1$, $0 \leq u_i \leq (z_i + 1)a_i - 1$, $i \geq 2$. Hence by Lemma 6.2.15, (2) is equivalent to (3) .

Then we now show that (1) is equivalent to (2) . By the second statement of Lemma 6.2.10, $E' = \iota^{(1)}(C', D')$. By Lemma 6.2.7, $D' = B'$. Therefore by the sixth statement of Lemma 6.2.12, we have

$$R_{E'}(2u_1 + 2ya_1 + 1, u_2, \ldots, u_n)$$
$$= R_{C'D'}(u_1 + ya_1, u_2, \ldots, u_n) + R_{D'C'}(u_1 + ya_1 + 1, u_2, \ldots, u_n)$$
$$= 0 \quad \text{(by Equation (6.26))} , \tag{6.27}$$

for all $0 \le u_1 \le a_1 - 1$, $y = 0$ or 1, $0 \le u_i \le (z_i + 1)a_i - 1$, $i \ge 2$.

Because of the fifth statement of Lemma 6.2.12 and $D' = B'$, we have

$$R_{E'}(2u_1 + 2ya_1, u_2, \ldots, u_n)$$
$$= R_{C'}(u_1 + ya_1, u_2, \ldots, u_n) + R_{D'}(u_1 + ya_1, u_2, \ldots, u_n)$$
$$= 2R_{A'}(u_1, \ldots, u_n) + R_{B'}(u_1 + ya_1, u_2, \ldots, u_n). \tag{6.28}$$

Letting, in Equation (6.28) , $u_1 = 0$ $y = 1$, $u_i = y_i a_i$, $0 \le y_i \le z_i$, $i \ge 2$, we have

$$R_{E'}(2a_1, y_2 a_2, \ldots, y_n a_n) = 2R_{A'}(0, y_2 a_2, \ldots, y_n a_n)$$
$$+ R_{B'}(a_1, y_2 a_2, \ldots, y_n a_n). \tag{6.29}$$

Suppose that (1) holds. Then from Equation (6.28) , for all $0 \le u_1 \le a_1 - 1$, $y = 0$ or 1, $0 \le u_i \le (z_i + 1)a_i - 1$, $i \ge 2$,

$$R_{E'}(2u_1 + 2ya_1, u_2, \ldots, u_n) \ne 0$$

only if

$$R_{A'}(u_1, \ldots, u_n) \ne 0 \quad or \quad R_{B'}(u_1 + ya_1, u_2, \ldots, u_n) \ne 0,$$

which occurs only if

$$u_i \equiv 0(\text{mod} a_i) \text{ for all } i \text{ and } y = 0 \text{ or } 1.$$

In the case $u_i \equiv 0(\text{mod} a_i)$ for all i and $y = 1$, because of Equation (6.29) and Lemma 6.2.8, for all $0 \le y_i \le z_i$, $i \ge 2$,

$$R_{E'}(2a_1, y_2 a_2, \ldots, y_n a_n)$$

$$= 2E_A(-1)^{\sum_{i \ge 2} y_i} \prod_{i \ge 2}(z_i + 1) + 2E_B(-1)^{1 + \sum_{i \ge 2} y_i} \prod_{i \ge 2}(z_i + 1)$$

$$= 0 \text{ (since } E_A = E_B).$$

Hence, by Equation (6.27) , the statement (2) holds.

Conversely, suppose that (2) holds. Then from Equation (6.28), for all $0 \leq u_1 \leq a_1 - 1, 0 \leq u_i \leq (z_i + 1)a_i - 1, i \geq 2$,

$$R_{E'}(2u_1, u_2, \ldots, u_n) = 2R_{A'}(u_1, \ldots, u_n) + R_{B'}(u_1, \ldots, u_n), \qquad (6.30)$$

$$R_{E'}(2u_1 + 2a_1, u_2, \ldots, u_n) = 2R_{A'}(u_1, \ldots, u_n) + R_{B'}(u_1 + a_1, u_2, \ldots, u_n)$$
$$2R_{A'}(u_1, \ldots, u_n) - R_{B'}(u_1, \ldots, u_n). \qquad (6.31)$$

Rearrange Equations (6.30) and (6.31) as

$$4R_{A'}(u_1, \ldots, u_n) = R_{E'}(2u_1, u_2, \ldots, u_n) + R_{E'}(2u_1 + 2a_1, u_2, \ldots, u_n)$$
$$2R_{B'}(u_1, \ldots, u_n) = R_{E'}(2u_1, u_2, \ldots, u_n) - R_{E'}(2u_1 + 2a_1, u_2, \ldots, u_n).$$

Putting $u_i = 0$ for all i, we have

$$4R_{A'}(0, \ldots, 0) = 2R_{B'}(0, \ldots, 0)$$

and so by Lemma 6.2.8, $E_A = E_B$. Also for all $0 \leq u_1 \leq a_1 - 1, 0 \leq u_i \leq (z_i + 1)a_i - 1, i \geq 2$,

$$R_{A'}(u_1, \ldots, u_n) \neq 0$$

only if

$$R_{E'}(2u_1, u_2, \ldots, u_n) \neq 0 \text{ or } R_{E'}(2u_1 + 2a_1, u_2, \ldots, u_n) \neq 0$$

which occurs only if $u_i \equiv 0 (\mathrm{mod} a_i)$ for all i. Applying a similar argument to $R_{B'}$, we have (1) .   **Q.E.D.**

Theorem 6.2.17 ([3], [10]) *Let A be an m-ary array of size $s \times t \times a_1 \times \ldots \times a_n$, $z = (z_1, \ldots, z_n)$ a type vector, and c an integer such that $ct \equiv s(\mathrm{mod} 2s)$. Then A is a GPA$(m; s, t, a_1, \ldots, a_n)$ of type $(1, 0, z)$ if and only if $\tau(A; c)$ is a GPA$(m; s, t, a_1, \ldots, a_n)$ of type $(1, 1, z)$.*

Proof. Let $A' = \epsilon(A; 1, 0, z)$, $B = \tau(A; c)$, $B' = \epsilon(B; 1, 1, z)$, and $D = \alpha^{(2)}(A', A')$. By the fifth statement of Lemma 6.2.11, $B' = \sigma(D; c)$. Therefore by the fourth statement of Lemma 6.2.13, by using the fifth statement of Lemma 6.2.12 and $ct \equiv s(\mathrm{mod} 2s)$, for all $0 \leq u \leq 2s - 1$, $0 \leq w \leq t - 1$, $y = 0$ or 1, $0 \leq u_i \leq (z_i + 1)a_i$,

$$R_{B'}(u, w + yt, u_1, \ldots, u_n) = R_D(u - c(w + yt), w + yt, u_1, \ldots, u_n)$$
$$= 2R_{A'}(u - cw - ys, w, u_1, \ldots, u_n).$$

The result follows from the definition of GPAs.   **Q.E.D.**

Corollary 6.2.3 ([3], [10]) *Let A be an m-ary array of size $s \times t \times a_1 \times \ldots \times a_n$, $z = (z_1, \ldots, z_n)$ be a type vector, and $t/\gcd(s,t)$ odd. Then A is a GPA$(m; s, t, a_1, \ldots, a_n)$ of type $(1, 0, z)$ if and only if $\tau(A; s/\gcd(s,t))$ is a GPA$(m; s, t, a_1, \ldots, a_n)$ of type $(1, 1, z)$.*

Proof. An integer c satisfies $ct \equiv s \pmod{2s}$ if and only if $t/\gcd(s,t)$ is odd and c is an odd multiple of $s/\gcd(s,t)$ **Q.E.D.**

Corollary 6.2.4 ([3], [10]) *Let $z = (z_1, \ldots, z_n) \neq (0, \ldots, 0)$ be a type vector. For $1 \leq i \leq n$, let $b_i \geq 0$ and let a_i be odd. Then there exists a GPA$(m; 2^{b_1} a_1, \ldots, 2^{b_n} a_n)$ of type z and energy E if and only if there exists a GPA$(m; 2^{b_1} a_1, \ldots, 2^{b_n} a_n)$ of type $(0^{(i'-1)}, 1, 0^{(n-i')})$ and energy E, where $z_{i'} = 1$ and for all $1 \leq i \leq n$, $b_{i'} \geq b_i$ whenever $z_i = 1$.*

Proof. It can be proved by using the last corollary.   **Q.E.D.**

   This corollary shows that the existence of a GPA of type $z \neq (0, \ldots, 0)$ implies the existence of a GPA of type $(1, 0, \ldots, 0)$ for some permutations of the original dimensions of the array.

Theorem 6.2.18 ([3], [10]) *Let A, B, C, and D be m-ary arrays of size $a_1 \times \ldots \times a_n$. Let $E = \alpha^{(1)}(I(A), I(B))$, and $F = \alpha^{(1)}(I(C), I(D))$. Then the following (1) and (2) hold if and only if (a) , (b) , (c) and (d) hold for all $0 \leq u_i \leq a_i - 1$:*

(1) $G = \alpha^{(1)}(I(E), I(F))$ *is a GPA$(m; 2, 2, a_1, \ldots, a_n)$ of type $(0, \ldots, 0)$ and of also simultaneously type $(1, 0, \ldots, 0)$;*

(2) $H = \iota^{(1)}(E, F)$ *is a GPA$(m; 4, a_1, \ldots, a_n)$ of type $(0, \ldots, 0)$.*

(a) $(R_A + R_B + R_C + R_D)(u_1, \ldots, u_n) \neq 0$ *only if $u_i = 0$ for all i;*

(b) $(R_{AB} + R_{BA} + R_{CD} + R_{DC})(u_1, \ldots, u_n) = 0$;

(c) $(R_{AC} + R_{BD})(u_1, \ldots, u_n) = 0$;

(d) $(R_{AD} + R_{BC})(u_1, \ldots, u_n) = 0$.

Proof. Note that $\epsilon(G; 0, \ldots, 0) = G$. Then by Lemma 6.2.14, (1) is equivalent to

$$P_G^{(1)}(u, w, u_1, \ldots, u_n) \neq 0, \text{ where } u, w = 0 \text{ or } 1, 0 \leq u_i \leq a_i - 1,$$

only if

$$u = w = u_i = 0 \text{ for all } i.$$

But by the twelfth statement of Lemma 6.2.12, for all $u, w = 0$ or 1, $0 \leq u_i \leq a_i - 1$,

$$P_G^{(1)}(u, w, u_1, \ldots, u_n)$$

$$= \begin{cases} (R_E + R_F)(w, u_1, \ldots, u_n) & \text{if } u = 0 \\ R_{EF}(w, u_1, \ldots, u_n) & \text{if } u = 1 \end{cases}$$

$$= \begin{cases} (R_A + R_B + R_C + R_D)(u_1, \ldots, u_n) & \text{if } (u, w) = (0, 0) \\ (R_{AB} + R_{BA} + R_{CD} + R_{DC})(u_1, \ldots, u_n) & \text{if } (u, w) = (0, 1) \\ (R_{AC} + R_{BD})(u_1, \ldots, u_n) & \text{if } (u, w) = (1, 0) \\ (R_{AD} + R_{BC})(u_1, \ldots, u_n) & \text{if } (u, w) = (1, 1) \end{cases}$$

by the tenth and eleventh statements of Lemma 6.2.12. Therefore (1) holds if and only if (a) , (b) , (c) and (d) hold for all $0 \leq u_i \leq a_i - 1$.

(2) holds if and only if $I(H)$ is a GPA$(m; 1, 4, a_1, \ldots, a_n)$ of type $(0, \ldots, 0)$. In fact, X is a GPA$(m; a_1, \ldots, a_n)$ of type z if and only if $I[A]$ is a GPA$(m; 1, a_1, \ldots, a_n)$ of type (z_0, z). Now using the fifth statement of Lemma 6.2.10, $I(H) = \iota^{(2)}(I(E), I(F))$, and since $I(E)$ and $I(F)$ each has size $1 \times 2 \times a_1 \times \ldots \times a_n$ we may write $G = \iota^{(1)}(I(E), I(F))$. Therefore if (1) is given, then by Lemma 6.2.15, (2) is equivalent to

$$R_{I(E)I(F)}(0, w, u_1, \ldots, u_n) = R_{I(E)I(F)}(0, w - 1, u_1, \ldots, u_n),$$

$$w = 0, 1, \ 0 \leq u_i \leq a_i - 1,$$

which, by the fourth statement of Lemma 6.2.12, is equivalent to

$$R_{EF}(w, u_1, \ldots, u_n) = R_{EF}(w - 1, u_1, \ldots, u_n), \ w = 0, 1, \ 0 \leq u_i \leq a_i - 1,$$

and by the tenth statement of Lemma 6.2.12 it is equivalent to

$$(R_{AC} + R_{BD})(u_1, \ldots, u_n) = (R_{AD} + R_{BC})(u_1, \ldots, u_n), \ 0 \leq u_i \leq a_i - 1.$$

Therefore (2) follows from (1) , (c) and (d) .   **Q.E.D.**

The following definition generalizes the concept of BSQ from the case of $m = 2$ to the case of general integer m.

Definition 6.2.14 ([3], [10]) *Let A, B, C, and D be m-ary arrays of size $a_1 \times \ldots \times a_n$. $\{A, B, C, D\}$ is called an $a_1 \times \ldots \times a_n$ m-ary supplementary quadruple of energy $E_A + E_B + E_C + E_D$ if the following two conditions are satisfied for all $0 \leq u_i \leq a_i - 1$:*

1. $(R_A + R_B + R_C + R_D)(u_1, \ldots, u_n) \neq 0$ *only if* $u_i = 0$ *for all i;*

2. $(R_{WX} + R_{YZ})(u_1, \ldots, u_n) = 0$ *for all* $\{W, X, Y, Z\} = \{A, B, C, D\}$.

Corollary 6.2.5 ([3], [10]) *Let $\{A, B, C, D\}$ be an $a_1 \times \ldots \times a_n$ m-ary supplementary quadruple. Then there exists a $GPA(m; 2, 2, u_1, \ldots, u_n)$ of type $(0, \ldots, 0)$ and also of type $(1, 0, \ldots, 0)$, and a $GPA(m; 4, u_1, \ldots, u_n)$ of type $(0, \ldots, 0)$, whose energy is $E_A + E_B + E_C + E_D$.*

Proof. It can be proved immediate by using Theorem 6.2.18 and Definition 6.2.14.   **Q.E.D.**

It is worth pointing that Corollary 6.2.5 implies Theorem 6.2.8 if $m = 2$.

Theorem 6.2.19 ([3], [10]) *Let $\{A_1, B_1, C_1, D_1\}$ and $\{A_2, B_2, C_2, D_2\}$ be respectively an $a_1 \times \ldots \times a_n$ m_1-ary and an $a_{n+1} \times \ldots \times a_{n+m}$ m_2-ary supplementary quadruple. Let*

$$A = \prod(A_1 + B_1, A_2) + \prod(A_1 - B_1, B_2),$$
$$B = \prod(A_1 + B_1, C_2) + \prod(A_1 - B_1, D_2),$$
$$C = \prod(C_1 + D_1, A_2) + \prod(C_1 - D_1, B_2),$$
$$D = \prod(C_1 + D_1, C_2) + \prod(C_1 - D_1, D_2).$$

Then $\{A, B, C, D\}$ is an $a_1 \times \ldots \times a_{n+m}$ $(1 + 4\lfloor m_1/2 \rfloor \lfloor m_2/2 \rfloor)$-ary supplementary quadruple. If $m_1 = m_2 = 2$, then the elements of A, B, C, D take only the values 2 or -2.

Proof. This theorem can be directly verified.   **Q.E.D.**

Theorem 6.2.20 ([3], [10]) *Let $z = (z_1, \ldots, z_n) \neq (0, \ldots, 0)$ be a type vector and let A be a $GPBA(a_1, \ldots, a_n)$ of type z with energy $E_A > 2$. Then*

1. $E_A \equiv 0 \pmod 4$;

2. *If $z_i = 1$ and a_i is even for some i, then a_j is even for some $j \neq i$.*

Proof. If $z_i = 0$ whenever a_i is even, then by Theorem 6.2.15 there exists a PBA$(a_1, \ldots, a_n)$ of energy $E_A = \prod_{i=1}^{n} a_i$, so $E_A \equiv 0 \pmod 4$. Assume for the rest of the proof that $z_i = 1$ and a_i is even, for some i. Then by Corollary 6.2.4, without loss of generality, we may assume that there exists a GPBA$(a_1, \ldots, a_n)$ of type $(1, 0, \ldots, 0)$, whose energy is E_A and a_1 is even. Call this array B and let $B' = \epsilon(B; 1, 0, \ldots, 0)$. By Lemma 6.2.7, $B' = \alpha^{(1)}(B, -B)$ and so putting $y = 0$ in Lemma 6.2.12, for all $0 \leq u_i \leq a_i - 1$,

$$(R_{B'}(u_1, \ldots, u_n))/2$$

$$= \begin{cases} P_B^{(1)}(u_1, \ldots, u_n) - P_B^{(1)}(a_1 - u_1, \ldots, a_n - u_n) & \text{if } u_1 \neq 0 \\ P_B^{(1)}(u_1, \ldots, u_n) & \text{if } u_1 = 0 \end{cases}$$

$$\equiv (a_1 - 2u_1) \prod_{i=2}^{n} a_i \pmod 4,$$

By the definition of GPAs we therefore have

$$(a_1 - 2u_1) \prod_{i=2}^{n} a_i \not\equiv 0 \pmod 4, \quad \text{only if } u_i = 0 \text{ for all } i. \tag{6.32}$$

Firstly note that

$$\text{if } \prod_{i=2}^{n} a_i \text{ is odd, then } a_1 = 2, \tag{6.33}$$

otherwise we may take $(u_1, \ldots, u_n) = (1, 0, \ldots, 0)$ and $(2, 0, \ldots, 0)$ in Equation (6.32), which gives a contradiction. Therefore

$$\prod_{i=2}^{n} a_i > 1, \tag{6.34}$$

otherwise $E_A = a_1 \prod_{i=2}^{n} a_i = 2$, contrary to hypothesis. Then using Equation (6.34), we may take $u_1 = 0$ and $(u_2, \ldots, u_n) \neq (0, \ldots, 0)$ in Equation (6.32), implying the Statement (1). Therefore $\prod_{i=2}^{n} a_i$ is even, otherwise by Equation (6.33) $E_A/2 = \prod_{i=2}^{n} a_i$ is odd, which contradict statement (1). This implies that s_j is even for some $j \geq 2$, and by assumption a_1 is even. Hence the statement (2) holds.   **Q.E.D.**

The following lemma gives conditions under which a GPA may be transformed into another GPA of the same type and size.

Lemma 6.2.16 ([3], [10]) *Let $z = (z_1, \ldots, z_n)$ be a type vector and $A = [A(a(1), \ldots, a(n))]$ an m-ary array of size $a_1 \times \ldots \times a_n$. For each of the following $a_1 \times \ldots \times a_n$ m-ary array $B = [B(b(1), \ldots, b(n))]$ of energy E_A, A is a $GPA(m; a_1, \ldots, a_n)$ of type z if and only if B is a $GPA(m; a_1, \ldots, a_n)$ of type z:*

1. $B(b(1), \ldots, b(n)) = A(b(1) + c, b(2), \ldots, b(n))$ *for all* $0 \le b(i) \le a_i - 1$, $0 \le c \le a_i - 1$, *and* $z_1 = 0$;

2. $B(b(1), \ldots, b(n)) = A(a_1 - b(1) - 1, b(2), \ldots, b(n))$ *for all* $0 \le b(i) \le a_i - 1$;

3. $B(b(1), \ldots, b(n)) = A(yb(1), b(2), \ldots, b(n))$ *for all* $0 \le b(i) \le a_i - 1$, *where* $\gcd(a_1, y) = 1$ *and* $z_1 = 0$;

4. $B = \phi^{(1)}(A)$, *where* a_1 *is even;*

5. $B = \sigma(A; c)$, *where* $ca_2 \equiv 0 (\bmod a_1)$ *and* $z_1 = 0$;

6. $B = \tau(A; c)$, *where* $ca_2 \equiv 0 (\bmod 2a_1)$ *and* $z_1 = 1$.

Theorem 6.2.21 ([3], [10]) *Let $u \ge 1$ be an integer. Then there exist binary arrays $M_1, \ldots, M_{10}$ satisfying the following properties:*

1. M_1 *is a* $GPBA(2)$ *of type* (1);

2. M_2 *is a* $GPBA(2, 2)$ *of type* $(0, 0)$ *and simultaneously type* $(1, 0)$;

3. M_3 *is a* $PBA(4)$;

4. M_4 *is a* $GPBA(2, 2, 3^{(2u)})$ *of type* $(0^{(2u+2)})$ *and simultaneously type* $(1, 0^{(2u+1)})$;

5. M_5 *is a* $PBA(4, 3^{(2u)})$;

6. M_6 *is a* $GPBA(2, 2, 2, 3^{(2u)})$ *of type* $(1, 0^{(2u+2)})$;

7. M_7 *is a* $GPBA(2, 4, 3^{(2u)})$ *of type* $(1, 0^{(2u+1)})$;

8. M_8 *is a* $GPBA(4, 2, 3^{(2u)})$ *of type* $(1, 0^{(2u+1)})$;

9. M_9 is a GPBA$(2, 3, 2, 3^{(2u-1)})$ of type $(1, 0^{(2u)})$;

10. M_{10} is a GPBA$(2, 3, 2, 2, 3^{(2u-1)})$ of type $(1, 0^{(2u+1)})$.

Proof. It is easy to check the following choices for M_1, M_2, and M_3:

$$M_1 = \begin{bmatrix} + \\ + \\ + \end{bmatrix}, \quad M_2 = \begin{bmatrix} + & + \\ + & - \end{bmatrix}, \quad M_3 = \begin{bmatrix} + \\ + \\ + \\ - \end{bmatrix}.$$

Let

$$A_1 = \begin{bmatrix} - & + & + \\ - & + & + \\ - & + & + \end{bmatrix}, \quad B_1 = \begin{bmatrix} + & + & + \\ - & - & - \\ - & - & - \end{bmatrix},$$

and

$$C_1 = \begin{bmatrix} + & - & - \\ - & - & + \\ - & + & - \end{bmatrix}, \quad D_1 = \begin{bmatrix} + & - & - \\ - & + & - \\ - & - & + \end{bmatrix}.$$

Then $\{A_1, B_1, C_1, D_1\}$ is a 3×3 binary supplementary quadruple of energy 4.3^2. Now by Theorem 6.2.19, if there exist $3 \times \ldots \times 3$ binary supplementary quadruple of energy 4.3^{2u} and $4.3^{2u'}$, then there exists a $3 \times \ldots \times 3$ binary supplementary quadruple of energy $4.3^{2(u+u')}$. Therefore there exists a $3 \times \ldots \times 3$ binary supplementary quadruple of energy 4.3^{2u} for all $u \geq 1$. Hence arrays M_4 and M_5 exist because of Corollary 6.2.5.

By Theorem 6.2.13 we may take $M_6 = \prod(M_1, M_4)$ and $M_7 = \prod(M_1, M_5)$. By Theorem 6.2.14 we may take $M_8 = \kappa(M_1, M_4)$. To form M_9, first use Theorem 6.2.12 to form a GPBA$(2, 2, 3, 3^{(2u-1)})$ of type $(1, 0, \ldots, 0)$ from M_4. Then use Corollary 6.2.3 to change the type to $(1, 1, 0, \ldots, 0)$ and then to $(0, 1, 0, \ldots, 0)$. Reordering dimensions gives M_9. To form M_{10}, first use Theorem 6.2.13 to form a GPBA$(2, 2, 3, 2, 3^{(2u-1)})$ of type $(1, 1, 0, \ldots, 0)$ as $\prod(M_1, M_9)$. Then use Corollary 6.2.3 to change the type to $(0, 1, 0, \ldots, 0)$ and reorder dimensions.   **Q.E.D.**

We shall refer to the arrays $M_1, \ldots, M_{10}$ freely from now on. In order to prove the recursive constructions, we introduce some more definitions here.

Definition 6.2.15 ([3], [10]) *A non-negative integer sequence* $s = (s_1, \ldots, s_n)$ *is called odd or positive according to the fact that* s_i *is respectively odd or positive for all* $1 \leq i \leq n$. *This vector is called empty if* $n = 0$. *For* $x \geq 1$, *denote by* A_x *the set of all positive vectors* $s = (s_1, \ldots, s_n)$ *satisfying* $\sum s_i = x$.

Definition 6.2.16 ([3], [10]) *Let* $s = (s_1, \ldots, s_n)$ *be a vector satisfying* $\sum s_i = x \geq 1$, *let* $a = (a_1, \ldots, a_m)$ *be a positive odd vector and let* $E \geq 1$. *Denote the set of all* $2^{s_1} \times \ldots \times 2^{s_n} \times a_1 \times \ldots \times a_m$ *k-ary arrays of energy* E *by* $S_x(E, k; s, a)$.

Lemma 6.2.17 ([3], [10]) *Let* $x \geq 1$, *and* $s = (s_1, \ldots, s_n) \in A_x$. *Let* $a = (a_1, \ldots, a_m)$ *be a positive odd vector. Let* $z = (z_1, \ldots, z_n)$ *and* $z' = (z'_1, \ldots, z'_m)$ *be type vectors.*

1. *Suppose that* $z_1 = 1$ *and that for all* $1 \leq i \leq n$, $s_1 \geq s_i$ *whenever* $z_i = 1$. *Then there exists a GPA of type* (z, z') *in* $S_x(E, k; s, a)$ *if and only if there exists a GPA of type* $(1, 0, \ldots, 0)$ *in* $S_x(E, k; s, a)$;

2. *Let* $x \geq 3$ *and* $n \geq 2$. *Then there exists a GPA of type* $(1, 0, \ldots, 0)$ *in* $S_x(E, k; s, a)$ *if there exist a GPA of type* $(1, 0, \ldots, 0)$ *and a GPA of type* $(1, 1, 0, \ldots, 0)$ *in* $S_{x-2}(E/4, k; b_1, b_2, s_3, \ldots, s_n; a)$, *where* $(b_1, b_2) = (s_1 - 1, s_2 - 1)$ *or* $(s_1, s_2 - 2)$;

3. *Let* $y \geq 2$ *and* $n \geq 2$. *Then there exists a GPA of type* $(0, \ldots, 0)$ *in* $S_{2y}(E, k; s, a)$ *if there exist a GPA of type* $(0, \ldots, 0)$ *and a GPA of type* $(1, 0, \ldots, 0)$ *in* $S_{2y-2}(E/4, k; b_1, b_2, s_3, \ldots, s_n; a)$, *where* $(b_1, b_2) = (s_1 - 1, s_2 - 1)$ *or* $(s_1 - 2, s_2)$;

4. *Let* $0 \leq N \leq n - 1$ *and let* $s_i = 1$ *for all* $n - N + 1 \leq i \leq n$. *If there exists a GPA of type* $(1, 0, \ldots, 0)$ *in* $S_{x-N}(2^{-N}E, k, s_1, \ldots, s_{n-N}; a)$, *then there exists a GPA of type* $(1, 0, \ldots, 0)$ *in* $S_x(E, k; s, a)$.

By the first statement of this lemma, the only GPAs we need to construct in $S_x(E, k; s, a)$ are those of type $(1, 0, \ldots, 0)$ or type $(0, \ldots, 0)$. We recursively construct these two types in the following propositions.

Proposition 6.2.1 ([3], [10]) *Let* a *be a positive odd vector. Let* $w \geq 1$ *and* $t \geq 0$ *be integers and let* $x = w + 2t$. *Then the following (1) implies* (2):

1. For each $s' = (s'_1, \ldots, s'_m) \in A_w$ there exists a GPA of type $(1, 0, \ldots, 0)$ in $S_w(E, k; s'; a)$ if $s'_1 \leq \lceil w/2 \rceil$ and $s'_i \leq \lceil w/2 \rceil + 1$ for all i.

2. For each $s = (s_1, \ldots, s_m) \in A_x$ there exists a GPA of type $(1, 0, \ldots, 0)$ in $S_x(2^{2t}E, k; s; a)$ if $s_1 \leq \lceil x/2 \rceil$ and $s_i \leq \lceil x/2 \rceil + 1$ for all i.

Proposition 6.2.2 ([3], [10]) *Let a be a positive odd vector. Let $w \geq 1$ and $t \geq 0$ be integers and let $y = w + t$. Then the following (1) and (2) imply (3) :*

1. For each $s' = (s'_1, \ldots, s'_m) \in A_{2w}$ there exists a GPA of type $(0, \ldots, 0)$ in $S_{2w}(E, k; s'; a)$ if $s'_1 \leq w + 1$ for all i.

2. For each integer w' satisfying $w \leq w' \leq y - 1$ and for each $s'' = (s''_1, \ldots, s''_{m'}) \in A_{2w'}$ there exists a GPA of type $(1, 0, \ldots, 0)$ in $S_{2w'}$ $\cdot(2^{2(w'-w)}E, k; s''; a)$ if $s''_1 \leq w'$ and $s''_i \leq w' + 1$ for all i. .

3. For each $s = (s_1, \ldots, s_n) \in A_{2y}$ there exists a GPA of type $(0, \ldots, 0)$ in $S_{2y}(2^{2t}E, k; s; a)$ if $s_i \leq y + 1$ for all i.

We wish to prove restrictions on $s = (s_1, \ldots, s_n)$ for the existence of a GPA of type $(1, 0, \ldots, 0)$ in $S_x(E, k; s; a)$. Suppose that for each y there exists a GPA of type $(0, \ldots, 0)$ in $S_{2y}(2^{2y-x}E, k; s'; a)$ if and only if $s'_i \leq y + 1$ for all i. We now prove that this implies $s_1 \leq \lceil x/2 \rceil$.

Proposition 6.2.3 ([3], [10]) *Let a be a positive odd vector and let $x \geq 1$ be an integer. Then the following (1) and (2) imply (3) :*

1. For each integer y satisfying $\lceil x/2 \rceil \leq y \leq x - 1$ and for each $s' = (s'_1, \ldots, s'_m) \in A_{2y}$ there exists a GPA of type $(0, \ldots, 0)$ in $S_{2y}(2^{2y-x}E, k; s'; a)$ if $s'_i \leq y + 1$ for all i;

2. For each integer y satisfying $\lceil x/2 \rceil \leq y \leq x - 1$ and for each $s'' = (s''_1, \ldots, s''_{m'}) \in A_{2y+2}$ there exists a GPA of type $(0, \ldots, 0)$ in $S_{2y+2}(2^{2y-x+2}E, k; s''; a)$ only if $s''_i \leq y + 2$ for all i;

3. For each $s = (s_1, \ldots, s_n) \in A_x$ there exists a GPA of type $(1, 0, \ldots, 0)$ in $S_x(E, k; s; a)$ only if $a_1 \leq \lceil x/2 \rceil$.

Theorem 6.2.22 ([3], [10]) *Let $n, u \geq 1$ and let $a_i \geq 1$ for $1 \leq i \leq n$. Let $z = (z_1, \ldots, z_n)$ and $z' = (z'_1, \ldots, z'_{2u})$ be type vectors,*

1. *There exists a* $\mathrm{PBA}(2^{a_1}, \ldots, 2^{a_n})$, *where* $\sum a_i = 2y \geq 2$, *if and only if* $a_i \leq y + 1$ *for all* i;

2. *There exists a* $\mathrm{GPBA}(2^{a_1}, \ldots, 2^{a_n})$ *of type* z, *where* $\sum a_i = x \geq 1$ *and* $z \neq (0, \ldots, 0)$,

   (a) *if* $a_i \leq \begin{cases} \lceil x/2 \rceil & \text{when } z_i = 1 (1 \leq i \leq n) \\ \lceil x/2 \rceil + 1 & \text{when } z_i = 0 (1 \leq i \leq n) \end{cases}$

   (b) *only if* $a_i \leq \lceil x/2 \rceil$ *when* $z_i = 1 (1 \leq i \leq n)$;

3. *There exists a* $\mathrm{PBA}(2^{a_1}, \ldots, 2^{a_n}, 3^{(2u)})$, *where* $\sum a_i = 2y \geq 2$, *if and only if* $a_i \leq y + 1$ *for all* i;

4. *There exists a* $\mathrm{GPBA}(2^{a_1}, \ldots, 2^{a_n}, 3^{(2u)})$ *of type* (z, z'), *where* $\sum a_i = x \geq 2$ *and* $z \neq (0, \ldots, 0)$,

   (a) *if* $a_i \leq \begin{cases} \lceil x/2 \rceil & \text{when } z_i = 1 (1 \leq i \leq n) \\ \lceil x/2 \rceil + 1 & \text{when } z_i = 0 (1 \leq i \leq n) \end{cases}$

   (b) *only if* $a_i \leq \lceil x/2 \rceil$ *when* $z_i = 1 (1 \leq i \leq n)$;

Proof. We first prove (1) and (2). By Definition 6.2.16, any $2^{a_1} \times \ldots \times 2^{a_n}$ binary array with $\sum a_i = x$ is a member of $S_x(2^x, 2; s; a)$, where a is empty. We shall apply Propositions 6.2.1, 6.2.2, and 6.2.3 with a being empty and $k = 2$.

We may satisfy the first condition of Proposition 6.2.1 when $(E, w) = (2, 1)$ and $(4, 2)$ using the existence of M_1 and M_2, respectively. Therefore

$$\text{there exists a GPBA}(2^{a_1}, \ldots, 2^{a_n}) \text{ of type } (1, 0, \ldots, 0),$$
$$\text{where } \sum a_i = x \geq 1 \tag{6.35}$$

if

$$a_1 \leq \lceil x/2 \rceil \text{ and } a_i \leq \lceil x/2 \rceil + 1 \text{ for all } i. \tag{6.36}$$

(2.a) follows from Lemma 6.2.17 (1).

We may then satisfy conditions (1) and (2) of Proposition 6.2.2 when $(E, w) = (4, 1)$ and $y \geq 1$ using respectively the existence of M_2, M_3 and the result that Equation (6.36) implies Equation (6.35). Therefore

$$\text{there exists a PBA}(2^{a_1}, \ldots, 2^{a_n}) \text{ where } \sum a_i = 2y \geq 2, \tag{6.37}$$

if

$$a_i \le y + 1 \text{ for all } i. \tag{6.38}$$

Equation (6.38) is also a necessary condition for Equation (6.37) , we have (1) .

Finally (1) implies conditions (1) and (2) of Proposition 6.2.3 when $E = 2^x$ and $x \ge 1$, and therefore there exists a GPBA$(2^{a_1}, \ldots, 2^{a_n})$ of type $(1, 0, \ldots, 0)$, where $\sum a_i = x \ge 1$, only if $a_1 \le \lceil x/2 \rceil$. (2.b) follows from Lemma 6.2.17.

We now outline the proof of (3) and (4) , which closely resembles that of (1) and (2). We apply Propositions 6.2.1, 6.2.2 and 6.2.3 with $a = 3^{(2u)}$ and $k = 2$. For $(E, w) = (4.3^{2u}, 2)$ and $(8.3^{2u}, 3)$ the respective existence of M_4, M_6, M_7, and M_8 satisfies the first statement of Proposition 6.2.1. For $(E, w) = (4.3^{2u}, 1)$ the existence of M_4 and M_5 satisfies the first statement of Proposition 6.2.2. We also require Proposition 6.2.3 and Lemma 6.2.17 to complete the proof.   **Q.E.D.**

We may follow a similar procedure to prove existence and nonexistence results for further families of GPBAs. We start the recursive constructions with the binary arrays M_9 and M_{10}.

Theorem 6.2.23 ([3], [10]) *Let $n, u \ge 1$ and let $a_i \ge 1$ for $1 \le i \le n$. Let $1 \le N \le min(n, 2u)$. Let $z = (z_1, \ldots, z_n) \ne (0, \ldots, 0)$ and $z' = (z'_{N+1}, \ldots, z'_{2u})$ be type vectors. Then there exists a GPBA$(2^{a_1}.3, \ldots, 2^{a_N}.3, 2^{a_{N+1}}, \ldots, 2^{a_n}, 3^{(2u-N)})$ of type (z, z'), where $\sum a_i = x \ge 2$,*

1. *if x is even and*

$$a_i \le \begin{cases} x/2 & \text{when } z_i = 1(1 \le i \le n) \\ x/2 + 1 & \text{when } z_i = 0(1 \le i \le n); \end{cases}$$

2. *if x is odd and either $a_i \le (x-1)/2$ for all $1 \le i \le n$ or*

$$a_i \le \begin{cases} (x-3)/2 & \text{when } z_i = 1(1 \le i \le n) \\ (x+1)/2 & \text{when } z_i = 0(1 \le i \le n); \end{cases}$$

3. *only if $a_i \le \lceil x/2 \rceil$ when $z_i = 1(1 \le i \le n)$.*

Proof. We may assume that $z_i = 1$ for all $1 \leq i \leq N$, otherwise by Theorem 6.2.12 we may transform to a GPBA of the same form but with smaller N. Reorder dimensions so that $a_1 = \max_{1 \leq i \leq N} a_i$. We may assume that for all $N + 1 \leq i \leq n$, $a_1 > a_i$ whenever $z_i = 1$, otherwise by Corollary 6.2.4 and Theorem 6.2.12 we may transform to a GPBA of the form already considered in the fourth statement of Theorem 6.2.22. Then by Corollary 6.2.4 and Theorem 6.2.12 we may transform to a GPBA$(2^{a_1}.3, 2^{a_2}, \ldots, 2^{a_n}, 3^{(2u-1)})$ of type $(1, 0, \ldots, 0)$.

1. *Following the method of Proposition 6.2.1, we use Theorem 6.2.16 to inductively construct the desired array. We begin with M_9, and make use of the arrays constructed in the fourth part of Theorem 6.2.22 for the induction step.*

2. *The method is similar, but a weaker inductive hypothesis is needed.*

3. *The method is similar to that of Proposition6.2.3 and make use the result of the third part of Theorem 6.2.22.*

Q.E.D

Up to now, we have proved several construction theorems for generalized perfect m-ary arrays in n-dimensions. By recursive application in the binary case we have the following ([3], [10])

1. *There exists a PBA$(2^{a_1}, \ldots, 2^{a_n}, 3^{(2u)})$, where $\sum a_i = 2y \geq 2$ and $u \geq 0$, if and only if $a_i \leq y + 1$ for all i;*

2. *There exists a GPBA$(2^{a_1}, \ldots, 2^{a_n}, 3^{(2u)})$ of type $(z_1, \ldots, z_n, z'_1, \ldots, z'_{2u})$, where $\sum a_i = x \geq 1$, $u \geq 0$, $(z_1, \ldots, z_n) \neq (0, \ldots, 0)$, and $x \geq 2$ whenever $u > 0$,*

   (a) *if $a_i \leq \begin{cases} \lceil x/2 \rceil & \text{when } z_i = 1(1 \leq i \leq n) \\ \lceil x/2 \rceil + 1 & \text{when } z_i = 0(1 \leq i \leq n) \end{cases}$*

   (b) *only if $a_i \leq \lceil x/2 \rceil$ when $z_i = 1(1 \leq i \leq n)$.*

3. *There exists a GPBA$(2^{a_1}.3, \ldots, 2^{a_N}.3, 2^{a_{N+1}}, \ldots, 2^{a_n}, 3^{(2u-N)})$ of type (z, z'), where $\sum a_i = x \geq 2$, $N \geq 1$, $z = (z_1, \ldots, z_n) \neq (0, \ldots, 0)$ and $z' = (z'_{N+1}, \ldots, z'_{2u})$,*

   (a) *if x is even and $a_i \leq \begin{cases} x/2 & \text{when } z_i = 1(1 \leq i \leq n) \\ x/2 + 1 & \text{when } z_i = 0(1 \leq i \leq n); \end{cases}$*

(b) *if x is odd and either $a_i \le (x-1)/2$ for all $1 \le i \le n$ or*

$$a_i \le \begin{cases} (x-3)/2 \text{ when } z_i = 1 (1 \le i \le n) \\ (x+1)/2 \text{ when } z_i = 0 (1 \le i \le n); \end{cases}$$

(c) *only if $a_i \le \lceil x/2 \rceil$ when $z_i = 1 (1 \le i \le n)$.*

6.3 Higher-Dimensional Hadamard Matrices Based on Orthogonal Designs ([17], [18], [19])

Besides the cases of $m = 1$ and $m = 2$, it has been proved that two-dimensional Hadamard matrices of order m can exist only if m is a multiple of four. The problem of finding the existence of at least one Hadamard matrix for all values of $m = 4t$ has been studied since 1892, and the most powerful recent results have depended entirely on the theory of orthogonal designs.

The first person to use an orthogonal design to find two-dimensional Hadamard matrices was J. Williamson in 1944. His work depended on using the orthogonal design of size 4×4 and type $(1, 1, 1, 1)$:

$$\begin{bmatrix} x & y & u & v \\ -y & x & v & -u \\ -u & -v & x & y \\ -v & u & -y & x \end{bmatrix} \tag{6.39}$$

and replacing the variables by four symmetric circulant (± 1)-valued matrices X, Y, U, V of order t that satisfy

$$XX^T + YY^T + UU^T + VV^T = 4tI_t \tag{6.40}$$

Subsequently, four symmetric circulant (± 1)-valued matrices satisfying Equation (6.40), or just four (± 1)-valued matrices satisfying Equation (6.40) and satisfying $MN^T = NM^T$, for $M, N \in \{X, Y, U, V\}$, have come to be known as Williamson matrices of order t. Williamson matrices of order t can be used in Equation (6.39) to obtain a two-dimensional Hadamard matrix of order $4t$. These matrices will be used later to obtain higher-dimensional Hadamard matrices.

6.3.1   Definitions of Orthogonality

For the details of this subsection the readers are recommended to the papers [1], [17].

Orthogonality for higher-dimensional matrices can be defined by several ways ([1], [17], [18], [19]). For example, the orthogonalities of higher-dimensional matrices can be quantified by defining an n-dimensional matrix to be orthogonal of property $(d_1, d_2, \ldots, d_n)$ with $2 \leq d_i \leq n$, where d_i indicates that in the i-th direction (i.e., the i-th coordinate), the (d_i-1)-st, d_i-th, (d_i+1)-st, $\ldots$, $(n-1)$-th dimensional layers are mutually uncorrelated, but the (d_i-2)-nd dimensional layer is not. $d_i = \infty$ means that not even the $(n-1)$-st dimensional layers are orthogonal. The two extreme definitions of orthogonalities can be described as follows:

Orthogonality by the first definition ([17]). In this case, the n-dimensional matrix has its two-dimensional layers, M, are orthogonal in all axis-normal directions, that is, if the inner product of their rows are pairwise zero, or equivalently if $MM^T = D_m$, a diagonal matrix of order m. In other words, an n-dimensional matrix is said to be orthogonal by the first definition if it has property $(2, 2, \ldots, 2)$.

Orthogonality by the second definition ([17]). In this case, the n-dimensional matrix has its $(n-1)$-dimensional sections normal to one coordinate axis are mutually uncorrelated but are not in themselves orthogonal in any sense; moreover, the sets of $(n-1)$-dimensional layers normal to other axes are neither mutually uncorrelated nor orthogonal. In other words, an n-dimensional matrix is said to be orthogonal by the second definition if it has property $(\infty, \infty, \ldots, \infty, n, \infty, \ldots, \infty)$.

When the term 'orthogonal' is used to describe an n-dimensional matrix without modifying the word orthogonal, it means that the orthogonality lies between the above extremes of the first and the second definitions of orthogonality. An absolutely proper n-dimensional Hadamard matrix is orthogonal of property $(2, 2, \ldots, 2)$; an absolutely improper n-dimensional Hadamard matrix is orthogonal of property $(n, n, \ldots, n)$.

Let a 2-dimensional matrix A be denoted by

$$A = \begin{bmatrix} A(1) \\ A(2) \\ \vdots \\ A(n) \end{bmatrix},$$

where $A(i) = (A(i,1), A(i,2), \ldots, A(i,n))$, $i = 1, 2, \ldots, n$. Similarly, let

$$B = \begin{bmatrix} B(1) \\ B(2) \\ \vdots \\ B(n) \end{bmatrix}.$$

The inner product of A and B will be meant

$$A \cdot B = A(1)B(1) + A(2)B(2) + \ldots A(n)B(n).$$

Alternatively, $A \cdot B$ can be written as the sum of the diagonal elements of AB^T, which is in fact $tr(AB^T)$. From the rows of a given orthogonal square matrix H, we construct layers of a three-dimensional matrix that is orthogonal by the second definition in the following way:

Let H be given by

$$H = \begin{bmatrix} H(1) \\ H(2) \\ \vdots \\ H(n) \end{bmatrix} = \begin{bmatrix} H(1,1) & H(1,2) & \ldots & H(1,n) \\ H(2,1) & H(2,2) & \ldots & H(2,n) \\ \vdots & \vdots & \vdots & \vdots \\ H(n,1) & H(n,2) & \ldots & H(n,n) \end{bmatrix},$$

where $H(p)H(q) = 0$, $p, q = 1, 2, \ldots, n$, $p \neq q$. Orthogonal layers $H(i,j)$ are obtained by taking the tensor products of the vectors $H(i)$, $H(j)$:

$$H(i,j) = H(i) \otimes (H(j))^T$$
$$= \begin{bmatrix} H(i,1)H(j,1) & H(i,1)H(j,2) & \ldots & H(i,1)H(j,n) \\ H(i,2)H(j,1) & H(i,2)H(j,2) & \ldots & H(i,2)H(j,n) \\ \vdots & \vdots & \vdots & \vdots \\ H(i,n)H(j,1) & H(i,n)H(j,2) & \ldots & H(i,n)H(j,n) \end{bmatrix}$$

$$= \begin{bmatrix} H(i,1)H(j) \\ H(i,2)H(j) \\ \vdots \\ H(i,n)H(j) \end{bmatrix}.$$

It is easy to check that $H(i,j)H(k,l) = 0$, for all $i, j, k, l = 1, 2, \ldots, n$, $j \neq l$. For

$$\begin{aligned} H(i,j)H(k,l) &= (H(i,1)H(j)) \cdot (H(k,1)H(l)) + (H(i,2)H(j))(H(k,2)H(l)) \\ &\quad + \ldots + (H(i,n)H(j)) \cdot (H(k,n)H(l)) \\ &= (H(i,1)H(k,1) + H(i,2)H(k,2) + \ldots \\ &\quad + H(i,n)H(k,n))(H(j) \cdot H(l)) \\ &= 0 \end{aligned}$$

since $H(j) \cdot H(l) = 0$, for all $l, j = 1, 2, \ldots, n$, $l \neq j$.

Fourth and higher-dimensional orthogonal matrices that are orthogonal by the second definition can be constructed by the analogous way. In fact,

Denote an n-dimensional matrix by $[M]_n$ and its $(n-1)$-dimensional layers by $[M_{n-1}]$. The inner product of two n-cube A_n and B_n is defined by the sum of the inner products of their respective $(n-1)$-layers parallel to the i-th coordinate plane $i = 1, 2, \ldots, n$, e.g.,

$$[A]_n \cdot [B]_n = \sum_{j=1}^{n} [A^i_j]_{n-1}[B^i_j]_{n-1},$$

where A^i_j denotes the j-th layer parallel to the i-th coordinate plane.

Then $[H]_n$ is said to be orthogonal if the inner product of the $(n-1)$-dimensional layers parallel to a coordinate hyperplane is pairwise equal to zero. Again we can construct such $(n-1)$-dimensional layers by taking the tensor products of the layers of an $(n-1)$-dimensional orthogonal cube.

The second definition of orthogonality of higher-dimensional matrices is quite in agreement with the orthogonality of the two-dimensional matrices from the point of view of dimensions. In a two-dimensional matrix M, the layers are the rows (or columns) of M that can be assumed as one-dimensional matrices, i.e., one-dimension lower than the dimension of M, so that in this case, too, orthogonality can be defined as the inner product of the parallel layer matrices pair-wise equal to zero.

Let $[H]_n$ be an n-cube of order h. We denote by $[H^i]_1$, $[H^i]_2$, ..., $[H^i]_{n-1}$ the 1-, 2-, ..., $(n-1)$-dimensional layer matrices of $[H]_n$, respectively. Where the i represents the layers embedded in the subspaces parallel to the i-th coordinate hyperplane. If the equation $[H^i]_1 \cdot [H^i]_1' = 0$ is satisfied by each pair of distinct 1-dimensional layers $[H^i]_1$ and $[H^i]_1'$ parallel to the i-th coordinate hyperplane, then $[H^i]_2 . [H^i]_2' = 0$ is also true for each distinct pair of 2-dimensional layers $[H^i]_2$ and $[H^i]_2'$. In general, it is easy to prove that ([17])

$$[H^i]_1 \cdot [H^i]_1' = 0 \implies [H^i]_2 . [H^i]_2' = 0$$
$$\implies , \ldots, \implies$$
$$\implies [H^i]_{n-1} . [H^i]_{n-1}' = 0.$$

It can be observed that $[H^i]_1 \cdot [H^i]_1' = 0$ is equivalent to the first definition of orthogonality, provided that i takes up all values from 1 to n. On the other hand, $[H^i]_{n-1} \cdot [H^i]_{n-1}' = 0$ is equivalent to the second definition of orthogonality. Thus the orthogonality by the first definition implies the orthogonality by the second definition ([17]) .

Between the first and second definitions of orthogonalities, there are variety of orthogonalities according to the dimension of the layers, which can vary from 1 to $n-1$, and the numbers of coordinate hyperplanes to which the layers can be parallel, which can go from 1 to n. If A and B are two n-cubes orthogonal of proprieties $(a_1, a_2, \ldots, a_n)$ and $(b_1, b_2, \ldots, b_n)$ and orders a and b, respectively, then the direct multiplication (or Kronecker product) $A \otimes B$ of A and B is an n-cube of order ab with orthogonal of property $(c_1, c_2, \ldots, c_n)$, where $c_i = \max(a_i, b_i)$, $i = 1, 2, \ldots, n$. A set $\mathcal{A} = \{A_1, A_2, \ldots, A_m\}$ of n-cubes orthogonal of proprieties $(a_{i1}, a_{i2}, \ldots, a_{in})$, where $a_{ij} \leq \max\{a_{i1}, a_{i2}, \ldots, a_{im}\}$ for $i = 1, 2, \ldots, m$ generates a monoid under the operation $\otimes$, i.e., it satisfies the following properties ([1], [17]) :

1. $A_i \otimes A_j$ is orthogonal;

2. $A_i(\otimes A_j \otimes A_k) = (A_i \otimes A_j) \otimes A_k$; and

3. $I_{1 \times 1} \in \mathcal{A}$ is the unit element.

Since $A_i \otimes A_j$ is equivalent to $A_j \otimes A_i$ by using an appropriate permutation of rows and columns, we could say that the monoid is 'combinatorially commutative.'

Example 1 ([1], [17], [18], [19]) : Let $A = [A(i,j,k)]$, $0 \leq i,j,k \leq 1$, be the 3-dimensional matrix of order 2 defined by

$$[A(i,0,k)] = \begin{bmatrix} -1 & 1 \\ -1 & -1 \end{bmatrix}, \text{ and } [A(i,1,k)] = \begin{bmatrix} -1 & 1 \\ 1 & 1 \end{bmatrix}.$$

These two faces are, as vectors $(-1,1,-1,-1)$ and $(-1,1,1,1)$, which are orthogonal, and so these faces(or parallel two-dimensional layers in the direction of y-axes) are said to be orthogonal. It can be verified that A is orthogonal of property $(2,2,3)$.

Let $B = [B(i,j,k)]$, $0 \leq i,j,k \leq 1$, be the 3-dimensional matrix of order 2 defined by

$$[B(i,0,k)] = \begin{bmatrix} 1 & 1 \\ 1 & -1 \end{bmatrix}, \text{ and } [B(i,1,k)] = \begin{bmatrix} -1 & 1 \\ 1 & 1 \end{bmatrix}.$$

Every two-dimensional face of B is an Hadamard matrix. So B is an absolutely 3-diemensional Hadamard matrix, which is orthogonal of property $(2,2,2)$ ([1], [17]) .

Example 2 (The Paley Cube [1], [17]) : Let $q \equiv 3 \pmod 4$ be a prime power and $z_0, z_1, \ldots, z_{q-1}$ be the elements of $GF(q)$, the Galois field. We define

$$p_{ij\ldots r} = \begin{cases} 1 & \text{if any of the subscripts is } q \\ \mathcal{X}(z_i + z_j + \ldots + z_r) & \text{otherwise,} \end{cases}$$

where each subscript runs from zero to q, $P = [p_{ij\ldots r}]$ is a $(q+1)$-dimensional Paley cube, and $\mathcal{X}(0) = -1$. Thus

$$\mathcal{X}(z) = \begin{cases} 1 & \text{if } z \text{ is a square in } GF(q) \\ -1 & \text{otherwise} \end{cases}$$

By the same reasoning as before, and using the two-dimensional properties of this matrix, we see that each two-dimensional face of the Paley cube except that one face containing all ones is an Hadamard matrix. So then, when $q \equiv 3 \pmod 4$ is a prime power, there is an almost Hadamard $(q+1)$-dimensional cube, called the Paley cube, of order $q+1$, which has one two-dimensional layer, in each direction, consisting of all ones with every other face being an Hadamard matrix. The Paley cube is orthogonal of property $(\infty, \infty, \ldots, \infty)$, but if the two-dimensional layer consisting of all ones is removed in one direction, the remaining n-dimensional matrix (note it is no longer a cube) has all two-dimensional layers in that direction orthogonal to each other.

6.3.2 Higher-Dimensional Orthogonal Designs ([1], [17], [18], [19])

Two-dimensional orthogonal designs of type $(s_1, s_2, \ldots, s_t)$ is defined by [20] as a square orthogonal matrices with entries from $\{0, \pm x_1, \pm x_2, \ldots, \pm x_t\}$, where $x_1, x_2, \ldots, x_t$ are commuting variables and s_j is the number of times $\pm x_j$ occurs in each row and column—that is, in which all distinct rows and columns have scalar product zero. Hence an $m \times m (= m^2)$ matrix $[d(i, j)]$ is an orthogonal design of type $(s_1, s_2, \ldots, s_t)$ if it has entries from $\{0, \pm x_1, \pm x_2, \ldots, \pm x_t\}$ and

$$\sum_{i=0}^{m-1} d(i, a)d(i, b) = \sum_{j=0}^{m-1} d(a, j)d(b, j) = \sum_{k=0}^{m-1} s_k x_k^2 \delta_{ab}. \tag{6.41}$$

Thus the two-dimensional Hadamard matrices are special cases in which the variables are from $\{\pm 1\}$ and $(s_1, s_2, \ldots, s_t)$ is (m). Therefore $[h(i, j)]$, $0 \le i, j \le m - 1$, is an Hadamard matrix if

$$\sum_{i=0}^{m-1} h(i, a)h(i, b) = \sum_{j=0}^{m-1} h(a, j)h(b, j) = m\delta_{ab}. \tag{6.42}$$

In general, a proper n-cube orthogonal design $D = [D(d(1), \ldots, d(n))]$, $0 \le d(i) \le m - 1$, of order m and type $(s_1, s_2, \ldots, s_t)^n$ is defined by [17] as such a cube in which all parallel two-dimensional layers, in any orientation parallel to a plane, are uncorrelated, which is equivalent to the requirement that $D(d(1), \ldots, d(n)) \in \{0, \pm x_1, \ldots, \pm x_t\}$, where $x_1, x_2, \ldots, x_t$ are commuting variables, and that

$$\sum_{d(1)} \cdots \sum_{d(r-1)} \sum_{d(r+1)} \cdots \sum_{d(n)} D(d(1), \ldots, d(r-1), a, d(r+1), \ldots, d(n))$$

$$\times D(d(1), \ldots, d(r-1), b, d(r+1), \ldots, d(n))$$

$$= \left(\sum_i s_i x_i^2 \right)^{n-1} \delta_{ab}, \tag{6.43}$$

where $(s_1, s_2, \ldots, s_t)$ are integers giving the occurrences of $\pm x_1, \ldots, \pm x_t$ in each row and column (called the type $(s_1, s_2, \ldots, s_t)^n$ by the first definition) , i.e., it is of property $(2, 2, \ldots, 2)$. In the fashion similar to the last subsection it is possible to define orthogonal designs according to the second definition or according to any other property of orthogonality.

Higher-dimensional orthogonal designs may be constructed by noting that if A is an n-cube orthogonal design of order a and type $(s_1, s_2, \ldots, s_t)^n$, and H is an n-dimensional Hadamard matrix of order h, then the direct multiplication $H \otimes A$ is an n-cube orthogonal design of order ah and type $(hs_1, hs_2, \ldots, hs_t)^n$. The property depends on the property of the matrices used ([17]) .

Example 1 ([17]). Let $H = [H(i, j, k)]$, $0 \leq i, j, k \leq 1$, be the 3-dimensional Hadamard matrix of order 2 defined by

$$[H(i,0,k)] = \begin{bmatrix} 1 & 1 \\ 1 & -1 \end{bmatrix}, \text{ and } [H(i,1,k)] = \begin{bmatrix} -1 & 1 \\ 1 & 1 \end{bmatrix}$$

and let $A = [A(i, j, k)]$, $0 \leq i, j, k \leq 1$, be the 3-cube orthogonal design of order 2 and type $(1, 1)^3$ defined by

$$[A(i,0,k)] = \begin{bmatrix} y & x \\ x & -y \end{bmatrix}, \text{ and } [A(i,1,k)] = \begin{bmatrix} -x & y \\ y & x \end{bmatrix}.$$

Then their direct multiplication $A \otimes H := C = [C(i, j, k)]$, $0 \leq i, j, k \leq 3$, is the 3-cube orthogonal design of order $2 \times 2 = 4$ and type $(2, 2)^3$ defined by

$$[C(i,0,k)] = \begin{bmatrix} -y & -x & -y & -x \\ -x & -y & -x & -y \\ -y & -x & y & x \\ -x & y & x & -y \end{bmatrix}; \quad [C(i,1,k)] = \begin{bmatrix} x & -y & x & -y \\ -y & -x & -y & -x \\ x & -y & -x & y \\ -y & -x & y & x \end{bmatrix};$$

and

$$[C(i,2,k)] = \begin{bmatrix} y & x & -y & -x \\ x & -y & -x & y \\ -y & -x & -y & -x \\ -x & y & -x & y \end{bmatrix}; \quad [C(i,3,k)] = \begin{bmatrix} -x & y & x & -y \\ y & x & -y & -x \\ x & -y & x & -y \\ -y & -x & -y & -x \end{bmatrix}.$$

Example 2 ([17]) . The following $B = [B(i, j, k)]$, $0 \le i, j, k \le 3$, is a 3-dimensional orthogonal design of order 4 and type $(1, 1, 1, 1)^3$:

$$[B(i,0,k)] = \begin{bmatrix} d & c & -b & -a \\ -c & d & -a & b \\ b & a & d & c \\ -a & b & c & -d \end{bmatrix} \; ; \quad [B(i,1,k)] = \begin{bmatrix} c & -d & a & -b \\ d & c & -b & -a \\ a & -b & -c & d \\ b & a & d & c \end{bmatrix} \; ;$$

and

$$[B(i,2,k)] = \begin{bmatrix} b & a & d & c \\ a & -b & -c & d \\ -d & -c & b & a \\ -c & d & -a & b \end{bmatrix} \; ; \quad [B(i,3,k)] = \begin{bmatrix} -a & b & c & -d \\ b & a & d & c \\ c & -d & a & -b \\ -d & -c & b & a \end{bmatrix} \; .$$

Theorem 6.3.1 ([1], [17], [18], [19]) *There exists an n-dimensional orthogonal design of order 2, type $(1, 1)^n$, and property $(2, 2, \ldots, 2)$.*

Proof. Let a and b be commuting variables. Let $H = [H(h(1), \ldots, h(n))]$, $0 \le h(i) \le 1$, $1 \le i \le n$, be defined by

$$H(h(1), \ldots, h(n)) = \begin{cases} a(-1)^{w/2+1} & w \equiv 0 \pmod 2 \\ b(-1)^{w/2-1} & w \equiv 1 \pmod 2, \end{cases}$$

where $w := \sum_{i=1}^{n} h(i)$, the weight of the subscripts.

In order to check the orthogonality of this H, we consider

$$H(0, 0, h(3), \ldots, h(n))H(0, 1, h(3), \ldots, h(n))$$
$$+ H(1, 0, h(3), \ldots, h(n))H(1, 1, h(3), \ldots, h(n)) \qquad (6.44)$$

and

$$H(0, 0, h(3), \ldots, h(n))H(1, 0, h(3), \ldots, h(n))$$
$$+ H(0, 1, h(3), \ldots, h(n))H(1, 1, h(3), \ldots, h(n)). \qquad (6.45)$$

Suppose $v = \sum_{i=3}^{n} h(i)$. Then we have the following four cases:

Case 1: If $v \equiv 0 \pmod 4$, then both Equations (6.44) and (6.45) become $-ab + ba = 0$;

Case 2: If $v \equiv 1 \pmod 4$, then both Equations (6.44) and (6.45) become $ba + a(-b) = 0$;

Case 3: If $v \equiv 2 \pmod 4$, then both Equations (6.44) and (6.45) become

$$a(-b) + (-b)(-a) = 0;$$

Case 4: If $v \equiv 3 \pmod 4$, then both Equations (6.44) and (6.45) become

$$(-b)(-a) + (-a)b = 0.$$

Similarly, it can be verified that all of the two-dimensional faces of H are orthogonal. So this H is the wanted n-dimensional orthogonal design of order 2, type $(1,1)^n$, and property $(2,2,\ldots,2)$.   **Q.E.D.**

Corollary 6.3.1 ([17], [18], [19]) *There exist n-dimensional orthogonal design of order 2^{t+1}, type $(2^t, 2^t)^n$, and property $(2,2,\ldots,2)$.*

Proof. The corollary follows the direct multiplication $A \otimes H$ of an absolutely proper n-dimensional Hadamard matrix A of order 2 and the n-dimensional orthogonal design H of order 2, type $(1,1)^n$, and property $(2,2,\ldots,2)$ constructed in Theorem 6.3.1.   **Q.E.D.**

An n-dimensional orthogonal design of type $(1,1,1,1)^n$ would be preserved under the following equivalence relations([17], [18], [19]) :

1. *Each variable is replaced throughout by its negative;*

2. *Rearrangement of the parallel k-dimensional hyper-planes;*

3. *Multiplication of every variable of one entire k-dimensional hyper-plane by -1.*

Horadam and Lin very recently proved in *J. Combin. Math. Comp.* in 1998 that there is only one equivalent $(1,1,1,1)^2$ designs of order 4 on the variables a, b, c, d, which is

$$\begin{bmatrix} a & b & c & d \\ -b & a & d & -c \\ -c & -d & a & b \\ -d & c & -b & a \end{bmatrix}.$$

Theorem 6.3.2 ([17], [18], [19]) *For each positive integer n, there exist n-dimensional orthogonal designs of order 4, type $(1,1,1,1)^n$, and property $(2,2,\ldots,2)$.*

Proof. We proceed inductively. First, define

$$a_1 = (-d, -c, b, a); \quad a_2 = (c, -d, a, -b); \quad a_3 = (b, a, d, c); \quad a_4 = (-a, b, c, -d).$$

and note that

$$a_i \cdot a_j = 0 \text{ for all } 1 \le i \ne j \le 4.$$

Now we can describe the faces of the $(1,1,1,1)^3$ design as

$$b_4^T = \begin{bmatrix} a_4 \\ a_3 \\ a_2 \\ a_1 \end{bmatrix}, \quad b_3^T = \begin{bmatrix} a_3 \\ -a_4 \\ a_1 \\ -a_2 \end{bmatrix}, \quad b_2^T = \begin{bmatrix} a_2 \\ -a_1 \\ -a_4 \\ a_3 \end{bmatrix}, \quad b_1^T = \begin{bmatrix} -a_1 \\ -a_2 \\ a_3 \\ a_4 \end{bmatrix},$$

or, equivalently,

$$\begin{bmatrix} b_4 \\ b_3 \\ b_2 \\ b_1 \end{bmatrix} = \begin{bmatrix} a_4 & a_3 & a_2 & a_1 \\ a_3 & -a_4 & a_1 & -a_2 \\ a_2 & -a_1 & -a_4 & a_3 \\ -a_1 & -a_2 & a_3 & a_4 \end{bmatrix}.$$

Now there are two inequivalent 2-dimensional $(1,1,1,1)^2$ orthogonal designs and both of them can be used to construct 3-dimensional $(1,1,1,1)^3$ orthogonal designs. Thus we have a $(1,1,1,1)^3$ design on the commuting variables a_1, a_2, a_3, a_4 and hence a $(1,1,1,1)^4$ design on the commuting variables a, b, c, d.

The orthogonality within the $(1,1,1,1)^3$ design is established by the construction. The orthogonality of the $(1,1,1,1)^4$ design is obtained by using the extra property $a_i . a_j = 0$ for all $1 \le i \ne j \le 4$.

To obtain the $(1,1,1,1)^{k+1}$ design, we assume the existence of the $(1,1,1,1)^j$ designs for all $j \le k$ made by the construction. Now we have a $(1,1,1,1)^k$ design whose hyper-rows, c_1, c_2, c_3, c_4, comprise objects which are the hyper-rows of the $(1,1,1,1)^{k-1}$ design. We now write down the hyper-rows of each of the four hyper-planes containing these rows as the columns of a 4×4 matrix, D. According to the construction, D is an orthogonal design of type $(1,1,1,1)$ whose objects are the c_i. Now we complete D to form a $(1,1,1,1)^3$ design, E, with objects c_i. Now E is a $(1,1,1,1)^{k+1}$ design whose orthogonality is guaranteed by the existence assumption.

In fact, at all stages of the above construction, the property was completely preserved. Hence the theorem follows.   **Q.E.D.**

6.3.3 Higher-Dimensional Hadamard Matrices from Orthogonal Designs

Let $z_1, z_2, \ldots, z_m$ be the elements of an abelian group G of order m. A type 2 or 1 matrix $A = [A(i,j)]$ is a matrix defined by ([17], [18], [19])

$$A(i,j) = \alpha(z_i \pm z_j),$$

where α is a map into a commutative ring. A circulant matrix of order m is a special case in that $G = Z_m$ (the cyclic group of order m), with $z_1 = 1$, $z_2 = 2$, $\ldots$, $z_m = m$, and so

$$
\begin{aligned}
A(i,j) &= \alpha(z_i \pm z_j) \\
&= \begin{cases} \alpha(z_i + z_j), \text{ for type 2 matrices,} \\ \alpha(z_i - z_j), \text{ for type 1 matrices.} \end{cases}
\end{aligned}
\tag{6.46}
$$

A set of t matrices X_1, X_2, $\ldots$, X_t of order m with $X_k = [X_k(i,j)]$ is called t-suitable matrices([17], [18], [19]) if

$$\sum_{k=1}^{t} \sum_{i=1}^{m} X_k(a,i) X_k(b,i) = f\delta(a,b) \tag{6.47}$$

and

$$\sum_{i=1}^{m} X_k(a,i) X_l(b,i) = \sum_{i=1}^{m} X_l(a,i) X_k(b,i), \tag{6.48}$$

where f is a constant or constant function. Thus the Williamson matrices of order m are 4-suitable matrices with entries ± 1 and $f = 4m$.

Suppose that $X_k = [X_k(i,j)]$, $k = 1, 2, \ldots, t$, are t-suitable matrices of type 2 and order m defined by([17], [18], [19])

$$X_k(i,j) = \psi_k(z_i + z_j).$$

Then from Equations (6.47) and (6.48), we have

$$
\sum_{k=1}^{t} \sum_{i=1}^{m} \psi_k(z_a + z_i) \psi_k(z_b + z_i) = f\delta(a,b)
$$

$$
= \sum_{k=1}^{t} \sum_{g \in G} \psi_k(z_a + g) \psi_k(z_b + g), \tag{6.49}
$$

and

$$\sum_{i=1}^{m} \psi_k(z_a + z_i)\psi_l(z_b + z_i) = \sum_{i=1}^{m} \psi_l(z_a + z_i)\psi_k(z_b + z_i). \tag{6.50}$$

We define the elements of the n-dimensional cube $X_k = [X_k(x(1), \ldots, x(n))]$, $1 \leq x(i) \leq m$, $1 \leq i \leq n$, for each k, by([17], [18], [19])

$$X_k(x(1), \ldots, x(n)) = \psi_k(z_{x(1)} + \ldots + z_{x(n)}) \tag{6.51}$$

To consider the inner product properties of the two-dimensional faces of this cube, we let the q-th coordinate take two values a and b, and the r-th coordinate run from 1 to m, all the other coordinates being constant. Then, with $y = \sum_{i=1}^{m} z_{x(i)} - z_{x(q)} - z_{x(r)}$, we have

$$\sum_{x(r)=1}^{m} X_k(x(1), \ldots, x(q-1), a, x(q+1), \ldots, x(n)).$$

$$X_k(x(1), \ldots, x(q-1), b, x(q+1), \ldots, x(n))$$

$$= \sum_{x(r)=1}^{m} \psi_k(y + z_a + z_{x(r)})\psi_k(y + z_b + z_{x(r)})$$

$$= \sum_{g \in G} \psi_k(y + z_a + g)\psi_k(y + z_b + g)$$

$$= \sum_{h \in G} \psi_k(z_a + h)\psi_k(z_b + h). \tag{6.52}$$

To find the inner product of the rows of the corresponding two-dimensional layers in different n-dimensional matrices X_k and X_l, we let the r-th coordinate sum from one to n, the q-th coordinate, take two values (a and b), and all other coordinates remain constant. Letting

$$y = \sum_{i=1}^{m} z_{x(i)} - z_{x(q)} - z_{x(r)},$$

we have([17], [18], [19])

$$\sum_{x(r)=1}^{m} X_k(x(1), \ldots, x(q-1), a, x(q+1), \ldots, x(n)).$$

$$X_l(x(1), \ldots, x(q-1), b, x(q+1), \ldots, x(n))$$

$$= \sum_{x(r)=1}^{m} \psi_k(y + z_a + z_{x(r)})\psi_l(y + z_b + z_{x(r)})$$

$$= \sum_{i=1}^{m} \psi_k(z_a + z_i)\psi_l(z_b + z_i)$$

$$= \sum_{i=1}^{m} \psi_l(z_a + z_i)\psi_k(z_b + z_i), \text{ (By Equation (6.50))}$$

$$= \sum_{x(r)=1}^{m} X_l(x(1), \ldots, x(q-1), a, x(q+1), \ldots, x(n))$$

$$X_k(x(1), \ldots, x(q-1), b, x(q+1), \ldots, x(n)). \tag{6.53}$$

Higher-dimensional Hadamard matrices may be constructed by replacing the variables of an n-cube orthogonal design of type $\overbrace{(1, 1, \ldots, 1)}^{t}{}^{n}$ by the t-suitable matrices([17], [18], [19]) .

Combining Equations (6.49) and (6.52) , we see that if the rows of t-suitable matrices are orthogonal, the rows of the n-dimensional matrices formed from these matrices, in any direction parallel to the axis, will also be orthogonal. The contribution of the different t-suitable matrices is cancelled out in the orthogonal design by using Equation (6.53) .

It is known that Williamson matrices of order m (or 4-suitable matrices in our present terminology) exist for the following orders ([20]) :

1. m: where $m \in \{1, 3, 5, \ldots, 29, 37, 43\}$;

2. $(p+1)/2$: $p \equiv 1 \pmod{4}$ a prime power;

3. 3^c, 7.3^{c-1}: c is a natural number;

4. $p^r(p+1)/2$: $p \equiv 1 \pmod{4}$ a prime power, r a natural number;

5. $s(4s-1)$: s is the order of a good matrix (see [21] for definition) ;

6. $s(4s+3)$: s is the order of a good matrix, and $4s+4$ is the order of a symmetric Hadamard matrix;

7. sv: s is the order of a good matrix, v is the order of an Abelian group G on which are defined a (v, k, λ) and a $(v, (v-1)/2, (v-s)/4)$ difference set $v - 4(k - \lambda) = 4s - 1$;

8. $3^{2r}(p_1^{r_1} \ldots p_n^{r_n})^4$: r, r_i are nonnegative integers, p_i are primes satisfying $p_i \equiv 3(\mathrm{mod}4)$ and $p_i > 3$.

The Williamson matrices are used to form n-dimensional cubes $X_1 = [X_1(x(1), \ldots, x(n))]$, $X_2 = [X_2(x(1), \ldots, x(n))]$, $X_3 = [X_3(x(1), \ldots, x(n))]$, and $X_4 = [X_4(x(1), \ldots, x(n))]$, which are used to replace the a, b, c, and d of the orthogonal design of order 4 and type $(1, 1, 1, 1)^3$ (see Example 2 of the last subsection). Because of the properties of these matrices, each of the faces parallel to the axes will be an Hadamard matrix of property $(2, 2, 2)$.

Thus when m is the order of 4 Williamson matrices, there is a 3-dimensional Hadamard matrix of order $4m$ and property $(2, 2, 2)$, i.e., there exists an absolutely proper 3-dimensional Hadamard matrix of order $4m$. In general, by Theorem 6.3.2 and the above construction, we have

Theorem 6.3.3 ([17], [18], [19]) *Let m be the order of 4 Williamson matrices. Then there is an n-dimensional Hadamard matrix of order $(4m)^m$ and property $(2, 2, \ldots, 2)$, i.e., there exists an absolutely proper n-dimensional Hadamard matrix of order $(4m)^m$.*

Definition 6.3.1 ([1], [17], [18], [19]) *Let $A = [A(a(1), \ldots, a(n))]$ and $B = [B(b(1), \ldots, b(n))]$, $0 \le a(i)$, $b(i) \le m - 1$, $1 \le i \le n$, be two n-dimensional matrices of order m. Then A and B are called anti-amicable if the following two conditions are satisfied*

1. *For each p, $1 \le p \le n$, and prefixed $0 \le a(1), \ldots, a(p-1), a(p+1), \ldots, a(n) \le m - 1$,*

$$\sum_{a(p)=0}^{m-1} A(a(1), \ldots, a(p-1), a(p), a(p+1), \ldots, a(n))$$
$$\times B(a(1), \ldots, a(p-1), a(p), a(p+1), \ldots, a(n)) = 0;$$

2. *For each pair of $p \ne q$, $1 \le p, q \le n$, prefixed $0 \le a(1), \ldots, a(p-1), a(p+1), \ldots, a(q-1), a(q+1), \ldots, a(n) \le m - 1$, and $0 \le x \ne z \le m - 1$, we have*

$$\sum_{a(p)=0}^{m-1} [A(a(1),\ldots,a(p-1),a(p),a(p+1),\ldots,a(q-1),x,$$
$$a(q+1),\ldots,a(n))B(a(1),\ldots,a(p-1),a(p),a(p+1),\ldots,$$
$$a(q-1),z,a(q+1),\ldots,a(n)) + A(a(1),\ldots,a(p-1),a(p),a(p+1),\ldots,$$
$$a(q-1),z,a(q+1),\ldots,a(n))B(a(1),\ldots,a(p-1),a(p),a(p+1),\ldots,$$
$$a(q-1),x,a(q+1),\ldots,a(n))] = 0.$$

In particular, two 2-dimensional matrices $A = \begin{bmatrix} a_{11} & a_{12} \\ a_{21} & a_{22} \end{bmatrix}$ and $B = \begin{bmatrix} b_{11} & b_{12} \\ b_{21} & b_{22} \end{bmatrix}$ are anti-amicable if

$$a_{11}b_{11} + a_{12}b_{12} = 0; \quad a_{21}b_{21} + a_{22}b_{22} = 0,$$

and

$$a_{11}b_{21} + a_{12}b_{22} + a_{21}b_{11} + a_{22}b_{12} = 0.$$

The following two 3-dimensional matrices $C = [C(i,j,k)]$ and $B = [B(i,j,k)]$ are 3-dimensional anti-amicable Hadamard matrices of order 2 and property $(2,2,2)$ ([1], [17], [18], [19]) , where

$$[C(i,j,0)] = \begin{bmatrix} 1 & 1 \\ -1 & 1 \end{bmatrix}, \quad [C(i,j,1)] = \begin{bmatrix} 1 & -1 \\ 1 & 1 \end{bmatrix}.$$

and

$$[D(i,j,0)] = \begin{bmatrix} 1 & -1 \\ 1 & 1 \end{bmatrix}, \quad [D(i,j,1)] = \begin{bmatrix} -1 & -1 \\ 1 & -1 \end{bmatrix}.$$

Lemma 6.3.1 ([1], [17], [18], [19]) *There exist n-dimensional anti-amicable Hadamard matrices of order 2 and property $(2,2,\ldots,2)$.*

Proof. Define two n-dimensional (± 1)-valued matrices $A = [A(a(1), \ldots, a(n))]$ and $B = [B(b(1), \ldots, b(n))]$ of order 2 by

$$A(a(1),\ldots,a(n)) = \begin{cases} (-1)^{s/2+1} & s \text{ even} \\ (-1)^{(s+1)/2} & s \text{ odd} \end{cases}$$

and

$$B(b(1),\ldots,b(n)) = \begin{cases} (-1)^{v/2} & v \text{ even} \\ (-1)^{(v+1)/2} & v \text{ odd}, \end{cases}$$

where

$$s := \sum_{i=1}^{n} a(i), \quad v := \sum_{i=1}^{n} b(i), \quad 0 \le a(i), \quad b(i) \le 1, \quad 1 \le i \le n.$$

It can be proved that these A and B are the wanted matrices of the lemma. In fact, each 2-dimensional section of A and B is of the form $\begin{bmatrix} a & b \\ b & -a \end{bmatrix}$ $(a, b = \pm 1)$, which is clearly an Hadamard matrix. Thus both A and B are absolutely proper n-dimensional Hadamard matrices of order 2.

In order to prove the anti-amicability, without loss of the generality, we consider the following four cases according to the values of $t = \sum_{i=3}^{n} a(i)$. Let $w = (a(3), \ldots, a(n))$. Then

Case 1: $t \equiv 0 \pmod 4$. Then

$$A(1,1,w)B(1,1,w) + A(1,2,w)B(1,2,w) = (-1)^2(-1) + (-1)^2(-1)^2 = 0,$$

$$A(2,1,w)B(2,1,w) + A(2,2,w)B(2,2,w) = (-1)^2(-1)^2 + (-1)(-1)^2 = 0$$

and

$$A(1,1,w)B(2,1,w) + A(1,2,w)B(2,2,w)$$
$$+A(2,1,w)B(1,1,w) + A(2,2,w)B(1,2,w)$$
$$= (-1)^2(-1)^2 + (-1)^2(-1)^2 + (-1)^2(-1) + (-1)(-1)^2$$
$$= 0.$$

Case 2: $t \equiv 1 \pmod 4$. Then

$$A(1,1,w)B(1,1,w) + A(1,2,w)B(1,2,w) = (-1)^2(-1)^2 + (-1)(-1)^2 = 0;$$

$$A(2,1,w)B(2,1,w) + A(2,2,w)B(2,2,w) = (-1)(-1)^2 + (-1)(-1) = 0;$$

and

$$A(1,1,w)B(2,1,w) + A(1,2,w)B(2,2,w)$$
$$+A(2,1,w)B(1,1,w) + A(2,2,w)B(1,2,w)$$
$$= (-1)^2(-1)^2 + (-1)(-1) + (-1)(-1)^2 + (-1)(-1)^2$$
$$= 0.$$

Case 3: $t \equiv 2(\mathrm{mod}4)$. Then

$$A(1,1,w)B(1,1,w) + A(1,2,w)B(1,2,w) = (-1)(-1)^2 + (-1)(-1) = 0,$$

$$A(2,1,w)B(2,1,w) + A(2,2,w)B(2,2,w) = (-1)(-1) + (-1)^2(-1) = 0$$

and

$$\begin{aligned} A(1,1,w)&B(2,1,w) + A(1,2,w)B(2,2,w) \\ &+A(2,1,w)B(1,1,w) + A(2,2,w)B(1,2,w) \\ &= (-1)(-1) + (-1)(-1) + (-1)(-1)^2 + (-1)^2(-1) \\ &= 0. \end{aligned}$$

Case 4: $t \equiv 3(\mathrm{mod}4)$. Then

$$A(1,1,w)B(1,1,w) + A(1,2,w)B(1,2,w) = (-1)(-1) + (-1)^2(-1) = 0,$$

$$A(2,1,w)B(2,1,w) + A(2,2,w)B(2,2,w) = (-1)^2(-1) + (-1)^2(-1)^2 = 0$$

and

$$\begin{aligned} A(1,1,w)&B(2,1,w) + A(1,2,w)B(2,2,w) \\ &+A(2,1,w)B(1,1,w) + A(2,2,w)B(1,2,w) \\ &= (-1)(-1) + (-1)^2(-1)^2 + (-1)^2(-1) + (-1)^2(-1) \\ &= 0. \end{aligned}$$

Q.E.D.

Definition 6.3.2 ([1], [17], [18], [19]) *Two 2-dimensional Hadamard matrices (or orthogonal designs) X and Y are said to be amicable if $XY' = YX'$.*

In general, two n-dimensional orthogonal designs $H = [H(h(1),\ldots, h(n))]$ and $G = [G(g(1),\ldots,g(n))]$, $0 \leq h(i), g(i) \leq m-1$, of order m are said to be amicable if for each pair $1 \leq p \neq q \leq n$, $0 \leq i, k \leq m-1$, and prefixed $(g(1),\ldots,g(p-1),g(p+1),\ldots,g(q-1),g(q+1),\ldots,g(n))$,

$$\sum_{g(p)=0}^{m-1} G(g(1),\ldots,g(p-1),g(p),g(p+1),\ldots,g(q-1),i,g(q+1),\ldots,g(n))$$

$$\times H(g(1),\ldots,g(p-1),g(p),g(p+1),\ldots,g(q-1),k,g(q+1),\ldots,g(n))$$

$$= \sum_{g(p)=0}^{m-1} G(g(1),\ldots,g(p-1),g(p),g(p+1),\ldots,g(q-1),k,g(q+1),\ldots,g(n))$$

$$\times H(g(1),\ldots,g(p-1),g(p),g(p+1),\ldots,g(q-1),i,g(q+1),\ldots,g(n))$$

$$(6.54)$$

If Equation (6.54) is satisfied except for some subscripts, we will say that H and G are amicable except for those subscripts.

After the try-and-fail search, it is found that

Theorem 6.3.4 ([1], [17], [18], [19]) *There exist no 3-dimensional amicable orthogonal designs of type* $(1,1)^3$ *and property* $(2,2,\ldots,2)$.

Let a, and b be commuting variables. Let $G = [G(g(1),\ldots,g(n))]$, $0 \le g(i) \le m-1$, $1 \le i \le n$, be the matrix defined by

$$G(g(1),\ldots,g(n)) = \begin{cases} (-1)^{w/2-g(n)}a & w \text{ even} \\ (-1)^{(w+1)/2-g(n)}b & w \text{ odd,} \end{cases} \qquad (6.55)$$

where $w := \sum_{i=1}^{n} g(i)$.

Lemma 6.3.2 ([1], [17], [18], [19]) *The matrix G constructed in Equation (6.55) is an n-dimensional orthogonal design of type* $(1,1)^n$ *and property* $(2,2,\ldots,2)$.

Proof. At first we prove the following two equations:

$$G(0,0,x)G(0,1,x) + G(1,0,x)G(1,1,x) = 0 \qquad (6.56)$$

and

$$G(0,0,x)G(1,0,x) + G(0,1,x)G(1,1,x) = 0, \qquad (6.57)$$

where $x := (g(3),\ldots,g(n))$ is the prefixed vector of length $n-2$.
Suppose $v := \sum_{i=3}^{n} g(i)$.

Case 1: If $v \equiv 0(\mathrm{mod}4)$, then the left hand sides of both Equations (6.56) and (6.57) become

$$(-1)^{-g(n)}a(-1)^{1-g(n)}b + (-1)^{1-g(n)}b(-1)^{1-g(n)}a = 0;$$

Case 2: If $v \equiv 1 (\mathrm{mod} 4)$, then the left hand sides of both Equations (6.56) and (6.57) become

$$(-1)^{1-g(n)} b(-1)^{1-g(n)} a + (-1)^{1-g(n)} a(-1)^{2-g(n)} b = 0;$$

Case 3: If $v \equiv 2 (\mathrm{mod} 4)$, then the left hand sides of both Equations (6.56) and (6.57) become

$$(-1)^{1-g(n)} a(-1)^{2-g(n)} b + (-1)^{2-g(n)} b(-1)^{2-g(n)} a = 0;$$

Case 4: If $v \equiv 3 (\mathrm{mod} 4)$, then the left hand sides of both Equations (6.56) and (6.57) become

$$(-1)^{2-g(n)} b(-1)^{-g(n)} a + (-1)^{-g(n)} a(-1)^{1-g(n)} b = 0;$$

Thus the face $A = [A(i,j)] := [G(i,j,x)]$ is a 2-dimensional orthogonal design of order 2.

Similarly, it can be proved that each 2-dimensional face of this matrix G is an orthogonal design of order 2.   **Q.E.D.**

Lemma 6.3.3 ([1], [17], [18], [19]) *There exist n-dimensional orthogonal designs of order 2 and type $(1,1)^n$ that are amicable except that one distinguished coordinate (the last) is constant.*

Proof. Let $G = [G(g(1), \ldots, g(n))]$ and $H = [H(h(1), \ldots, h(n))]$ be two matrices produced by Equation (6.55) starting from the commuting variables (a_1, b_1) and (a_2, b_2), respectively. Then by the same proof as that of Lemma 6.3.2, it can be proved that G and H are the wanted designs. **Q.E.D.**

Lemma 6.3.4 ([1], [17], [18], [19]) *There exist n-dimensional amicable Hadamard matrices of order 2 and property $(2, 2, \ldots, 2)$.*

Proof. It can be proved by replacing the variables a_i, b_i by ± 1.   **Q.E.D.**

For more details about the higher-dimensional Hadamard matrices the readers are recommended to the papers [22-31].

Bibliography

[1] P.J. Shlichta, *Higher-Dimensional Hadamard Matrices*, IEEE Trans. On Inform. Theory, Vol.IT-25, No.5, pp.566-572, 1979.

[2] Y. X. Yang and X.D. Lin, *Coding and Cryptography* , PPT Press, Beijing, 1992.

[3] J. Jedwab, *Perfect Arrays, Barker Arrays and Difference Sets*, PhD Thesis, University of London, 1991.

[4] H.D. Luke, L. Bomer and M.Antweiler, *Perfect Binary Arrays*, Signal Processing, Vol.17, No.1, pp69-80, 1989.

[5] H.D. Luke, *Sequences and Arrays with Perfect Periodic Correlation*, IEEE Trans. AES, Vol.24, No.2, pp287-294, 1988.

[6] J.A.Davis and J.Jedwab, *A Summary of Menon Difference Sets*, Congressus Numerantium, 93(1993), pp203-207.

[7] W.Launey, *A Note on N-Dimensional Hadamard Matrices of Order 2^t and Reed-Muller Codes*, IEEE Trans. on Inform. Theory, Vol.27, No.3, pp664-667, 1991.

[8] K.T.Arasu, J.A.Davis, J.Jedwab and S.K. Sehgal, *New Constructions of Menon Difference Sets*, J. Combin. Theory (A), Vol.64, No.2, pp329-336, 1993.

[9] J.Jedwab, C.J.Mitchell, *Constructing New Perfect Binary Arrays*, Electron. Lett. 1988; 24(11):650-652.

[10] J.Jedwab, *Generalized Perfect Arrays and Menon Difference Sets*, Designs, Codes and Cryptography, Vol.2, pp19-68, 1992.

[11] M.Y.Xia, *Some Infinite Classes of Special Williamson Matrices and Difference Sets*, J. Combin. Theory (A), Vol.61, pp230-242, 1992.

[12] Y.X.Yang, *Quasi-Perfect Binary Arrays*, Chinese J. of Electronics, Vol.20, No.4, pp37-44, 1992.

[13] J.Jedwab, *Nonexistence Of Perfect Binary Arrays*, Electron. Lett. Vol.27, No.14, pp1252-1253, 1991.

[14] Kopilovich, *On Perfect Binary Arrays*, Electron. Lett. 1988; 24(9):566-567.

[15] P.Wild, *Infinite Families of Perfect Binary Arrays*, Electron. Lett. 1988; 24(14):845-847.

[16] W.K.Chan, M.K.Siu, and P.Tong, *Two-Dimensional Binary Arrays With Good Autocorrelation*, Inform. and Control, 1979; 42, 125-130.

[17] J.Hammer and J.Seberry, *Higher-Dimensional Orthogonal Designs And Applications*, IEEE Trans. Inform. Theory, Vol.27, No.6, pp772-779, 1981.

[18] J. Seberry, *Higher-Dimensional Orthogonal Designs And Hadamard Matrices*, Combinatorics VII: Proc. Seventh Australian Conf., Lecture Notes in Mathematics, Springer-Verlag, New York, 1980.

[19] J. Hammer and J. Seberry, *Higher-Dimensional Orthogonal Designs And Hadamard Matrices (II)*, Proc. Ninth Conf. on Numerical Mathematics, Congressus Numerantium, Utilities Mathematics, Winnipeg, pp23-29, 1979.

[20] A.V. Geramita and J. Seberry, *Orthogonal Designs: Quadratic Forms and Hadamard Matrices*, New York: Marcel Dekker, 1979.

[21] J.Seberry Wallis, *On the Existence of Hadamard Matrices* , J. Combinatorial Theory, Ser. A, Vol.21, pp. 188-195, 1976.

[22] Y.X. Yang, *Proofs of Some Conjectures on Higher Dimensional Hadamard Matrices*, Chinese Science Bulletin, Vol.31, No.24, pp1662-1667, 1986.

[23] Y.X.Yang, *Existence, Construction Methods and Enumeration of Higher-Dimensional Hadamard Matrices*, IEEE ISIT'90, U.S.A, 1990.

[24] Y.X.Yang, *On the Perfect Binary Arrays*, J. of Electronics (China), Vol.7, No.2, pp175-181, 1990.

[25] Y.X.Yang, *Dyadic Methods in Communication Theory*, Advances in Mathematics, Vol.19, No.2, pp254-255, 1990.

[26] S.Q.Li, and Y.X.Yang, *On the Circulant Hadamard Conjecture*, Select Papers for the J. of BUPT, pp80-83, 1990.

[27] Y.X. Yang, *New Proofs for the Conjectures of Higher-Dimensional Hadamard Matrices*, J. Systems Science and Mathematical Sciences, Vol.8, No.1, pp52-55, 1988.

[28] S.Q. Li, and Y.X.Yang, *The Analysis of Array Sampling and Folding*, J. of Beijing Univ. of Posts and Telecomm, Vo.12, No.1, pp28-34, 1989.

[29] Y.X. Yang, *On the Constructions of Higher Dimensional Hadamard Matrices*, J. of Beijing Univ. of Posts and Telecomm, Vo.11, No.2, pp31-38, 1988.

[30] Y. X. Yang, *Applications of Higher-Dimensional Matrices to Cryptography*, J. of Beijing Univ. of Posts and Telecomm, Vo.12, No.4, pp41-46, 1989.

[31] D. Calabro, J.K.Wolf, *On The Synthesis of Two-Dimensional Arrays With Desirable Correlation Properties*, Information and Control, 1968; 11: 537-560.

Concluding Questions

The topic of (higher-) dimensional Hadamard matrices is a developing one. As the first book concerning about the topic, this book has investigated many problems of constructions, existences, enumeration, transforms and fast algorithms, whilst there are still many open problems in both areas of theory and practice. In order to motivate more research we list here some interesting research problems.

1. (Hadamard Conjecture) There exists an Hadamard matrix of order $4t$ for each positive integer t;

2. (Wallis Conjecture) There exists a weighting matrix $W(4t, k)$, $0 \leq k \leq 4t$, for each positive integer t; (This conjecture is clearly a generalization of the Hadamard conjecture.)

3. It is conjectured that there exist amicable Hadamard matrices of each order $n \equiv 0 \pmod 4$.

4. (Conjecture) Let x, y, q, m, and t be non-negative integers satisfying $m = 2^t q$ and $0 \leq x+y \leq m$. When t is sufficiently large, there always exists an orthogonal design of type $(x, y, m - x - y)$ and order m.

5. Prove or disprove the existence of three-dimensional Hadamard matrices of orders $4k + 2 \neq 2.3^b$, $k > 1$, $b \geq 0$.

6. Construct more three-dimensional Hadamard matrices of orders $4k + 2$, $k > 1$.

7. The three-dimensional Hadamard matrices can be constructed from two-dimensional ones. Could the two-dimensional Hadamard matrices be constructed by three- or higher-dimensional Hadamard matrices? If the answer is positive, it is possible that more new two-dimensional Hadamard matrices would be discovered. In general, could the lower-dimensional Hadamard matrices be constructed by decomposing the higher-dimensional Hadamard matrices?

8. What is the enumeration of n-dimensional Hadamard matrices of order two (or equivalently the H–Boolean functions in n-variables)?

9. What is the enumeration of the Boolean functions satisfying the SAC of order $k \leq n-4$? Construct more Boolean functions satisfying the SAC of order k.

10. Construct and enumerate Bent functions.

11. Construct and enumerate Boolean functions satisfying propagation criterion of degree k and order m.

12. Let $n \geq 4$. Is there an n-dimensional Hadamard matrix of order $2t$, for each $t \geq 1$?

13. What are the necessary and sufficient conditions for the existence of $PBA(a_1,\ldots,a_n)$? How can one construct as many PBAs as possible?

14. What are the necessary and sufficient conditions for the existence of higher-dimensional orthogonal designs(HDOD)? How can one construct as many HDODs as possible?

15. How can one use the higher-dimensional arrays introduced in this book to construct families of secure crypto-systems?

16. Try to develop more applications of higher-dimensional Walsh and Hadamard matrices to engineering areas, e.g., signal processing, telecommunications (especially in mobile, optical, and data communications), image processing, EMS *et al.*.

Remark. After the manuscript of this book was finished, the author learnt of many new wonderful works on higher-dimensional Hadamard and perfect binary arrays by Yu Qing Chen, De Launey, K.J.Horadam and Cantian Lin *et al.*.

Index